T0282706

LONDON MATHEMATICAL SOCIETY LECTURE NOTE SERIES

The books in the series listed below are available from booksellers, or, in case of difficulty, from Cambridge University Press.

34 Representation theory of Lie groups, M.F. ATIYAH *et al*
36 Homological group theory, C.T.C. WALL (ed)
39 Affine sets and affine groups, D.G. NORTHCOTT
40 Introduction to H_p spaces, P.J. KOOSIS
46 p-adic analysis: a short course on recent work, N. KOBLITZ
49 Finite geometries and designs, P. CAMERON, J.W.P. HIRSCHFELD & D.R. HUGHES (eds)
50 Commutator calculus and groups of homotopy classes, H.J. BAUES
57 Techniques of geometric topology, R.A. FENN
59 Applicable differential geometry, M. CRAMPIN & F.A.E. PIRANI
66 Several complex variables and complex manifolds II, M.J. FIELD
69 Representation theory, I.M. GELFAND *et al*
74 Symmetric designs: an algebraic approach, E.S. LANDER
76 Spectral theory of linear differential operators and comparison algebras, H.O. CORDES
77 Isolated singular points on complete intersections, E.J.N. LOOIJENGA
79 Probability, statistics and analysis, J.F.C. KINGMAN & G.E.H. REUTER (eds)
80 Introduction to the representation theory of compact and locally compact groups, A. ROBERT
81 Skew fields, P.K. DRAXL
82 Surveys in combinatorics, E.K. LLOYD (ed)
83 Homogeneous structures on Riemannian manifolds, F. TRICERRI & L. VANHECKE
86 Topological topics, I.M. JAMES (ed)
87 Surveys in set theory, A.R.D. MATHIAS (ed)
88 FPF ring theory, C. FAITH & S. PAGE
89 An F-space sampler, N.J. KALTON, N.T. PECK & J.W. ROBERTS
90 Polytopes and symmetry, S.A. ROBERTSON
91 Classgroups of group rings, M.J. TAYLOR
92 Representation of rings over skew fields, A.H. SCHOFIELD
93 Aspects of topology, I.M. JAMES & E.H. KRONHEIMER (eds)
94 Representations of general linear groups, G.D. JAMES
95 Low-dimensional topology 1982, R.A. FENN (ed)
96 Diophantine equations over function fields, R.C. MASON
97 Varieties of constructive mathematics, D.S. BRIDGES & F. RICHMAN
98 Localization in Noetherian rings, A.V. JATEGAONKAR
99 Methods of differential geometry in algebraic topology, M. KAROUBI & C. LERUSTE
100 Stopping time techniques for analysts and probabilists, L. EGGHE
101 Groups and geometry, ROGER C. LYNDON
103 Surveys in combinatorics 1985, I. ANDERSON (ed)
104 Elliptic structures on 3-manifolds, C.B. THOMAS
105 A local spectral theory for closed operators, I. ERDELYI & WANG SHENGWANG
106 Syzygies, E.G. EVANS & P. GRIFFITH
107 Compactification of Siegel moduli schemes, C-L. CHAI
108 Some topics in graph theory, H.P. YAP
109 Diophantine analysis, J. LOXTON & A. VAN DER POORTEN (eds)
110 An introduction to surreal numbers, H. GONSHOR
111 Analytical and geometric aspects of hyperbolic space, D.B.A. EPSTEIN (ed)

113 Lectures on the asymptotic theory of ideals, D. REES
114 Lectures on Bochner-Riesz means, K.M. DAVIS & Y-C. CHANG
115 An introduction to independence for analysts, H.G. DALES & W.H. WOODIN
116 Representations of algebras, P.J. WEBB (ed)
117 Homotopy theory, E. REES & J.D.S. JONES (eds)
118 Skew linear groups, M. SHIRVANI & B. WEHRFRITZ
119 Triangulated categories in the representation theory of finite-dimensional algebras, D. HAPPEL
121 Proceedings of *Groups - St Andrews 1985*, E. ROBERTSON & C. CAMPBELL (eds)
122 Non-classical continuum mechanics, R.J. KNOPS & A.A. LACEY (eds)
124 Lie groupoids and Lie algebroids in differential geometry, K. MACKENZIE
125 Commutator theory for congruence modular varieties, R. FREESE & R. MCKENZIE
126 Van der Corput's method of exponential sums, S.W. GRAHAM & G. KOLESNIK
127 New directions in dynamical systems, T.J. BEDFORD & J.W. SWIFT (eds)
128 Descriptive set theory and the structure of sets of uniqueness, A.S. KECHRIS & A. LOUVEAU
129 The subgroup structure of the finite classical groups, P.B. KLEIDMAN & M.W.LIEBECK
130 Model theory and modules, M. PREST
131 Algebraic, extremal & metric combinatorics, M-M. DEZA, P. FRANKL & I.G. ROSENBERG (
132 Whitehead groups of finite groups, ROBERT OLIVER
133 Linear algebraic monoids, MOHAN S. PUTCHA
134 Number theory and dynamical systems, M. DODSON & J. VICKERS (eds)
135 Operator algebras and applications, 1, D. EVANS & M. TAKESAKI (eds)
136 Operator algebras and applications, 2, D. EVANS & M. TAKESAKI (eds)
137 Analysis at Urbana, I, E. BERKSON, T. PECK, & J. UHL (eds)
138 Analysis at Urbana, II, E. BERKSON, T. PECK, & J. UHL (eds)
139 Advances in homotopy theory, S. SALAMON, B. STEER & W. SUTHERLAND (eds)
140 Geometric aspects of Banach spaces, E.M. PEINADOR and A. RODES (eds)
141 Surveys in combinatorics 1989, J. SIEMONS (ed)
142 The geometry of jet bundles, D.J. SAUNDERS
143 The ergodic theory of discrete groups, PETER J. NICHOLLS
144 Introduction to uniform spaces, I.M. JAMES
145 Homological questions in local algebra, JAN R. STROOKER
146 Cohen-Macaulay modules over Cohen-Macaulay rings, Y. YOSHINO
147 Continuous and discrete modules, S.H. MOHAMED & B.J. MÜLLER
148 Helices and vector bundles, A.N. RUDAKOV et al
149 Solitons, nonlinear evolution equations and inverse scattering, M.J. ABLOWITZ & P.A. CLARKSON
150 Geometry of low-dimensional manifolds 1, S. DONALDSON & C.B. THOMAS (eds)
151 Geometry of low-dimensional manifolds 2, S. DONALDSON & C.B. THOMAS (eds)
152 Oligomorphic permutation groups, P. CAMERON
153 L-functions and arithmetic, J. COATES & M.J. TAYLOR (eds)
154 Number theory and cryptography, J. LOXTON (ed)
155 Classification theories of polarized varieties, TAKAO FUJITA
156 Twistors in mathematics and physics, T.N. BAILEY & R.J. BASTON (eds)
157 Analytic pro-*p* groups, J.D. DIXON, M.P.F. DU SAUTOY, A. MANN & D. SEGAL
158 Geometry of Banach spaces, P.F.X. MÜLLER & W. SCHACHERMAYER (eds)
159 Groups St Andrews 1989 volume 1, C.M. CAMPBELL & E.F. ROBERTSON (eds)
160 Groups St Andrews 1989 volume 2, C.M. CAMPBELL & E.F. ROBERTSON (eds)
161 Lectures on block theory, BURKHARD KÜLSHAMMER
162 Harmonic analysis and representation theory for groups acting on homogeneous trees, A. FIGA-TALAMANCA & C. NEBBIA
163 Topics in varieties of group representations, S.M. VOVSI
166 Surveys in combinatorics, 1991, A.D. KEEDWELL (ed)
167 Stochastic analysis, M.T. BARLOW & N.H. BINGHAM (eds)
175 Adams memorial symposium on algebraic topology 1, N. RAY & G. WALKER (eds)
176 Adams memorial symposium on algebraic topology 2, N. RAY & G. WALKER (eds)

LONDON MATHEMATICAL SOCIETY LECTURE NOTE SERIES

Managing Editor: Professor J.W.S. Cassels, Department of Pure Mathematics and Mathematical Statistics, University of Cambridge, 16 Mill Lane, Cambridge CB2 1SB, England

The books in the series listed below are available from booksellers, or, in case of difficulty, from Cambridge University Press.

34 Representation theory of Lie groups, M.F. ATIYAH *et al*
36 Homological group theory, C.T.C. WALL (ed)
39 Affine sets and affine groups, D.G. NORTHCOTT
40 Introduction to H_p spaces, P.J. KOOSIS
46 p-adic analysis: a short course on recent work, N. KOBLITZ
49 Finite geometries and designs, P. CAMERON, J.W.P. HIRSCHFELD & D.R. HUGHES (eds)
50 Commutator calculus and groups of homotopy classes, H.J. BAUES
57 Techniques of geometric topology, R.A. FENN
59 Applicable differential geometry, M. CRAMPIN & F.A.E. PIRANI
66 Several complex variables and complex manifolds II, M.J. FIELD
69 Representation theory, I.M. GELFAND *et al*
74 Symmetric designs: an algebraic approach, E.S. LANDER
76 Spectral theory of linear differential operators and comparison algebras, H.O. CORDES
77 Isolated singular points on complete intersections, E.J.N. LOOIJENGA
79 Probability, statistics and analysis, J.F.C. KINGMAN & G.E.H. REUTER (eds)
80 Introduction to the representation theory of compact and locally compact groups, A. ROBERT
81 Skew fields, P.K. DRAXL
82 Surveys in combinatorics, E.K. LLOYD (ed)
83 Homogeneous structures on Riemannian manifolds, F. TRICERRI & L. VANHECKE
86 Topological topics, I.M. JAMES (ed)
87 Surveys in set theory, A.R.D. MATHIAS (ed)
88 FPF ring theory, C. FAITH & S. PAGE
89 An F-space sampler, N.J. KALTON, N.T. PECK & J.W. ROBERTS
90 Polytopes and symmetry, S.A. ROBERTSON
91 Classgroups of group rings, M.J. TAYLOR
92 Representation of rings over skew fields, A.H. SCHOFIELD
93 Aspects of topology, I.M. JAMES & E.H. KRONHEIMER (eds)
94 Representations of general linear groups, G.D. JAMES
95 Low-dimensional topology 1982, R.A. FENN (ed)
96 Diophantine equations over function fields, R.C. MASON
97 Varieties of constructive mathematics, D.S. BRIDGES & F. RICHMAN
98 Localization in Noetherian rings, A.V. JATEGAONKAR
99 Methods of differential geometry in algebraic topology, M. KAROUBI & C. LERUSTE
100 Stopping time techniques for analysts and probabilists, L. EGGHE
101 Groups and geometry, ROGER C. LYNDON
103 Surveys in combinatorics 1985, I. ANDERSON (ed)
104 Elliptic structures on 3-manifolds, C.B. THOMAS
105 A local spectral theory for closed operators, I. ERDELYI & WANG SHENGWANG
106 Syzygies, E.G. EVANS & P. GRIFFITH
107 Compactification of Siegel moduli schemes, C-L. CHAI
108 Some topics in graph theory, H.P. YAP
109 Diophantine analysis, J. LOXTON & A. VAN DER POORTEN (eds)
110 An introduction to surreal numbers, H. GONSHOR
111 Analytical and geometric aspects of hyperbolic space, D.B.A. EPSTEIN (ed)

113 Lectures on the asymptotic theory of ideals, D. REES
114 Lectures on Bochner-Riesz means, K.M. DAVIS & Y-C. CHANG
115 An introduction to independence for analysts, H.G. DALES & W.H. WOODIN
116 Representations of algebras, P.J. WEBB (ed)
117 Homotopy theory, E. REES & J.D.S. JONES (eds)
118 Skew linear groups, M. SHIRVANI & B. WEHRFRITZ
119 Triangulated categories in the representation theory of finite-dimensional algebras, D. HAPPEL
121 Proceedings of *Groups - St Andrews 1985*, E. ROBERTSON & C. CAMPBELL (eds)
122 Non-classical continuum mechanics, R.J. KNOPS & A.A. LACEY (eds)
124 Lie groupoids and Lie algebroids in differential geometry, K. MACKENZIE
125 Commutator theory for congruence modular varieties, R. FREESE & R. MCKENZIE
126 Van der Corput's method of exponential sums, S.W. GRAHAM & G. KOLESNIK
127 New directions in dynamical systems, T.J. BEDFORD & J.W. SWIFT (eds)
128 Descriptive set theory and the structure of sets of uniqueness, A.S. KECHRIS & A. LOUVEAU
129 The subgroup structure of the finite classical groups, P.B. KLEIDMAN & M.W.LIEBECK
130 Model theory and modules, M. PREST
131 Algebraic, extremal & metric combinatorics, M-M. DEZA, P. FRANKL & I.G. ROSENBERG (
132 Whitehead groups of finite groups, ROBERT OLIVER
133 Linear algebraic monoids, MOHAN S. PUTCHA
134 Number theory and dynamical systems, M. DODSON & J. VICKERS (eds)
135 Operator algebras and applications, 1, D. EVANS & M. TAKESAKI (eds)
136 Operator algebras and applications, 2, D. EVANS & M. TAKESAKI (eds)
137 Analysis at Urbana, I, E. BERKSON, T. PECK, & J. UHL (eds)
138 Analysis at Urbana, II, E. BERKSON, T. PECK, & J. UHL (eds)
139 Advances in homotopy theory, S. SALAMON, B. STEER & W. SUTHERLAND (eds)
140 Geometric aspects of Banach spaces, E.M. PEINADOR and A. RODES (eds)
141 Surveys in combinatorics 1989, J. SIEMONS (ed)
142 The geometry of jet bundles, D.J. SAUNDERS
143 The ergodic theory of discrete groups, PETER J. NICHOLLS
144 Introduction to uniform spaces, I.M. JAMES
145 Homological questions in local algebra, JAN R. STROOKER
146 Cohen-Macaulay modules over Cohen-Macaulay rings, Y. YOSHINO
147 Continuous and discrete modules, S.H. MOHAMED & B.J. MÜLLER
148 Helices and vector bundles, A.N. RUDAKOV et al
149 Solitons, nonlinear evolution equations and inverse scattering, M.J. ABLOWITZ & P.A. CLARKSON
150 Geometry of low-dimensional manifolds 1, S. DONALDSON & C.B. THOMAS (eds)
151 Geometry of low-dimensional manifolds 2, S. DONALDSON & C.B. THOMAS (eds)
152 Oligomorphic permutation groups, P. CAMERON
153 L-functions and arithmetic, J. COATES & M.J. TAYLOR (eds)
154 Number theory and cryptography, J. LOXTON (ed)
155 Classification theories of polarized varieties, TAKAO FUJITA
156 Twistors in mathematics and physics, T.N. BAILEY & R.J. BASTON (eds)
157 Analytic pro-*p* groups, J.D. DIXON, M.P.F. DU SAUTOY, A. MANN & D. SEGAL
158 Geometry of Banach spaces, P.F.X. MÜLLER & W. SCHACHERMAYER (eds)
159 Groups St Andrews 1989 volume 1, C.M. CAMPBELL & E.F. ROBERTSON (eds)
160 Groups St Andrews 1989 volume 2, C.M. CAMPBELL & E.F. ROBERTSON (eds)
161 Lectures on block theory, BURKHARD KÜLSHAMMER
162 Harmonic analysis and representation theory for groups acting on homogeneous trees, A. FIGA-TALAMANCA & C. NEBBIA
163 Topics in varieties of group representations, S.M. VOVSI
166 Surveys in combinatorics, 1991, A.D. KEEDWELL (ed)
167 Stochastic analysis, M.T. BARLOW & N.H. BINGHAM (eds)
175 Adams memorial symposium on algebraic topology 1, N. RAY & G. WALKER (eds)
176 Adams memorial symposium on algebraic topology 2, N. RAY & G. WALKER (eds)

LONDON MATHEMATICAL SOCIETY LECTURE NOTE SERIES

Managing Editor: Professor J.W.S. Cassels, Department of Pure Mathematics and Mathematical Statistics, University of Cambridge, 16 Mill Lane, Cambridge CB2 1SB, England

The books in the series listed below are available from booksellers, or, in case of difficulty, from Cambridge University Press.

34 Representation theory of Lie groups, M.F. ATIYAH *et al*
36 Homological group theory, C.T.C. WALL (ed)
39 Affine sets and affine groups, D.G. NORTHCOTT
40 Introduction to H_p spaces, P.J. KOOSIS
46 p-adic analysis: a short course on recent work, N. KOBLITZ
49 Finite geometries and designs, P. CAMERON, J.W.P. HIRSCHFELD & D.R. HUGHES (eds)
50 Commutator calculus and groups of homotopy classes, H.J. BAUES
57 Techniques of geometric topology, R.A. FENN
59 Applicable differential geometry, M. CRAMPIN & F.A.E. PIRANI
66 Several complex variables and complex manifolds II, M.J. FIELD
69 Representation theory, I.M. GELFAND *et al*
74 Symmetric designs: an algebraic approach, E.S. LANDER
76 Spectral theory of linear differential operators and comparison algebras, H.O. CORDES
77 Isolated singular points on complete intersections, E.J.N. LOOIJENGA
79 Probability, statistics and analysis, J.F.C. KINGMAN & G.E.H. REUTER (eds)
80 Introduction to the representation theory of compact and locally compact groups, A. ROBERT
81 Skew fields, P.K. DRAXL
82 Surveys in combinatorics, E.K. LLOYD (ed)
83 Homogeneous structures on Riemannian manifolds, F. TRICERRI & L. VANHECKE
86 Topological topics, I.M. JAMES (ed)
87 Surveys in set theory, A.R.D. MATHIAS (ed)
88 FPF ring theory, C. FAITH & S. PAGE
89 An F-space sampler, N.J. KALTON, N.T. PECK & J.W. ROBERTS
90 Polytopes and symmetry, S.A. ROBERTSON
91 Classgroups of group rings, M.J. TAYLOR
92 Representation of rings over skew fields, A.H. SCHOFIELD
93 Aspects of topology, I.M. JAMES & E.H. KRONHEIMER (eds)
94 Representations of general linear groups, G.D. JAMES
95 Low-dimensional topology 1982, R.A. FENN (ed)
96 Diophantine equations over function fields, R.C. MASON
97 Varieties of constructive mathematics, D.S. BRIDGES & F. RICHMAN
98 Localization in Noetherian rings, A.V. JATEGAONKAR
99 Methods of differential geometry in algebraic topology, M. KAROUBI & C. LERUSTE
100 Stopping time techniques for analysts and probabilists, L. EGGHE
101 Groups and geometry, ROGER C. LYNDON
103 Surveys in combinatorics 1985, I. ANDERSON (ed)
104 Elliptic structures on 3-manifolds, C.B. THOMAS
105 A local spectral theory for closed operators, I. ERDELYI & WANG SHENGWANG
106 Syzygies, E.G. EVANS & P. GRIFFITH
107 Compactification of Siegel moduli schemes, C-L. CHAI
108 Some topics in graph theory, H.P. YAP
109 Diophantine analysis, J. LOXTON & A. VAN DER POORTEN (eds)
110 An introduction to surreal numbers, H. GONSHOR
111 Analytical and geometric aspects of hyperbolic space, D.B.A. EPSTEIN (ed)

113 Lectures on the asymptotic theory of ideals, D. REES
114 Lectures on Bochner-Riesz means, K.M. DAVIS & Y-C. CHANG
115 An introduction to independence for analysts, H.G. DALES & W.H. WOODIN
116 Representations of algebras, P.J. WEBB (ed)
117 Homotopy theory, E. REES & J.D.S. JONES (eds)
118 Skew linear groups, M. SHIRVANI & B. WEHRFRITZ
119 Triangulated categories in the representation theory of finite-dimensional algebras, D. HAPPEL
121 Proceedings of *Groups - St Andrews 1985*, E. ROBERTSON & C. CAMPBELL (eds)
122 Non-classical continuum mechanics, R.J. KNOPS & A.A. LACEY (eds)
124 Lie groupoids and Lie algebroids in differential geometry, K. MACKENZIE
125 Commutator theory for congruence modular varieties, R. FREESE & R. MCKENZIE
126 Van der Corput's method of exponential sums, S.W. GRAHAM & G. KOLESNIK
127 New directions in dynamical systems, T.J. BEDFORD & J.W. SWIFT (eds)
128 Descriptive set theory and the structure of sets of uniqueness, A.S. KECHRIS & A. LOUVEAU
129 The subgroup structure of the finite classical groups, P.B. KLEIDMAN & M.W.LIEBECK
130 Model theory and modules, M. PREST
131 Algebraic, extremal & metric combinatorics, M-M. DEZA, P. FRANKL & I.G. ROSENBERG (
132 Whitehead groups of finite groups, ROBERT OLIVER
133 Linear algebraic monoids, MOHAN S. PUTCHA
134 Number theory and dynamical systems, M. DODSON & J. VICKERS (eds)
135 Operator algebras and applications, 1, D. EVANS & M. TAKESAKI (eds)
136 Operator algebras and applications, 2, D. EVANS & M. TAKESAKI (eds)
137 Analysis at Urbana, I, E. BERKSON, T. PECK, & J. UHL (eds)
138 Analysis at Urbana, II, E. BERKSON, T. PECK, & J. UHL (eds)
139 Advances in homotopy theory, S. SALAMON, B. STEER & W. SUTHERLAND (eds)
140 Geometric aspects of Banach spaces, E.M. PEINADOR and A. RODES (eds)
141 Surveys in combinatorics 1989, J. SIEMONS (ed)
142 The geometry of jet bundles, D.J. SAUNDERS
143 The ergodic theory of discrete groups, PETER J. NICHOLLS
144 Introduction to uniform spaces, I.M. JAMES
145 Homological questions in local algebra, JAN R. STROOKER
146 Cohen-Macaulay modules over Cohen-Macaulay rings, Y. YOSHINO
147 Continuous and discrete modules, S.H. MOHAMED & B.J. MÜLLER
148 Helices and vector bundles, A.N. RUDAKOV et al
149 Solitons, nonlinear evolution equations and inverse scattering, M.J. ABLOWITZ & P.A. CLARKSON
150 Geometry of low-dimensional manifolds 1, S. DONALDSON & C.B. THOMAS (eds)
151 Geometry of low-dimensional manifolds 2, S. DONALDSON & C.B. THOMAS (eds)
152 Oligomorphic permutation groups, P. CAMERON
153 L-functions and arithmetic, J. COATES & M.J. TAYLOR (eds)
154 Number theory and cryptography, J. LOXTON (ed)
155 Classification theories of polarized varieties, TAKAO FUJITA
156 Twistors in mathematics and physics, T.N. BAILEY & R.J. BASTON (eds)
157 Analytic pro-*p* groups, J.D. DIXON, M.P.F. DU SAUTOY, A. MANN & D. SEGAL
158 Geometry of Banach spaces, P.F.X. MÜLLER & W. SCHACHERMAYER (eds)
159 Groups St Andrews 1989 volume 1, C.M. CAMPBELL & E.F. ROBERTSON (eds)
160 Groups St Andrews 1989 volume 2, C.M. CAMPBELL & E.F. ROBERTSON (eds)
161 Lectures on block theory, BURKHARD KÜLSHAMMER
162 Harmonic analysis and representation theory for groups acting on homogeneous trees, A. FIGA-TALAMANCA & C. NEBBIA
163 Topics in varieties of group representations, S.M. VOVSI
166 Surveys in combinatorics, 1991, A.D. KEEDWELL (ed)
167 Stochastic analysis, M.T. BARLOW & N.H. BINGHAM (eds)
175 Adams memorial symposium on algebraic topology 1, N. RAY & G. WALKER (eds)
176 Adams memorial symposium on algebraic topology 2, N. RAY & G. WALKER (eds)

London Mathematical Society Lecture Note Series. 175

Adams Memorial Symposium on Algebraic Topology: 1

Manchester 1990

Edited by
N. Ray and G. Walker
Department of Mathematics, University of Manchester

Published by the Press Syndicate of the University of Cambridge
The Pitt Building, Trumpington Street, Cambridge CB2 1RP
40 West 20th Street, New York, NY 10011-4211, USA
10 Stamford Road, Oakleigh, Victoria 3166, Australia

First published 1992

Library of Congress cataloguing in publication data available

British Library cataloguing in publication data available

ISBN 0 521 42074 1 paperback

Transferred to digital printing 2004

Contents of Volume 1

Preface

The international Symposium on algebraic topology which was held in Manchester in July 1990 was originally conceived as a tribute to Frank Adams by mathematicians in many countries who admired and had been influenced by his work and leadership. Preparations for the meeting, including invitations to the principal speakers, were already well advanced at the time of his tragic death in a car accident on 7 January 1989, at the age of 58 and still at the height of his powers.

Those members of the Symposium, and readers of these volumes, who had the good fortune to know Frank as a colleague, teacher and friend will need no introduction here to the qualities of his intellect and personality. Others are referred to Ioan James's article, published as *Biographical Memoirs of Fellows of the Royal Society*, Vol. 36, 1990, pages 3–16, and to the Memorial Address and the Reminiscences written by Peter May and published in *The Mathematical Intelligencer*, Vol. 12, no. 1, 1990, pages 40–44 and 45–48.

We, the editors of these proceedings, were both research students of Frank's during his years at Manchester, As might be imagined, this was a remarkable and unforgettable experience. There was inspiration in plenty, and, on occasion, humble pie to be eaten as well. The latter became palatable as we learned to appreciate that the vigour of Frank's responses was never directed at us as individuals, but rather towards the defence of mathematics. In fact we both discovered that when suitably prompted, Frank was astonishly willing to repeatedly explain arguments that we had bungled. He also provided warm and understanding support, friendship and guidance far beyond his role as research supervisor.

This was an exciting period for Manchester, where Frank's influence was admirably complemented by Michael Barratt, and for algebraic topology in general. When he came to Manchester in 1962, Frank had just developed the K-theory operations he used to solve the problem of vector fields on spheres. In the years that followed he developed his series of papers on $J(X)$, and regularly lectured on subject matter which eventually became his Chicago Lecture Notes volume "Stable homotopy and generalised homology".

Our opening article, by Peter May, describes these and other achievements in more detail, and forms in a sense an introduction to the whole of the book. Although some attempt has been made to group papers according to the themes which May identifies, we cannot pretend that anything very systematic has been attained, or is even desirable. Most of the contributions here are based on talks given at the Symposium, as the reader will see by consulting lists on pages xi–xii. Aside from this, we feel it sufficient to remark that all the articles have been refereed, and that every attempt has been made to attain a mathematical standard worthy of association with the name of J. F. Adams – with what success we must leave the reader to judge.

We also hope that the Symposium itself might be seen as a significant tribute to his philosophy and powers. In keeping with his views on the value of mathematics in transcending political and geographical boundaries, we were fortunate to attract a large number of participants from many countries, including Eastern Europe and the Soviet Union.

In conclusion, we would like to thank the many organisations and individuals who made possible both the Symposium and these volumes.

The bulk of the initial funding was provided by the Science and Engineering Research Council, with substantial additions being made by the London Mathematical Society and the University of Manchester Research Support Fund. Support for important peripherals was given by the NatWest Bank, Trinity College Cambridge, and the University of Manchester Mathematics Department and Vice-Chancellor's Office. We would especially like to thank John Easterling and Mark Shackleton in this context.

During the Symposium our sanity would not have survived intact without the able assistance of all our Manchester students and colleagues in algebraic topology, and most significantly, the fabulous organisational and front-desk skills of the Symposium Secretary, Jackie Minshull. And the high point of the Symposium, an ascent of Tryfan (Frank's favourite Welsh Peak), would have been far less enjoyable without the presence of Manchester guide Bill Heaton.

Mrs Grace Adams and her family were most helpful in providing photographs and other information, and were very supportive of the Symposium despite their bereavement.

The production of these volumes was first conceived by the Cambridge University Press Mathematical Editor David Tranah, and their birth pangs were considerably eased by his laid-back skills. Our referees rose to the task of supplying authoritative reports within what was often a tight deadline. We should also thank those authors who offered a manuscript which we have not had space to publish.

Finally, we both owe a great debt to our respective families, for sustaining us throughout the organisation of the Symposium, and for continuing to support us as its ripples spread downwards through the following months. Therefore, to Sheila Kelbrick and our daughter Suzanne, and to Wendy Walker, thank you.

These volumes are dedicated to Frank's memory.

Nigel Ray Grant Walker

University of Manchester
September 1991

Contents of Volume 2

Programme of one-hour invited lectures

A. K. Bousfield	On K_*-local stable homotopy theory.
G. E. Carlsson	Applications of bounded K-theory.
F. W. Clarke	Cooperations in elliptic homology.
M. C. Crabb	The Adams conjecture and the J map.
E. S. Devinatz	Duality in stable homotopy theory.
W. G. Dwyer	Construction of a new finite loop space.
P. Goerss	Projective and injective Hopf algebras over the Dyer-Lashof algebra.
M. J. Hopkins	p-adic interpolation of stable homotopy groups.
J. R. Hubbuck	Fields of spaces.
S. Jackowski	Maps between classifying spaces revisited.
J. D. S. Jones	Morse theory and classifying spaces.
N. J. Kuhn	A representation theoretic approach to the Steenrod algebra.
J. Lannes	The Segal conjecture from an unstable viewpoint.
M. E. Mahowald	On the tertiary suspension.
J. P. May	The work of J.F. Adams.
M. Mimura	Characteristic classes of exceptional Lie groups.
S. A. Mitchell	Harmonic localization and the Lichtenbaum-Quillen conjecture.
G. Nishida	p-adic Hecke algebra and $\mathrm{Ell}_*(X_\Gamma)$.
S. B. Priddy	The complete stable splitting of BG.
D. C. Ravenel	The telescope conjecture.
C. A. Robinson	Ring spectra and the new cohomology of commutative rings.
Y. Rudjak	Orientability of bundles and fibrations and related topics.
V. Vershinin	The Adams spectral sequence as a method of studying cobordism theories.
C. W. Wilkerson	Lie groups and classifying spaces.

Programme of contributed lectures

A. J. Baker	MSp from a chromatic viewpoint.
M. Bendersky	v_1-periodic homotopy groups of Lie groups — II.
C.-F. Bödigheimer	Homology operations for mapping class groups.
B. Botvinnik	Geometric properties of the Adams-Novikov spectral sequence.
D. M. Davis	v_1-periodic homotopy groups of Lie groups.
B. I. Gray	Unstable periodicity.
J. P. C. Greenlees	Completions, dimensionality and local cohomology.
J. Harris	Lannes' T functor on summands of $H^*(B(Z/p)^n)$.
H.-W. Henn	Refining Quillen's description of $H^*(BG; F_p)$.
K. Hess	The Adams-Hilton model for the total space of a fibration.
J. R. Hunton	Detruncating Morava K-theory.
S. Hutt	A homotopy theoretic approach to surgery on Poincaré spaces.
A. Jeanneret	Topological realisation of certain algebras associated to Dickson algebras.
K. Y. Lam	The geometric dimension problem according to J.F. Adams.
J. R. Martino	The dimension of a stable summand of BG.
J. McCleary	Hochschild homology and the cobar construction.
J. McClure	Integral homotopy of $THH(bu)$ — an exercise with the Adams spectral sequence.
N. Minami	The stable splitting of $BSL_3(Z)$.
J. Morava	The most recent bee in Ed Witten's bonnet.
F. Morel	The representability of mod p homology after one suspension.
E. Ossa	Vector bundles over loop spaces of spheres.
M.M. Postnikov	Simplicial sets with internal symmetries.
H. Sadofsky	The Mahowald invariant and periodicity.
R. Schwänzl	Hermitian K-theory of A_∞-rings.
K. Shimomura	On a spectrum whose BP_*-homology is $(BP_*/I_n)[t_1, \ldots, t_k]$.
V. P. Snaith	Adams operations and the determinantal congruence conjecture of M.J. Taylor.
M. C. Tangora	The theorems of Poisson, Euler and Bernoulli on the Adams spectral sequence.
C. B. Thomas	Characteristic classes of modular representations.
R. M. W. Wood	The boundedness conjecture for the action of the Steenrod algebra on polynomials.

Programme of Posters

D. Arlettaz: The Hurewicz homomorphism in dimension 2.

M. Beattie: Proper suspension and stable proper homotopy groups.

T. Bisson: Covering spaces as geometric models of cohomology operations.

D. Blanc Operations on resolutions and the reverse Adams spectral sequence.

J. M. Boardman: Group cohomology and gene splitting.

P. Booth: Cancellation and non-cancellation amongst products of spherical fibrations.

C. Casacuberta and M. Pfenniger: On orthogonal pairs in categories and localization.

S. Edwards: Complex manifolds with c_1 non-generating.

V. Franjou: A short proof of the $\mathcal{U}$-injectivity of H^*RP^∞.

V. G. Gorbunov: Symplectic bordism of projective spaces and its application.

T. Hunter: On Steenrod algebra module maps between summands in $H^*((Z/2)^s; F_2)$.

K. Ishiguro Classifying spaces of compact simple Lie groups and p-tori.

N. Iwase: Generalized Whitehead spaces with few cells.

M. Kameko: Generators of $H^*(RP^\infty \times RP^\infty \times RP^\infty)$.

S. Kochman: Lambda algebras for generalized Adams spectral sequences.

I. Leary and N. Yagita: p-group counterexamples to Atiyah's conjecture on filtration of $R_C(G)$.

A. T. Lundell Concise tables of homotopy of classical Lie groups and homogeneous spaces.

G. Moreno: Lower bounds for the Hurewicz map and the Hirzebruch Riemann-Roch formula.

R. Nadiradze Adams spectral sequence and elliptic cohomology.

N. Oda: Localisation of the homotopy set of the axes of pairings.

A. A. Ranicki: Algebraic L-theory assembly.

N. Ray: Tutte algebras of graphs and formal groups.

J. Rutter: The group of homotopy self-equivalence classes of non-simply connected spaces: a method for calculation.

C. R. Stover On the structure of $[\Sigma\Omega\Sigma X, Y]$, described independently of choice of splitting $\Sigma\Omega\Sigma X \longrightarrow \bigvee_{n=1}^{\infty} \Sigma X^{(n)}$.

P. Symonds: A splitting principle for group representations.

Z. Wojtkowiak: On 'admissible' maps and their applications.

K. Xu: Representing self maps.

Participants in the Symposium

Jaume Aguadé *(Barcelona)*
Sadoon Al-Musawe *(Birmingham)*
Dominique Arlettaz *(Lausanne)*
Peter Armstrong *(Edinburgh)*
Tony Bahri *(Rider Coll, New Jersey)*
Andrew Baker *(Manchester)*
Michael Barratt *(Northwestern)*
Malcolm Beattie *(Oxford)*
Martin Bendersky *(CUNY)*
Terence Bisson *(Buffalo)*
David Blanc *(Northwestern)*
Michael Boardman *(Johns Hopkins)*
C.-F. Bodigheimer *(Göttingen)*
Imre Bokor *(Zurich)*
Peter Booth *(Newfoundland)*
Boris Botvinnik *(Khabarovsk)*
Pete Bousfield *(UIC)*
Ronnie Brown *(Bangor)*
Shaun Bullett *(QMWC, London)*
Mike Butler *(Manchester)*
David Carlisle *(Manchester)*
Gunnar Carlsson *(Princeton)*
Carles Casacuberta *(Barcelona)*
Francis Clarke *(Swansea)*
Fred Cohen *(Rochester)*
Michael Crabb *(Aberdeen)*
Don Davis *(Lehigh)*
Ethan Devinatz *(Chicago)*
Albrecht Dold *(Heidelberg)*
Emmanuel Dror-Farjoun *(Jerusalem)*
Bill Dwyer *(Notre Dame)*
Peter Eccles *(Manchester)*
Steven Edwards *(Indiana)*
Michael Eggar *(Edinburgh)*
Sam Evens *(Rutgers)*
Vincent Franjou *(Paris)*
Paul Goerss *(Washington)*
Marek Golasinski *(Toru, Poland)*
Vassily Gorbunov *(Novosibirsk)*
Brayton Gray *(UIC)*
David Green *(Cambridge)*
John Greenlees *(Sheffield)*
J. Gunawardena *(Hewlett-Packard)*
Derek Hacon *(Rio de Janeiro)*
Keith Hardie *(Cape Town)*
John Harris *(Toronto)*
Adam Harrison *(Manchester)*
Philip Heath *(Newfoundland)*
Hans-Werner Henn *(Heidelberg)*
Matthias Hennes *(Bonn)*
Kathryn Hess *(Stockholm)*
Peter Hilton *(SUNY, Binghamton)*
Peter Hoffman *(Waterloo, Canada)*
Mike Hopkins *(MIT)*
John Hubbuck *(Aberdeen)*
Reinhold Hübl *(Regensburg)*
Tom Hunter *(Kentucky)*
John Hunton *(Manchester)*
Steve Hutt *(Edinburgh)*
Kenshi Ishiguro *(Purdue)*
Norio Iwase *(Okayama)*
Stefan Jackowski *(Warsaw)*
Jan Jaworowski *(Indiana)*
Alain Jeanneret *(Neuchâtel)*
David Johnson *(Kentucky)*
Keith Johnson *(Halifax, Nova Scotia)*
John Jones *(Warwick)*
Masaki Kameko *(Johns Hopkins)*
Klaus Heiner Kamps *(Hagen)*
Nondas Kechagias *(Queens, Ont)*
John Klippenstein *(Vancouver)*
Karlheinz Knapp *(Wuppertal)*
Stan Kochman *(York U, Ont)*
Akira Kono *(Kyoto)*

Piotr Krason *(Virginia)*
Nick Kuhn *(Virginia)*
Kee Yuen Lam *(Vancouver)*
Peter Landweber *(Rutgers)*
Jean Lannes *(Paris)*
Ian Leary *(Cambridge)*
Kathryn Lesh *(Brandeis)*
Al Lundell *(Boulder, Col)*
Maria Luisa Sa Magalheas *(Porto)*
Zafer Mahmud *(Kuwait)*
Mark Mahowald *(Northwestern)*
Howard Marcum *(Ohio)*
John Martino *(Yale)*
Tadeusz Marx *(Warsaw)*
Yoshihoru Mataga *(UMDS, Japan)*
Honoré Mavinga *(Wisconsin)*
Peter May *(Chicago)*
John McCleary *(Vassar Coll)*
Jim McClure *(Kentucky)*
Chuck McGibbon *(Wayne State)*
Haynes Miller *(MIT)*
Mamoru Mimura *(Okayama)*
Norihiko Minami *(MSRI)*
Bill Mitchell *(Manchester)*
Steve Mitchell *(Washington)*
Jack Morava *(Johns Hopkins)*
Fabien Morel *(Paris)*
Guillermo Moreno *(Mexico)*
Fix Mothebe *(Manchester)*
Roin Nadiradze *(Tbilisi)*
Goro Nishida *(Kyoto)*
Nobuyuki Oda *(Fukuoka)*
Bob Oliver *(Aarhus)*
Erich Ossa *(Wuppertal)*
Akimou Osse *(Neuchâtel)*
John Palmieri *(MIT)*
Markus Pfenniger *(Zurich)*
Mikhail Postnikov *(Moscow)*
Stewart Priddy *(Northwestern)*
Andrew Ranicki *(Edinburgh)*
Douglas Ravenel *(Rochester)*
Nige Ray *(Manchester)*
Alan Robinson *(Warwick)*
Yuli Rudjak *(Moscow)*
John Rutter *(Liverpool)*
Hal Sadofsky *(MIT)*
Brian Sanderson *(Warwick)*
Pepe Sanjurjo *(Madrid)*
Bill Schmitt *(Memphis, Tenn)*
Roland Schwanzl *(Osnabruck)*
Lionel Schwartz *(Paris)*
Graeme Segal *(Oxford)*
Paul Shick *(John Carroll Univ)*
Don Shimamoto *(Swarthmore Coll)*
Katsumi Shimomura *(Tottori)*
Hubert Shutrick *(Karlstad, Sweden)*
Raphael Sivera *(Valencia)*
Vic Snaith *(McMaster, Ont)*
Richard Steiner *(Glasgow)*
Christopher Stover *(Chicago)*
Chris Stretch *(Ulster)*
Neil Strickland *(Manchester)*
Michael Sunderland *(Oxford)*
Wilson Sutherland *(Oxford)*
Peter Symonds *(McMaster)*
Martin Tangora *(UIC)*
Charles Thomas *(Cambridge)*
Rob Thompson *(Northwestern)*
Japie Vermeulen *(Cape Town)*
Vladimir Vershinin *(Novosibirsk)*
Rainer Vogt *(Osnabruck)*
Grant Walker *(Manchester)*
Andrsej Weber *(Warsaw)*
Clarence Wilkerson *(Purdue)*
Steve Wilson *(Johns Hopkins)*
Zdzislaw Wojtkowiak *(Bonn)*
Reg Wood *(Manchester)*
Lyndon Woodward *(Durham)*
Xu Kai *(Aberdeen)*
Nobuaki Yagita *(Tokyo)*

Addresses of Contributors

J P May

Department of Mathematics
University of Chicago
5734 University Avenue
Chicago, Illinois 60637, U.S.A.

Kathryn P Hess

Département de Mathématiques
Ecole Polytechnique Fédérale de Lausanne
CH-1015 Lausanne, Switzerland

J D S Jones

Mathematics Institute
University of Warwick
Coventry CV4 7AL, U.K.

J McCleary

Department of Mathematics
Vassar College
Poughkeepsie, New York 12601, U.S.A.

Z Fiedorowicz

Department of Mathematics
Ohio State University
231 West 18th Avenue
Columbus, Ohio 43210-1174, U.S.A.

R. Schwanzl

Fachbereich Mathematik/Informatik
Universität Osnabrück
45 Osnabrück, Postfach 4469
Germany.

R. Vogt

Fachbereich Mathematik/Informatik
Universität Osnabrück
45 Osnabrück, Postfach 4469
Germany.

Dominique Arlettaz	Institut de Mathematiques Université de Lausanne CH-1015 Lausanne, Switzerland.
K. Y. Lam	Department of Mathematics University of British Columbia Vancouver, B.C. V6T 1Y4, Canada
D. Randall	Department of Mathematics Loyola University New Orleans, Louisiana 70018, U.S.A.
Mamoru Mimura	Department of Mathematics Faculty of Science, Okayama University Okayama 700, Japan
C. A. McGibbon	Department of Mathematics Wayne State University Detroit, Michigan 48202, U.S.A.
J. M. Møller	Mathematisk Institut Universitetsparken 5 DK-2100 København Ø, Denmark
David Blanc	Department of Mathematics The Hebrew University Givat Ram Campus 91 000 Jerusalem, Israel
Christopher Stover	Department of Mathematics University of Chicago 5734 University Avenue Chicago, Illinois 60637, U.S.A.

Nobuyuki Oda
Department of Applied Mathematics
Faculty of Science
Fukuoka University 8.19.1
Nanakuma Jonan-ku
Fukuoka 814-01, Japan

I. M. James
Mathematical Institute
24-29 St Giles
Oxford OX1 3LB.

Ronald Brown
School of Mathematics
University of Wales
Dean Street, Bangor LL57 1UT, U.K.

Carles Casacuberta
SFB 170, Mathematisches Institut
Universität Göttingen
3400 Göttingen, Germany

Georg Peschke
Department of Mathematics
University of Alberta
Edmonton T6G 2G1, Canada

Markus Pfenniger
School of Mathematics
University of Wales
Dean Street, Bangor LL57 1UT, U.K.

Peter Hilton
Department of Mathematical Sciences
SUNY at Binghamton
Binghamton, New York 13901, U.S.A.

Victor P. Snaith
Department of Mathematics
McMaster University
Hamilton, Ontario L8S 4K1, Canada

M. C. Crabb
Department of Mathematics
University of Aberdeen
The Edward Wright Building
Dunbar Street, Aberdeen AB9 2TY.

J. R. Hubbuck
Department of Mathematics
University of Aberdeen
The Edward Wright Building
Dunbar Street, Aberdeen AB9 2TY.

Kai Xu
present address unknown
Please send c/o J. R. Hubbuck
at University of Aberdeen, see above

Zdzisław Wojtkowiak
Département de Mathématiques
Université de Nice
Parc Valrose, F-06034 Nice, France

Kenshi Ishiguro
SFB 170, Mathematisches Institut
Universität Göttingen
3400 Göttingen, Germany

John Martino
Department of Mathematics
University of Virginia
Charlottesville, Virginia 22903, U.S.A.

Stewart Priddy
Department of Mathematics
Northwestern University
Evanston, Illinois 60208, U.S.A.

Douglas C. Ravenel
Department of Mathematics
University of Rochester
Rochester, New York 14627, U.S.A.

A. K. Bousfield — Department of Mathematics
University of Illinois at Chicago
Chicago, Illinois 60680, U.S.A.

John R. Hunton — DPMMS
University of Cambridge
16 Mill Lane, Cambridge CB2 1SB

N. P. Strickland — Department of Mathematics
University of Manchester
Manchester M13 9PL, U.K.

Donald M. Davis — Department of Mathematics
Lehigh University
Bethlehem, Pennsylvania 18015, U.S.A.

Mark Mahowald — Department of Mathematics
Northwestern University
Evanston, Illinois 60208, U.S.A.

Martin Bendersky — Department of Mathematics
CUNY, Hunter College
New York 10021, U.S.A.

Dianne Barnes — Department of Mathematics
Northwestern University
Evanston, Illinois 60208, U.S.A.

David Poduska — Department of Mathematics
Case Western Reserve University
Cleveland
Ohio 44106, U.S.A.

Paul Shick
Department of Mathematics
John Carroll University
Cleveland, Ohio 44118, U.S.A.

Goro Nishida
Research Institute for Mathematical Sciences
Kyoto University
Kitashirakawa Sakyo-ku, Kyoto 606, Japan

Francis Clarke
Department of Mathematics
University College of Swansea
Singleton Park, Swansea SA2 8PP, U.K.

Keith Johnson
Department of Mathematics
Dalhousie University
Halifax, Nova Scotia B3H 3J5, Canada

J.P.C. Greenlees
Department of Mathematics
University of Sheffield
Hicks Building, Hounsfield Road
Sheffield S3 7RH

Martin C. Tangora
Department of Mathematics
University of Illinois at Chicago
Chicago, Illinois 60680, U.S.A.

Alain Jeanneret
Institut de Mathématique
Université de Neuchatel
Chantemerle 20
Neuchatel, CH-2000, Switzerland.

D. P. Carlisle
Department of Computer Science
University of Manchester
Manchester M13 9PL, U.K.

R. M. W. Wood
Department of Mathematics
University of Manchester
Manchester M13 9PL, U.K.

Mohamed Ali Alghamdi,
Department of Mathematics
Faculty of Science
King Abdulaziz University
PO Box 9028, Jeddah 21413, Saudi Arabia

Nicholas J. Kuhn
Department of Mathematics
University of Virginia
Charlottesville, Virginia 22903, U.S.A.

Andrew Baker
Department of Mathematics
University of Glasgow
University Gardens, Glasgow G12 8QW, U.K.

V. G. Gorbunov
Department of Mathematics
University of Manchester
Manchester M13 9PL, U.K.

V. V. Vershinin
Institute of Mathematics
Universitetskii Pr. 4
Novosibirsk, USSR 630090

Jack Morava
Department of Mathematics
Johns Hopkins University
Baltimore, Maryland 21218, U.S.A.

The Work of J. F. Adams[1]

I first met Frank here in Manchester in 1964, when this building was being planned. I remember from the first feeling that he was a far more impressive man than the anecdotes of his exploits had led me to expect, and a far nicer one. I also felt humbled by the sheer amount of mathematics that he knew and perhaps more so by the amount that he somehow assumed I knew. I feel a little the same way now, faced with this audience and this topic. Still, I don't want to spend much time in reminiscence.[2] I want rather to give a quick guided tour through Frank's work, largely letting it speak for itself.

I should say that Frank's collected works are to be published in the near future by the Cambridge University Press. Like this talk, the collected works are organized by subject matter rather than by strict chronology. However, I will begin not quite at the beginning of his work with a sequence of four papers submitted between 1955 and 1958. All dates cited are dates of submission, not necessarily of appearance.

A. The cobar construction, the Adams spectral sequence, higher order cohomology operations, and the Hopf invariant one problem

1. *On the chain algebra of a loop space* (1955, with Peter Hilton) [**5**][3]
2. *On the cobar construction* (1956) [**6**]

Let K be a CW-complex with trivial 1-skeleton. In the first paper, a DGA-algebra $A(K)$ is constructed whose homology is the Pontryagin algebra $H_*(\Omega K)$; as an algebra, $A(K)$ is free on generators in bijective correspondence with the cells of K (other than the vertex). As Kathryn Hess explained in her talk a few hours ago, this Adams-Hilton model is small enough to be of concrete value for computations and is still being used and studied today. In the second paper, a larger, but functorial, DGA is given whose homology is $H_*(\Omega K)$, namely the cobar construction $F(C_*(K))$. This construction was discussed in John McCleary's talk on Hochschild homology. Nowadays, an obvious and trivial next step after the introduction of

[1] Reconstruction and expansion of the talk given at the conference, most of which was not written out beforehand

[2] A more personal tribute has been published in The Mathematical Intelligencer, Vol. 12, No. 1, 1990, 40–48.

[3] Details of publication of Adams' works discussed here can be found in the complete bibliography which follows this paper.

the cobar construction would be to filter it and so arrive at what is called the Eilenberg-Moore spectral sequence for the computation of $H_*(\Omega K)$. In fact, Moore and Adams were already in contact before this paper was written, and it was cited by Eilenberg and Moore as an important precursor to their work.

3. *On the structure and applications of the Steenrod algebra* (1957) **[9]**

Adams viewed this paper as a step towards the solution of the Hopf invariant one problem. The main theorem states that if $\pi_{2n-1}(S^n)$ and $\pi_{4n-1}(S^{2n})$ both contain elements of Hopf invariant one, then $n \leq 4$. It is now chiefly celebrated for the introduction of the Adams spectral sequence converging from $\mathrm{Ext}_A^{s,t}(H^*(X), Z_p)$ to ${}_p\pi_*^s(X)$. Products are defined in the spectral sequence when $X = S^0$, and the sub-Hopf algebras A_r are used to compute products of the elements h_i inductively, where h_i corresponds to Sq^{2^i}. The basic argument runs as follows. Let $n = 2^m$, $m \geq 3$. Assuming that h_m is a permanent cycle, $h_0(h_m)^2$ would survive to E_∞ if $d_2 h_{m+1} = 0$. This would contradict the fact that, in $\pi_*^s(S^0)$, $2x^2 = 0$ if $\deg(x)$ is odd. This seems straightforward enough today, but it was revolutionary at the time. The idea of reducing such a fundamental topological problem as Hopf invariant one to the non-triviality of a particular differential in a spectral sequence was quite new and unexpected.

Adams was curiously modest about the Adams spectral sequence. He always referred to it as a formalization of the Cartan-Serre method of killing homotopy groups. I think we all see it as something very much more than that. Its introduction was a watershed, and it substantially raised the level of algebraic sophistication of our subject.

4. *On the non-existence of elements of Hopf invariant one* (1958) **[14]**

If $\pi_{2n-1}(S^n)$ contains an element of Hopf invariant one, then $n = 1, 2, 4,$ or 8. The proof is based on showing that Sq^{2^m} decomposes in terms of secondary cohomology operations if $m > 3$. The paper contains definitive homological algebra for the study of Ext_A, including minimal resolutions and the cobar construction with its $\smile$ and $\smile_1$ products. It uses Milnor's description of A^* to redo the calculations in the previous paper. It gives a detailed study of stable secondary cohomology operations via universal examples, which are generalized two-stage Postnikov systems. The results include axioms for the operations, existence and uniqueness theorems, the relationship between the operations and Tor_A^2, and a Cartan formula. Particular operations are studied via homological algebra, and a key computation in CP^∞ is used to start the induction which shows that the undetermined constants in the decompositions of the Sq^{2^m} are non-zero.

Adams was a problem solver. He introduced exactly the tools he needed to solve the problems he studied, and he had relatively little interest in Bourbaki style analysis of the foundations or in systematic calculations. He had an extraordinary talent for proving important and easily formulated conceptual theorems through a mix of new ideas, new foundational constructions, and adroit calculations. The solution of the Hopf invariant one problem was the first of many such successes.

B. Applications of K-theory

1. *Vector fields on spheres* (1961) **[23]**

Having so spectacularly solved the Hopf invariant one problem, Adams turned next to the vector fields problem. It was natural for him to try cohomology operations here too. A 1960 note **[20]** gave a partial result, and he was still working in cohomology in July, 1961, when he gave a series of lectures in Berkeley. When the solution came, however, it used K-theory and Adams wrote of his cohomological efforts: "The author's work on this topic may be left in decent obscurity, like the bottom nine-tenths of an iceberg." Write $n = (2a+1)2^b$ and $b = c + 4d$ and let $\rho(n) = 2^c + 8d$. Hurwicz-Radon and Eckman had shown that there exist $\rho(n) - 1$ linearly independent vector fields on S^{n-1}. Adams proved that there do not exist $\rho(n)$ such fields. It suffices to show that the truncated projective space $\mathbf{R}P^{m+\rho(m)}/\mathbf{R}P^{m-1}$ is not coreducible (the bottom cell is not a retract up to homotopy) for any m. He introduced what are now called the Adams operation ψ^k into real and complex K-theory, he calculated the K-theory of truncated projective spaces, with their Adams operations, and he showed that there is no splitting of their real K-theory which is compatible with the operations. All of Adams' papers are well written, but the exposition in this classic paper is especially lovely.

For background, James, in part, and Atiyah had shown that the bundle $O(n)/O(n-k) \rightarrow S^{n-1}$ admits a cross-section if and only if n is a multiple of the order of the image of the canonical line bundle in $\widetilde{J}(\mathbf{R}P^{k-1})$, and analogously in the complex and quaternionic cases. Curiously, it was left to Atiyah and Bott to observe that Adams' calculations actually imply that $\widetilde{KO}(\mathbf{R}P^k) \cong \widetilde{J}(\mathbf{R}P^k)$. This group is cyclic of order $2^{\varphi(k)}$, where $\varphi(k)$ is the number of j such that $0 < j \leq k$ and $j \equiv 0, 1, 2$, or $4 \bmod 8$. Mark Mahowald discussed the significance of this calculation in his talk.

2. *On complex Stiefel manifolds* (1964, with Grant Walker) **[29]**

It is shown that $U(n)/U(n-k) \rightarrow S^{2n-1}$ admits a cross-section if and only if M_k divides n; here $\nu_p(M_k) = \sup\{r+\nu_p(r)\,|\,1 \le r \le (k-1)/(p-1)\}$ if $p \le k$ and $\nu_p(M_k) = 0$ if $p > k$. Atiyah and Todd had shown that the condition is necessary, and they had conjectured that it is sufficient. As already noted, Atiyah had reduced the problem to a calculation in $\tilde{J}(CP^{k-1})$, and this paper analyzes $\tilde{J}(CP^n)$ by the methods of $J(X)$-I,II. It gives a worked example of the general study in those papers.

3. *On the groups* $J(X)$*-I* (1963), *II* (1963), *III* (1963), *IV* (1965) ([**25**], [**28**], [**31**], [**35**])

The program in this fundamentally important cycle of papers is to give effective means for computing the group $J(X) = \tilde{J}(X) \oplus Z$ of fiber homotopy equivalence classes of stable vector bundles over a finite CW-complex X. The basic idea is to give computable upper and lower bounds $J''(X)$ and $J'(X)$ for $J(X)$ and to show that the two bounds coincide. Thus $J(X)$ would be captured in the diagram of epimorphisms

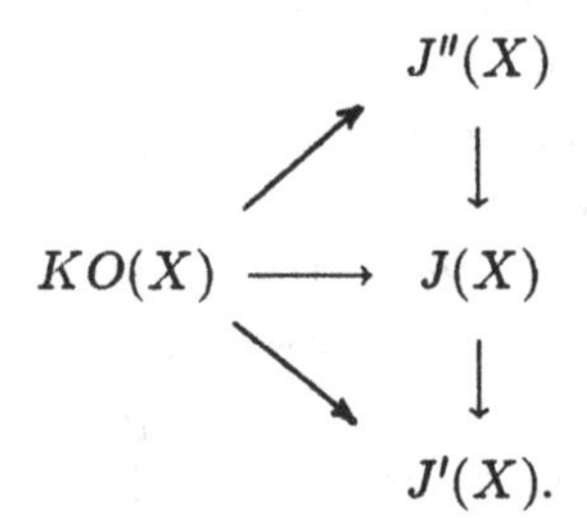

That $J''(X)$ really is an upper bound depends on the celebrated Adams conjecture: "If k is an integer, X is a finite CW-complex and $y \in KO(X)$, then there exists a non-negative integer $e = e(k,y)$ such that $k^e(\psi^k - 1)y$ maps to zero in $J(X)$."

As Michael Crabb explained in his talk, it is now possible to give a fairly elementary proof of the Adams conjecture. It is fortunate that such an argument was not discovered early on. The proofs of the Adams conjecture by Sullivan and Quillen led to a veritable cornucopia of new mathematics, including localizations and completions of spaces and the higher algebraic K-groups of rings.

J(X)-I. The Adams conjecture is proven if y is a linear combination of $O(1)$ and $O(2)$ bundles or if $X = S^{2n}$ and y is a complex bundle. The proof is based on the Dold theorem mod k: if there is a fiberwise map $E_\xi \to E_\eta$ of degree $\pm k$ on each fiber, then $k^e\xi$ and $k^e\eta$ are fiber homotopy equivalent for some $e > 0$.

J(X)-II. The group $J''(X)$ is specified as $KO(X)/W(X)$, where $W(X)$ is the subgroup generated by all elements $k^{e(k)}(\psi^k - 1)y$ for a suitable function e (independent of y). The cannibalistic classes ρ^k are defined by the formula $\rho^k(\xi) = \varphi^{-1}\psi^k\varphi(1)$ on Spin(8n)-bundles ξ, and it is shown that they can be defined more generally after localization. If ξ and η are fiber homotopy equivalent, then $\rho^k(\xi) = \rho^k(\eta)[\psi^k(1+y)/(1+y)]$ for some $y \in \widetilde{K}(X)$ (independent of k). $J'(X)$ is specified as $KO(X)/V(X)$, where $V(X)$ is the subgroup of those x such that $\rho^k(x) = \psi^k(1+y)/(1+y)$ in $KO(X) \otimes Z[1/k]$ for all $k \neq 0$ and some $y \in \widetilde{K}O(X)$. Explicit computations give the groups $KO(\mathbf{R}P^n) = J''(\mathbf{R}P^n) = J'(\mathbf{R}P^n)$ and $J''(S^n) = J'(S^n)$. The latter calculations imply that $J(\pi_{8n+i}(SO)) = Z_2$ if $i = 0$ or 1 and that $J(\pi_{4n-1}(SO))$ is cyclic of order $m(2n)$, where $m(2n)$ is the denominator of $B_n/4n$, although Adams was left with an ambiguity when n is even because he only had the complex and not the real Adams conjecture for bundles over spheres. Of course, these basic calculations are essential to the understanding of the stable homotopy groups of spheres.

J(X)-III. The main theorem of the series is proven: $J'(X) = J''(X)$. This is based on the fundamental commutative diagram

$$\begin{array}{ccc}
\sum_k \widetilde{K}SO(X) & \xrightarrow{\sum k^{e(k)}(\psi^k - 1)} & \widetilde{K}SO(X) \\
\downarrow{\scriptstyle \sum \vartheta^k} & & \downarrow{\scriptstyle \prod \rho^\ell} \\
1 + \widetilde{K}SO(X) & \xrightarrow{\prod \psi^\ell/1} & \prod_\ell 1 + \widetilde{K}SO(X) \otimes Z[1/\ell]
\end{array}$$

The diagram is obtained by summing individual diagrams for pairs (k, ℓ), and the ϑ^k are constructed in the course of the character theoretic proof. The main theorem follows from the fact that this diagram is a weak pullback. The paper also explains and exploits the modular periodicity of the Adams operations.

The paper has a tantalizing last section. It asks for a theory Sph(X) of stable spherical fibrations in which $J(X)$ is a direct summand mapped to by $KO(X)$; Sph(X) should be represented by $BF \times Z$, where F is the monoid of homotopy equivalences of spheres. It also asks for a theory Sph$(X; kO)$ of kO-oriented stable spherical fibrations and gives a number of probable consequences. With characteristic honesty, Adams wrote of this discussion "I will not call the results "theorems", since the underlying assumptions have not been stated precisely enough." This section makes vividly clear just how prescient this whole series of papers was. Many relevant and now standard tools were unavailable to Adams, but he foresaw much that would later be formulated and proven with them.

For example, an alternative version of the diagram above can be constructed conceptually by exploiting localized classifying spaces rather than representation theory. At $p = 2$, the relevant diagram is:

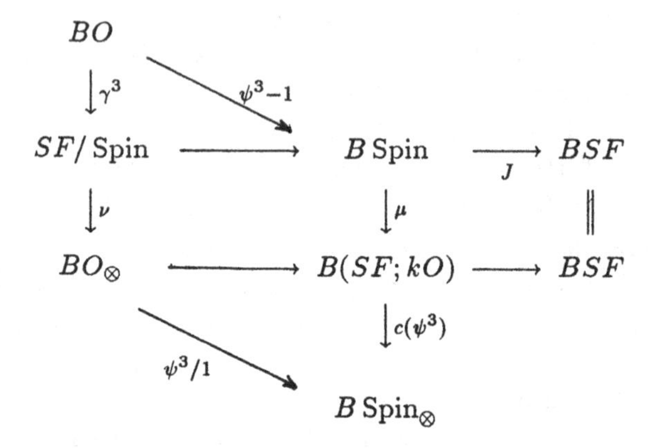

Here $B(SF; kO)$ classifies $\mathrm{Sph}(F; kO)$, $c(\psi^3)$ is the universal cannibalistic class determined by ψ^3, and μ is given by the Atiyah-Bott-Shapiro orientation. The rows are fibration sequences, so μ determines ν and the Adams conjecture determines γ^3. The composite $c(\psi^3)\circ\mu$ is ρ^3, the composite $\nu\circ\gamma^3$ can be taken as ϑ^3, and these two maps are 2-local equivalences.[4] I discussed this approach with Frank, who had envisioned something of the sort. He very much liked it, but he rightly emphasized that you can't proceed this way before you have the Adams conjecture.

J(X)-IV. The results of I–III are applied to computations in the stable homotopy groups of spheres. The starting point is an abstract analysis of the "d and e invariants" of a half exact functor k from the homotopy category to an abelian category $\mathcal{A}$. For $f: X \to Y$, $d(f) = f^* \in \mathrm{Hom}(k(Y), k(X))$. If $d(f) = 0$ and $d(\Sigma f) = 0$, then $e(f)$ is the class of

$$0 \to k(\Sigma X) \to k(Cf) \to k(Y) \to 0$$

in $\mathrm{Ext}^1(k(Y), k(\Sigma X))$. In the applications, $\mathcal{A}$ consists of finitely generated Abelian groups with Adams operations, k is taken to be $\widetilde{K}$ or $\widetilde{KO}$, and X and Y are taken to be spheres or Moore spaces, for which the Hom and Ext target groups are readily computed.

[4]For details, see Chapter V of [J. P. May (with contributions by Nigel Ray, Frank Quinn, and J. Tornehave) E_∞ ring spaces and E_∞ ring spectra. Springer Lecture Notes in Mathematics Vol 577, 1977].

Calculations of these invariants on the groups π_r^s for $r > 1$ are related to $J : \pi_r(SO) \to \pi_r^s$. Here $d = 0$ except in the real case with $r \equiv 1$ or 2 mod 8, when d detects a direct summand $Z_2 \not\subset \mathbf{Im}\, J$ generated by elements μ_r. For any r, the real e invariant detects $\mathbf{Im}\, J$ as a direct summand, the case $r \equiv 7$ mod 8 being incomplete in the paper since the full Adams conjecture was not yet available. Various composition products and Toda brackets are detected by means of Yoneda and Massey products in the target groups. This seems a little like a magical boot strap operation since the Hom and Ext calculations involved are fairly elementary. The conception is a marvelous example of algebraic modeling of topological phenomena.

The complex e-invariant is determined by the Chern character, and, via Adams' paper on the Chern character, to be discussed shortly, this leads to a proof by K-theory of the Hopf invariant one problem for any prime p. Finally, the e-invariant is used to prove that if Y is the mod p^f Moore space for an odd prime p (with bottom cell in a suitable odd dimension) and if $r = 2(p-1)p^{f-1}$, then there is a map $A : \Sigma^r Y \to Y$ which induces an isomorphism on $\widetilde{K}$, so that all of the iterates of A are essential. You have heard about these vitally important periodicity maps in several talks, for example those of Katsumi Shimomura, Doug Ravenel, and especially that of Pete Bousfield.

4. *K-theory and the Hopf Invariant* (1964, with Michael Atiyah) **[34]**

This paper gives the beautiful and definitive K-theoretic proof of the Hopf invariant one result for all primes p. For $p = 2$, it is based on the relation $\psi^2\psi^3 = \psi^3\psi^2$ applied in the obvious 2-cell complex. Alain Jeanneret showed us that this trick can still be used to good effect to obtain new results.

5. *Geometric dimension of bundles over* $\mathbf{R}P^n$ (1974) **[51]**

Since Kee Y. Lam's talk gave a rather complete summary of the results in this nice paper, I will not discuss its main thrust. However, in view of the current interest in periodicity maps, I want to mention an addendum that it gives to the discussion of Moore spaces in J(X)-IV: for $n \geq 5$, there is a map $\Sigma^8 Y_n \to Y_n$ which induces an isomorphism on $\widetilde{K}Sp$, where Y_n is the mod 2 Moore space with bottom cell in dimension n.

C. Characteristic classes and calculations in K-theory and cobordism

1. *On formulae of Thom and Wu* (1961) **[19]**

In this beautiful early example of modern algebraic modeling, Adams shows that any Poincaré duality algebra over the Steenrod algebra has "Wu classes" and thus Stiefel Whitney classes which satisfy all of the same formulas which relate these classes in the cohomology of differential manifolds. The proof is based on the construction and analysis of a suitable universal left and right A-algebra.

2. *On Chern characters and the structure of the unitary group* (1960) **[18]**

Using Bott periodicity to study the Postnikov system of $BU[2q, \ldots, \infty)$, Adams defines characteristic classes $ch_{q,r}(\xi) \in H^{2q+2r}(X; \mathbf{Z})$ for stable bundles ξ over (2q-1)-connected spaces. If ch_r is the rth component of the Chern character, then $ch_{q,r}(\xi)$ rationalizes to $m(r)ch_{q+r}(\xi)$, and the $ch_{q,r}(\xi)$ relate appropriately to Steenrod operations when reduced mod p. Yuli Rudjak noted that some of the ideas Adams introduced here are relevant to the study of the orientability of various kinds of bundles.

3. *Chern characters revisited* (1971) **[47]**

This gives a more modern and sophisticated approach to the Chern character. The image of $H_*(bu; \mathbf{Z})$ in $H_*(bu; \mathbf{Q}) = \mathbf{Q}[u]$, $\deg u = 2$, is shown to be the subgroup generated by $\{u^r/m(r) \mid r \geq 0\}$. The elegant one prime at a time proof is based on the simple A-module structure of $H^*(bu; \mathbf{Z}_p)$. Viewing ch_r as a map $bu \to K(\mathbf{Q}, 2r)$, it follows that the image of $m(r)ch_r : bu_n(X) \to H_{n-2r}(X; \mathbf{Q})$ is integral for any X.

4. *The Hurewicz homomorphism for MU and BP* (1970, with Arunas Liulevicius) **[45]**

This paper gives a nice proof via the Adams spectral sequence of an interpretation of the Hattori-Stong theorem: $\pi_*(BP) \to \pi_*(k \wedge BP)$ is a split monomorphism, and similarly for MU, where k represents connective K-theory.

This is one of the very few papers in which Adams allowed his coauthor to do the actual writing. Adams preferred to hold pen in hand himself, although he paid careful attention to the suggestions of his collaborators.

The following five papers can be viewed as a series in which Adams applied to K-theory the algebraic foundations that he established for the calculational study of generalized homology and cohomology theories.

5. *Hopf algebras of cooperations for real and complex K-theory* (1970, with Albert Harris and Robert Switzer) **[42]**

Using the stable Adams operations in localized K-theory to obtain integrality conditions, $K_*(K)$ is computed as a subring of $K_*(K) \otimes \mathbf{Q}$, which is a ring of finite Laurent series on two variables; $KO_*(KO)$ is also determined. Francis Clarke showed us how to relate this to the study of the ring of cooperations in elliptic homology.

6. *Operations of the Nth kind in K-theory* (1972) [**48**]

In a report on work with David Baird, Adams indicates that, for K-theory localized at an odd prime p, $\mathrm{Ext}^{s,t}_{K_*(K)}(\widetilde{K}_*(X), \widetilde{K}_*(Y)) = 0$ for all finite X and Y and all t when $s \geq 3$. He then speculates about stable homotopy theory "seen through the spectacles of K-theory". Pete Bousfield's beautiful talk on the structure of stable homotopy theory localized at K showed us how these speculations have come to fruition.

7. *Operations on K-theory of torsion-free spaces* (1975, with Peter Hoffman) [**54**]

For integers $n < m$, this paper computes the ring of those operations $K(X) \to K(X)$ which are defined and natural for CW-spectra X such that $\pi_r(X) = 0$ for $r < 2n$, $H_r(X)$ is free for all r, and $H_r(X) = 0$ for $r > 2m$.

8. *Stable operations on complex K-theory* (1976, with Francis Clarke) [**59**]

It is observed that although linear combinations of ψ^1 and ψ^{-1} are the only obvious stable operations, $K^0(K)$ is actually uncountable.

9. *Primitive elements in the K-theory of BSU* (1975) [**56**]

The kernel and cokernel of $PK^0(BU; k) \to PK^0(BSU; k)$ are computed for any commutative ring k. In particular, somewhat surprisingly, it is found that $PK^0(BU; \widehat{Z}_p) \to PK^0(BSU; \widehat{Z}_p)$ is an isomorphism.

D. Stable homotopy and generalized homology

Especially during his last few years at Manchester, Adams made frequent extended trips to the United States. His usual destination was Chicago, a place where he always felt very comfortable and at home. Some of his most influential writing is in notes prepared for delivery in lecture series at Chicago (1967, 1970, and 1971) and at a conference in Seattle (1968). According to Nigel Ray, he tried out some of these lectures on people at Manchester.

1967 *S. P. Novikov's work on operations on complex cobordism* * [**49**][5]

Novikov's work in question was only available in Russian at the time, and it was quite difficult reading even for those who knew Russian. Adams' clear exposition allowed the quick assimilation of this material into the main stream of algebraic topology in the West.

1968 *Lectures on generalized homology* (Seattle) [**39**]

1. *The universal coefficient theorem and the Künneth theorem*

[5]The three lecture notes denoted * are in the "Chicago blue book" (titled after the 1971 lectures). The University of Chicago Press will keep it in print, and I would like to be told if anybody has trouble obtaining a copy.

This classic account shows that the four "UCT's" imply the four "KT's" by specialization, that two of the UCT's imply the other two by duality, and that one UCT can be viewed as a special case of the ASS (Adams' preferred abbreviation for the Adams spectral sequence). It gives a treatment of the remaining UCT, by Atiyah's method in K-theory, that still seems to represent the state of the art. The account is applied to the Conner-Floyd theorem and to other relations between K and MU.

2. *The Adams spectral sequence*

This account of the generalized ASS shows much progress beyond earlier tries at generalization. The now generally accepted preference for homology over cohomology is expounded. Convergence is not studied here.

3. *Hopf algebra and comodule structure*

Definitive foundations are given for the algebra used to describe E_2 of the generalized ASS in terms of homology. The material here was taken for granted in quite a few talks at this conference, for example those of Doug Ravenel, Katsumi Shimomura, and Vladimir Vershinin.

4. *Splitting generalized cohomology theories with coefficients*

This gives a splitting of KU and a parallel splitting of MU via idempotents; the former is still the standard reference, but the latter was soon superseded by Quillen's approach via formal group laws.

5. *Finiteness theorems*

A systematic generalized treatment of coherent rings is given. One application gives that, for finite CW-complexes X, $MU^*(X)$ admits a finite MU^*-resolution by finitely generated free modules (as was also shown by Conner and Smith). Another application (due to J. Cohen) shows that a space Y with non-trivial reduced mod p cohomology has infinitely many non p-trivial stable homotopy groups.

1970 *Quillen's work on formal groups and complex cobordism* * [**49**]

Just as the 1967 lectures allowed the rapid assimilation of Novikov's work, so these lectures allowed the rapid assimilation of Quillen's work. I remember these lectures as great fun. The first eight strictly alternated algebra and topology, giving a connected development of the theory of formal groups on the one hand and a clear exposition of their role in topology on the other. The calculations of $MU^*(MU)$ and $BP^*(BP)$ of Novikov and Quillen were reworked as explicit calculations of $MU_*(MU)$ and $BP_*(BP)$. These calculations have been cited in several talks here.

An interesting survey, *Algebraic topology in the last decade* [**43**], based on a lecture given at a conference at the University of Wisconsin, also dates from 1970.

1971 *Stable homotopy and generalized homology* * [**49**]

This classic lecture series is still the best introduction to its topics, although many of us prefer more idealistic approaches to the construction of the stable homotopy category. Its treatment of convergence of the ASS marked a major improvement over earlier work. The important idea of studying the stable homotopy category by means of localizations at spectra was first introduced here.

E. Lectures on Lie groups (1969) [38]

This excellent exposition of the basics of Lie theory, with emphasis on the representation theory of compact Lie groups, is based on lectures given at Manchester.[6] Working tools for many of the papers in the next two groups can be found here.

F. Finite H-spaces and compact Lie groups

1. *The sphere considered as an H-space mod p* (1960) [**16**]

The very first appearance of the localization of a space appeared here, presaging the study of finite H-spaces via localization. Adams gave the slogan "an odd sphere is an H-space mod p" as early as 1956.

2. *H-spaces with few cells* (1960) [**21**]

It is shown that if $H^*(G; Z_2) \cong E(x_q, x_n)$, $q \leq n$, then q and n are both one of 1, 3, 7, or else $(q,n) = (1,2), (3,5), (7,11)$, or $(7,15)$. Adams left open the question of realizability of the last two, but that was settled in the negative in John Hubbuck's thesis.

3. *Finite H-spaces and algebras over the Steenrod algebra* (1978, with Clarence Wilkerson) [**63**]

Explicit necessary and sufficient algebraic conditions are given on an unstable A-algebra P which ensure that $P \cong H^*(BT)^W$ for a torus T and finite group W. These conditions hold if P is a polynomial algebra on generators of even degree prime to p, and W is then a p-adic generalized reflection group; moreover, there is a space X such that $P \cong H^*(X)$. The program carried out in this paper is due to Wilkerson, and it is one that Frank particularly liked. The paper contains lovely algebra. Here is a nice quote; it follows a description of a proof based on appeal to Bezout's theorem: "We hope that the reader finds something appealing in this strategy; we shall

[6] It was reprinted by the University of Chicago Press in 1982. Again, I would like to be told if anybody has trouble obtaining it.

not risk spoiling this impression by giving the details. The proof we shall present is our second."

4. *Finite H-spaces and Lie groups* (1980) [**66**]

This "unsatisfactory report" explains the difficulties involved in trying to obtain Adams-Wilkerson type results in K-theory. The requisite algebraic models would require a filtration; the natural choice is not the obvious choice but the rational filtration, defined in terms of the Chern character, and one must relate the Adams operations to it. This leads to a notion of good integrality which is satisfied in the absence of torsion and in some but not all classical cases with torsion. An amusing "letter from E_8" explains that he needs so much torsion in his integral cohomology in order to ensure the absence of torsion in his K-theory.

This report was given in a conference in honor of Saunders MacLane in Aspen, Colorado. Frank and I went there together, and we had agreed to entertain the category theorists with silly lectures—my own started with a warning that it would contain a deliberate categorical blunder that would destroy the validity of all of the results. We spent the afternoons climbing in the Rockies. Going up from Independence Pass, at 11,000 feet, to the nearest peak, at over 13,000 feet, entailed quite a bit of crawling on hands and knees over snow and ice, but the view was well worth the effort. Frank and I seemed to do some such daft thing at all of the conferences we attended together.

Frank had great affection for the low-dimensional and exceptional Lie groups. Their study was a lifetime hobby, and the next three papers are "Snippets of a book I have in preparation." It is hoped that his unpublished notes will be edited and published in the not too distant future; appropriate people at this conference have agreed to do the work.

5. *Spin(8), triality, F_4 and all that* (1981) [**67**]

This paper gives a nice conceptual explanation of the exceptional symmetry of D_4 = Spin(8) given by the fact that Out(Spin(8)) $\cong \Sigma_3$. The exceptional 27-dimensional Jordan algebra J is constructed as the sum of $\mathbf{R}^3$ and the three irreducible representations of Spin(8), together with a linear, a bilinear, and a trilinear form. Then F_4 is the group of $\mathbf{R}$-linear maps $J \to J$ which preserve all three forms, and Spin(8) is the subgroup which fixes the elements of $\mathbf{R}^3$. The subgroup of F_4 which preserves the preferred basis elements of $\mathbf{R}^3$ maps onto Σ_3 with kernel Spin(8), giving a clear and attractive proof that the outer automorphisms of Spin(8) are specified by conjugation by elements of F_4.

6. *The fundamental representations of E_8* (1985) [**73**]

The object here is to describe explicit polynomial generators for $R(E_8)$. Let α be the adjoint representation. Then $R(E_8)$ is generated by $\lambda^i\alpha$ for $1 \leq i \leq 5$, β, $\lambda^1\beta$, and γ, and explicit concrete descriptions of β and γ are given. In this area, Frank expected a lot from his readers; he thought that everybody should understand Lie groups as well as he did. Here is the argument for one fairly obscure lemma: "Sketch proof. Take the obvious steps yourself—it's quicker than going to the library."

7. *2-tori in E_8* (1986) [**77**]

The phenomena uncovered here are interesting and intricate. Maximal 2-tori in E_8 fall into two conjugacy classes, one of rank 9 and one of rank 8, and explicit constructions are given. Any 2-torus in E_8 is conjugate to one in $Ss^+(16)$, and maximal 2-tori in $Ss^+(16)$ remain maximal in E_8. However, maximal 2-tori in $Ss^+(16)$ fall into four conjugacy classes, two of rank 9 and two of rank 8, while maximal 2-tori of Spin(16) fall into two conjugacy classes, both of rank 9. For n even, the maximal 2-tori in $PO(n)$ are determined. For $n \equiv 0 \mod 8$ and $n > 8$, the maximal 2-tori of $PSO(n)$ and $Ss^+(n)$ are determined .

This kind of explicit information is obviously relevant to Quillen's work on the mod 2 cohomology of compact Lie groups, and it will be necessary to the program Mimura described to us of understanding the mod 2 cohomology of exceptional Lie groups in terms of 2-tori.

G. Maps between classifying spaces of compact Lie groups

1. *Maps between classifying spaces* (1975, with Zafer Mahmud) [**55**]

Stefan Jackowski gave us a summary of this important paper (abbreviated MBCS below) in his talk on the same topic, so I shall say relatively little. Let G and G' be compact connected Lie groups with maximal tori T and T' and Weyl groups W and W'. Let $\vartheta : H^*(BG';\mathbf{Q}) \to H^*(BG;\mathbf{Q})$ be a homomorphism of $\mathbf{Q}$-algebras which "commutes with Steenrod operations for all sufficiently large primes p". There is an "admissible map" $\varphi : \pi_1(T)\otimes \mathbf{Q} \to \pi_1(T') \otimes \mathbf{Q}$ such that the following diagram commutes:

$$\begin{array}{ccc} H^*(BG';\mathbf{Q}) & \xrightarrow{\vartheta} & H^*(BG;\mathbf{Q}) \\ \downarrow & & \downarrow \\ H^*(BT';\mathbf{Q}) & \xrightarrow{\varphi^*} & H^*(BT;\mathbf{Q}) \end{array}$$

Two choices of φ differ by composition by an element of W'. Moreover, any

ϑ is induced by a map f defined after finite localization. Conversely, given an admissible map φ, there is a unique homomorphism ϑ such that the diagram commutes, and ϑ can be induced by a map f defined after finite localization. Thus there is a bijective correspondence between homomorphisms ϑ induced by maps f defined after finite localization and W'-equivalence classes of admissible maps φ. For any $f : BG \to BG'$, $f^* : K(BG') \to K(BG)$ carries $R(G')$ into $R(G)$. There is a detailed Lie theoretic analysis of admissible maps, and there are lots of concrete calculational examples and case by case calculations.

This paper initiated serious work on this topic, long before others were interested. The talks of Jackowski, Stewart Priddy, and others made clear that this is now a thriving area of algebraic topology.

2. *Maps between classifying spaces II* (1978) **[62]**

In this fascinating and relatively neglected paper, G and G' are compact, but not necessarily connected, Lie groups. The group $FF(X)$ of "formally finite elements of $K(X)$" is defined as the group of differences of elements annihilated by all but finitely many λ^i. Clearly $f^* : K(X') \to K(X)$ carries $FF(X')$ into $FF(X)$ for any $f : X \to X'$. Let $\alpha : R(G) \to K(BG)$ be the natural map. With $X = BG$, $\mathbf{Im}(\alpha) \subset FF(BG)$. Adams proves that equality holds if G is finite or if $\pi_0(G)$ is the union of its Sylow subgroups. In these cases, $f^* : K(BG') \to K(BG)$ carries $R(G')$ into $R(G)$ for any map $f : BG \to BG'$, but this is not true in general. However, for any G there is an $n > 0$ such that $nFF(BG) \subset \mathbf{Im}(\alpha)$. If G is monogenic and $x \in FF(BG)$, then $x = \alpha(\rho)$ for an honest representation ρ of G, but there is a finite p-group G and an element $x \in FF(BG)$ such that x is not of the form $\alpha(\rho)$ for any representation ρ. For finite G, the crux is that $x \in \mathbf{Im}(\alpha)$ if the total exterior power $\lambda_t(x)$ is a rational function of t. However, there is a G such that $\pi_0(G)$ is a p-group and an x such that $\lambda_t(x)$ is a rational function of x and yet $nx \notin \mathbf{Im}(\alpha)$ for any $n > 0$. The elements of formal dimension 2 in $K(BSL(2,5))$ are analyzed in detail.

This paper is independent of MBCS, and I think that its ideas can be exploited further. The basic method is to approximate G by its finite subgroups. At the time, this was a novel idea. In particular, finite approximations of toral extensions over finite groups first appear here. There is also an attractive theory of characters $\widehat{\chi}_g$ defined on elements of $K(BG, \widehat{Z}_p)$, and it is observed that $K(BG)$ injects into the product of the $K(BG, \widehat{Z}_p)$. If $\widehat{\chi}_g(x) = 0$ for all g of prime power order, then $x = 0$. These characters take values in $\widehat{Z}_p \otimes \mathbf{C}$, but they lie in $\mathbf{C}$ if $\lambda_t(x)$ is a rational function of

t. Another pleasant result, which should have been known before, is that a continuous class function $G \to \mathbf{C}$ which restricts to a virtual character for every finite subgroup of G is a virtual character.

3. *Maps between classifying spaces III* (1983, with Mahmud) [**71**]

As in MBCS, G and G' are compact connected Lie groups. For any $f : BG \to BG'$, $f^* : R(G') \to R(G)$ carries $RO(G')$ to $RO(G)$ and $RSp(G')$ to $RSp(G)$. There is a canonical, easily determined, "Dynkin element" $\delta \in Z(G)$, defined up to multiplication by squares of elements of order 4 in $Z(G)$, whose behavior on self-conjugate representations distinguishes real from symplectic representations. Under suitable hypotheses, necessary to rule out counterexamples, it is shown that the admissible map associated to f^* in MBCS preserves Dynkin elements (up to indeterminacy).

4. *Maps between p-completed classifying spaces* (1988, with Zdzislaw Wojtkowiak) [**81**]

Working in the context of MBCS, complete all spaces at p. It is observed that a result of Dwyer and Zabrodsky on maps $B\pi \to BG'$ for finite p-groups π implies that, given $f : BG \to BG'$, there is a compatible map $BT \to BT'$, unique up to the action of W'. It follows that all of the results about admissible maps in MBCS carry over to "admissible maps" $\pi_1(T) \otimes \hat{Z}_p \to \pi_1(T') \otimes \hat{Z}_p$, since these maps are obtained by extension of scalars from p-local admissible maps.

H. Modules over the Steenrod algebra and their Ext groups

1. *A finiteness theorem in homological algebra* (1960) [**17**]

This gives a general vanishing result of the form $\mathrm{Ext}_A^{s,t}(k,k) = 0$ in a range $s\varepsilon < t < sq$. When A is the mod p Steenrod algebra, $\varepsilon = 1$ and $q = 2(p-1)$, which is not best possible.

2. *A periodicity theorem in homological algebra* (1961, 1965) [**26**], [**36**]

The 1965 paper gives details of results sketched in Adams' 1961 Berkeley notes. Consider $H^{s,t}(A) = \mathrm{Ext}_A^{s,t}(\mathbf{Z}_2, \mathbf{Z}_2)$, where A is the mod 2 Steenrod algebra. It is shown that $H^{s,t}(A) = 0$ for $0 < s < t < U(s)$, where $U(s)$ is approximately $3s$ and is best possible. It is also shown that, in small regions near the vanishing line, there are periodicity isomorphisms $\pi_r : H^{s,t}(A) \cong H^{s+2^r, t+3\cdot 2^r}(A)$ given by $\pi_r(x) = \langle h_{r+1}, (h_0)^{2^r}, x\rangle$.

From his earliest work in stable homotopy theory, Adams was very concerned with the possibility of discerning systematic periodicity phenomena in stable homotopy groups. Pertaining to the algebraic periodicity just mentioned, he asked in his 1961 Berkeley lecture notes: "What geometric phenomena can one find which show a periodicity and which on passing to algebra give the sort of periodicity encountered in the last lecture?". His persistent interest in this kind of question has perhaps helped shape the present intense focus on periodicity phenomena in stable homotopy theory that was evidenced, for example, in the talks of Mark Mahowald, Jack Morava, Ethan Devinatz, Mike Hopkins, Doug Ravenel, Don Davis, and Martin Bendersky.

3. *Modules over the Steenrod algebra* (1970, with Harvey Margolis) [**41**]

This gives Adams' clarification of interesting results of Margolis. Let B be a sub Hopf algebra of the mod 2 Steenrod algebra. A bounded below B-module M is free if and only if $H(M; P_t^s) = 0$ for all $P_t^s \in B$ with $s < t$. If M is free, then M is injective. If $f : M \to N$ induces an isomorphism on homology for all P_t^s, then M and N are stably equivalent, in the sense that they become isomorphic after adding suitable free modules to each.

4. *Sub-Hopf-algebras of the Steenrod algebra* (1973, with Margolis) [**50**]

All sub Hopf algebras of the mod p Steenrod algebra are constructed.

5. *What we don't know about* $\mathbf{R}P^\infty$ (1972) [**48**]

Consideration of truncated projective spaces led Adams to several conjectures which, in retrospect, can be viewed as versions of the Segal conjecture for the group $\mathbf{Z}_2$. One can form an A-algebra $P = \mathbf{Z}_2[x, x^{-1}]$ which, as Adams observes, is certainly not the cohomology of any spectrum. He conjectures that $\mathrm{Ext}_A(P, \mathbf{Z}_2) \cong \mathrm{Ext}_A(\Sigma^{-1}\mathbf{Z}_2, \mathbf{Z}_2)$, and he explains how this might come about in terms of the sub Hopf algebras A_r.

6. *Calculation of Lin's Ext groups* (1979, with W. H. Lin, Don Davis, and Mark Mahowald) [**64**]

This paper gives Adams' simplification of Davis and Mahowald's simplification of Lin's proof of Adams' Ext conjecture just stated. Exactly as Adams envisioned, the proof is by inductive use of the A_r. This implies the Segal conjecture for $\mathbf{Z}_2$, although the topology is not presented here.

7. *The Segal conjecture for elementary abelian p-groups* (1984, with Jeremy Gunawardena and Haynes Miller) [**76**]

This gives the Ext calculation which was used in the original proof of the Segal conjecture for elementary Abelian p-groups, although, again, the topology is not presented here. Let V be an elementary p-group of rank n. Let $H^*(V)_{\mathrm{loc}}$ be the localization obtained by inverting βx for all non-zero elements $x \in H^1(V)$. The quotient map $H^*(V)_{\mathrm{loc}} \to \mathbf{Z}_p \otimes_A H^*(V)_{\mathrm{loc}}$ induces an isomorphism on $\mathrm{Tor}_A(\mathbf{Z}_p, ?)$ and $\mathbf{Z}_p \otimes_A H^*(V)_{\mathrm{loc}}$ is 0 except in degree $-n$, where it has rank $p^{n(n-1)/2}$ and can be identified with the Steinberg representation of $\mathrm{Aut}(V) = \mathrm{GL}(n, \mathbf{Z}_p)$. For $p = 2$ and $n = 1$, this is the result of the paper above; for $p > 2$ and $n = 1$, it is Gunawardena's thesis. The proof is based on use of the Singer construction. Various other Ext calculations are also given, including as a very special case the isomorphism $\mathbb{Z}_p\{\mathrm{Hom}\,\mathbb{Z}_p(U,V)\} \cong \mathrm{Hom}_A(H^*(V), H^*(U))$ that Nick Kuhn took as the starting point of his talk.

The three authors of this paper originally planned a sequel giving the topology, but it was superseded by a sequel by three other authors that gave an efficient use of their Ext calculations to prove the Segal conjecture for elementary Abelian p-groups, together with a version of Carlsson's reduction of the conjecture for all finite p-groups to this case.[7]

Several people at the conference asked me how this proof of the Segal conjecture for p-groups compares to the beautiful new proof that Jean Lannes presented to us. Elementary Abelian p-groups play no special role in Lannes' direct induction, and this is a line of attack that Adams himself pursued without success. Lannes uses his magic functor T (about which more later) as a substitute for the Singer construction. However, his proof gives the nonequivariant form of the Segal conjecture, and, to generalize from p-groups to all finite groups, one requires the original equivariant form. Both arguments should be of lasting interest .

I. Miscellaneous papers in homotopy and cohomology theory

1 *An example in homotopy theory* (1957) [**7**]

2. *An example in homotopy theory* (1963, with Grant Walker) [**27**]

The first of these notes displays two CW-complexes which are of the same n-type for all n but are not homotopy equivalent. The second displays a phantom map from a CW-complex to a finite-dimensional CW-complex. Each example was the first of its kind.

3. *A variant of E. H. Brown's representability theorem* (1970) [**40**]

A group-valued contravariant functor on the homotopy category of finite CW-complexes is shown to be representable if it satisfies the wedge and Mayer-Vietoris axioms. Brown showed this for functors taking countable sets as values; thus the result eliminates the annoying countability assumption at the price of insisting on group values. A consequence is that any homology theory which satisfies the direct limit axiom is representable.

4. *Idempotent functors in homotopy theory* (1973) [**52**]

The notion of an idempotent functor is advertised in connection with localization and completion theory. Axioms are given on a set S of maps

[7] J. Caruso, J. P. May, and S. B. Priddy. The Segal conjecture for elementary Abelian p-groups II. p-adic completions in equivariant cohomology. Topology 26(1987), 413–433.

in the homotopy category of connected CW-complexes which ensure, by Brown's theorem, that S is the set of equivalences for an idempotent functor. The key axiom, which is hard to verify in practice, is an analog of the solution set condition in the adjoint functor theorem.

In fact, this paper results from a failed swindle presented in 1973 lectures at Chicago. Adams presented the theory without the solution set analog, thereby constructing localizations and completions without doing a shred of work. Pete Bousfield, who is not generally an excitable chap, was sitting behind me, and I vividly remember him tapping me on the shoulder and asking "How does he know it is a set?" However, even Frank's rare mistakes had good effects. Pete was inspired by these lectures to develop the theory of localizations of spaces and spectra at generalized homology theories. As the talks of Devinatz, Hopkins, Bousfield, Greenlees, and others made clear, such localizations are of fundamental importance in homotopy theory.

5. *The Kahn-Priddy theorem* (1972) [**46**]

In this semi-expository paper, it is shown, among other things, that the 2-local stable map $\mathbf{R}P^{\infty} \to S^0$ which induces an epimorphism on homotopy groups is essentially unique, and similarly for odd primes.

6. *Uniqueness of BSO* (1975, with Stewart Priddy) [**58**]

With all spaces and spectra localized or completed at a prime p, if X is a connective spectrum whose zeroth space is equivalent to BSO, then X is equivalent to bso. That is, the space BSO is the zeroth space of only one connective spectrum. The same is true for BSU, and a variant is true for BO and BU. The proof of this remarkable result is well worth summarizing. First, using ideas from Adams' work on the Chern character, it is shown that the Postnikov system argument for the calculation of the mod p cohomology of bso works equally well for X, so that there is an isomorphism of A-modules $H^*(X) \cong H^*(\text{bso})$. Second, the theory of stable equivalence of modules over A_1 (for $p = 2$) and $E[x, y]$ (for $p > 2$) introduced by Adams and Margolis allows sufficient calculation of the groups $\text{Ext}_A(H^*(Y), H^*(X))$ to see that it looks to the eyes of E_∞ of the ASS as if there ought to be an equivalence $X \to \text{bso}$. Finally, an approximation of X in the form $X \simeq W \wedge M$, where W has finite skeleta and M is the Moore spectrum for $Z_{(p)}$ or $\widehat{Z}_{(p)}$, gets around the convergence problem for the ASS. As important motivating examples, it follows that $BSO_{\otimes}$ and, when $p > 2$, F/PL are equivalent to BSO as infinite loop spaces.

Stewart Priddy told us something about his collaboration with Frank on this project. Actually, I proposed the problem to both authors, but my own ideas on the subject led nowhere.

7. *A generalization of the Segal conjecture* (1986, with Jean-Pierre Haeberly, Stefan Jackowski, and J. P. May) [**78**]

For a finite group G, a multiplicative subset S of the Burnside ring $A(G)$, and an ideal I in $A(G)$, the cohomology theory $S^{-1}(\pi_G^*(?))_I^\wedge$ is $\mathcal{H}$-invariant, where $\mathcal{H} = \bigcup\{\mathrm{Supp}(P) \mid P \cap S = \varphi, P \supset I\}$. That is, a G-map $f : X \to Y$ which restricts to an equivalence $f^H : X^X \to Y^H$ for $H \in \mathcal{H}$ induces an isomorphism $S^{-1}(\pi_G^*(f))_I^\wedge$. The proof reduces the general case to Carlsson's equivariant version of the original Segal conjecture, which is the special case in which G is a p-group, S is empty, and $I = (p)$.

The paper developed in a typically competitive manner, with successive generalizations leading to the present elegant, but perhaps hard to assimilate, conclusion; needless to say, the ultimate formulation was due to Adams.

8. *A generalization of the Atiyah-Segal completion theorem* (1986, with Haeberly, Jackowski, and May) [**79**]

For a compact Lie group G, a multiplicative subset S of the representation ring $R(G)$, and an ideal I in $R(G)$, the cohomology theory $S^{-1}(K_G^*(?))_I^\wedge$ is $\mathcal{H}$-invariant, where $\mathcal{H} = \bigcup\{\mathrm{Supp}(P) \mid P \cap S = \varphi, P \supset I\}$, and similarly for real K-theory. The proof is by a surprisingly easy direct reduction to quotation of equivariant Bott periodicity. The case in which S is empty and I is the augmentation ideal is the original Atiyah-Segal completion theorem, which is thus given a new proof. The case $I = 0$ is the Segal localization theorem. With S empty, the theorem is due to the second and third authors; the proof given is a simplification of Jackowski's.

I got to hold pen in hand on this one, Frank on the previous one.

9. *Atomic spaces and spectra* (1988, with Nick Kuhn) [**80**]

This last work is a quintessential example of Adams' characteristic algebraic modeling of topological phenomena. If X is a p-complete space or spectrum of finite type, then $[X, X]$ is a profinite monoid with zero; in the spectrum case, it is a profinite ring. As a matter of algebra, if M is a profinite monoid with zero, then either M contains a non-trivial idempotent or M is"good", in the sense that every element of M is either invertible or topologically nilpotent; if M is a profinite ring and is good, it is local with radical as maximal ideal and $M/\operatorname{rad}(M)$ is a finite field. If X is indecomposable, then $[X, X]$ is good. If $[X, X]$ is good, then X is atomic and prime. There is a p-local spectrum of finite type which is indecomposable and for which $[X, X]$ is not good. Every finite field of characteristic p arises as $[X, X]/\operatorname{rad}[X, X]$ for an indecomposable stable summand X of the classifying space of some finite p-group; every finite field of characteristic 2 so arises from some finite spectrum X. John Hubbuck told us more about this topological realizability of finite fields.

J. Infinite loop spaces(1975 IAS lecture notes) [61]

If you want to get graduate students interested in algebraic topology, this is a very good book to have them read. It contains capsule introductions to a variety of topics, none of which have yet made it to any text book. The exposition is delightful, despite the elephantine jokes. "I am very grateful to J. P. May, B. J. Sanderson and S. B. Priddy for reading the first draft of this book $\cdots$; I have benefited greatly from their comments. It goes without saying that I accept the responsibility for any jokes which remain." It was deliberately thin on details, except for its discussion of the transfer and its quite full treatment of the Madsen, Snaith, Tornehave theorem that passage to maps of zeroth spaces gives an injection $[bso, bso] \to [BSO, BSO]$.

K. Two unpublished expository papers

There are two expository papers of Adams that were distributed in handwritten form but not published during his lifetime; they will appear in his collected works.

1. *Two theorem of J. Lannes* (1986) [**83**]

Lannes introduced a functor T which is characterized by an adjunction $\text{Hom}_{\mathcal{U}}(T(M), P) \cong \text{Hom}_{\mathcal{U}}(M, H^*(B\mathbf{Z}_p) \otimes P)$, where $\mathcal{U}$ is the category of unstable A-modules. He proved that T is exact and that it preserves tensor products. In the case $p = 2$, Adams gives an alternative approach to these two theorems "which aims to make $T(M)$ more accessible to direct calculation". Nick Kuhn told us about some of the useful formulas to be found in this paper.

2. *The work of M. J. Hopkins* (1988) [**84**]

This gives a concise and clear discussion of the theorems of Ethan Devinatz, Mike Hopkins, and Jeff Smith. It explains the three versions of their nilpotence theorem and also their theorem about the existence and essential uniqueness of v_n-self maps, and it raises interesting questions about "generators" and "Euler characteristics" for the classes of finite spectra that arise in the proof of the latter theorem. This capsule summary was the culmination of a series of lectures at Cambridge which gave a rather complete exposition of the nilpotence theorems.

Perhaps the first paragraph of the paper provides a fitting close to this discussion of Frank's work.

"The work I shall report has the following significance. At one time it seemed as if homotopy theory was utterly without system; now it is almost proved that systematic effects predominate."

In fact, this happy state of affairs is a testament to Adams' own contributions, which led the way and paved the ground. The real tribute to Frank at this conference was the mathematical content of the talks. Our subject is thriving, and nowhere more so than in the directions that he himself pioneered.

Beyond his published work, Adams contributed to the development of our subject in many other ways. He was a good friend and mentor to many of us. His correspondence with mathematicians all over the world was extraordinary. He wrote me well over 1,000 handwritten pages, and I was only one of many regular correspondents. He was especially generous and helpful to young mathematicians just starting out. There are many proofs attributed to Adams in the work of others, and there are many proofs and some essentially complete rewrites that come from his "anonymous" referee's reports. He wrote many perspicacious comments on papers for Mathematical Reviews. He was a force to be reckoned with at any occasion such at this, and his guiding presence and astute questions enlivened many a conference.

As the two unpublished papers illustrate, Frank throughout his career made it his business to learn and assimilate fully all of the most important new developments in algebraic topology. Moreover, he worked hard to ease the assimilation of these developments by others. When Frank started out in 1955, it was perfectly possible for one man to understand thoroughly all that there was to know in algebraic topology. This was still more or less possible when I met him in 1964. But our field has by now expanded and developed in so many directions that no one person can expect to master all of it, as Frank tried so hard to do. Many of its most vital directions were set in course by Frank Adams, and his influence will permeate the subject for the foreseeable future. His untimely death is a great loss to our subject, and a great personal loss to all of us who knew him well. He will long be missed, and long be remembered. His place in the history of mathematics is assured.

Bibliography of J. F. Adams

1. On decompositions of the sphere, J. London Math. Soc. 29 (1954), 96–99.

2. A new proof of a theorem of W. H. Cockcroft, J. London Math. Soc. 30 (1955), 482–488.

3. Four applications of the self-obstruction invariants, J. London Math. Soc. 31 (1956), 148–159.

4. On products in minimal complexes, Trans. Amer. Math. Soc. 82 (1956), 180–189.

5. (with P. J. Hilton) On the chain algebra of a loop space, Comment. Math. Helv. 30 (1956), 305–330.

6. On the cobar construction, Proc. Nat. Acad. Sci. USA 42 (1956), 409–412.

   (Also in Colloque de Topologie Algébrique, Louvain 1956, Georges Thone 1957, 81–87.)

7. An example in homotopy theory, Proc. Cambridge Philos. Soc. 53 (1957), 922–923.

8. Une relation entre groupes d'homotopie et groupes de cohomologie, C. R. Acad. Sci. Paris 245 (1957), 24–26.

9. On the structure and applications of the Steenrod algebra. Comment. Math. Helv. 32 (1958), 180–214.

10. On the non-existence of elements of Hopf invariant one mod p, Notices of the Am. Math. Soc. 5 (1958), 25.

11. On the non-existence of elements of Hopf invariant one, Bull. Amer. Math. Soc. 64 (1958), 279–282.

12. Théorie de l'homotopie stable, Bull. Soc. Math. France 87 (1959), 277–280.

13. Exposé sur les travaux de C. T. C. Wall sur l'algèbre de cobordisme Ω, Bull. Soc. Math. France 87 (1959), 281–284.

14. On the non-existence of elements of Hopf invariant one, Ann. of Math. 72 (1960), 20–104.

15. Appendix to a paper of E. R. Reifenberg, Acta Math. 104 (1960), 76–91.

16. The sphere, considered as an H-space mod p, Quart. J. Math. Oxford Ser. (2), 12 (1961), 52–60.

17. A finiteness theorem in homological algebra, Proc. Cambridge Philos. Soc. 57 (1961), 31–36.

18. On Chern characters and the structure of the unitary group, Proc. Cambridge Philos. Soc. 57 (1961), 189–199.

19. On formulae of Thom and Wu, Proc. London Math. Soc. (3) 11 (1961), 741–752.

20. Vector fields on spheres, Topology 1 (1962), 63–65.

21. H-spaces with few cells, Topology 1 (1962), 67–72.

22. Vector fields on spheres, Bull. Amer. Math. Soc. 68 (1962), 39–41.

23. Vector fields on spheres, Ann. of Math. 75 (1962), 603–632.

24. Applications of the Grothendieck-Atiyah-Hirzebruch functor $K(X)$, Proc. Internat. Congr. Math. 1962, 435–441.

(Also in Proc. Colloq. Algebraic Topology Aarhus 1962, 104–113.)

25. On the groups $J(X)$—I, Topology 2 (1963), 181–195.

26. Stable Homotopy Theory, Lecture Notes in Math. 3, Springer 1964.

(Second edition 1966; third edition 1969.)

27. (with G. Walker) An example in homotopy theory, Proc. Cambridge Philos. Soc. 60 (1964), 699–700.

28. On the groups $J(X)$—II, Topology 3 (1965), 137–171.

29. (with G. Walker) On complex Stiefel manifolds, Proc. Cambridge Philos. Soc. 61 (1965), 81–103.

30. On the groups $J(X)$, in Differential and Combinatorial Topology, Princeton Univ. Press 1965, 121–143.

31. On the groups $J(X)$—III, Topology 3 (1965), 193–222.

32. (with P. D. Lax and R. S. Phillips) On matrices whose real linear combinations are non-singular, Proc. Amer. Math. Soc. 16 (1965), 318–322; with a correction, ibid. 17 (1966), 945–947.

33. A spectral sequence defined using K-theory, Proc. Colloq. de Topologie, Bruxelles 1964, Gauthier-Villars 1966, 149–166.

34. (with M. F. Atiyah) K-theory and the Hopf invariant, Quart. J. Math. Oxford Ser. (2) 17 (1966), 31–38.

35. On the groups $J(X)$—IV, Topology 5 (1966), 21–71; with a correction, ibid. 7 (1968), 331.

36. A periodicity theorem in homological algebra, Proc. Cambridge Philos. Soc. 62 (1966), 365–377.

37. A survey of homotopy theory, Proc. Internat. Congr. Math. 1966, MIR, Moscow 1968, 33–43.

38. Lectures on Lie groups, W. A. Benjamin Inc., New York 1969. (Reprinted by Univ. of Chicago Press 1982.)

39. Lectures on generalised cohomology, in Category Theory, Homology Theory and their Applications III, Lecture Notes in Math. 99, Springer 1969, 1–138.

40. A variant of E. H. Brown's representability theorem, Topology 10 (1971), 185–198.

41. (with H. R. Margolis) Modules over the Steenrod algebra, Topology 10 (1971), 271–282.

42. (with A. S. Harris and R. M. Switzer) Hopf algebras of co-operations for real and complex K-theory, Proc. London Math. Soc. (3) 23 (1971), 385–408.

43. Algebraic topology in the last decade, in Proc. Sympos. Pure Math. 22, American Mathematical Society 1971, 1–22.

44. Algebraic Topology: a Student's Guide, London Math. Soc. Lecture Note Ser. 4, Cambridge Univ. Press 1972.

45. (with A. Liulevicius) The Hurewicz homomorphism for MU and BP, J. London Math. Soc. (2) 5 (1972), 539–545.

46. The Kahn-Priddy theorem, Proc. Cambridge Philos. Soc. 73 (1973), 45–55.

47. Chern characters revisited, Illinois J. Math. 17 (1973), 333–336; with an addendum, ibid. 20 (1976), 372.

48. Operations of the nth kind in K-theory, and what we don't know about RP^{∞}, in New Developments in Topology, London Math. Soc. Lecture Note Ser. 11, Cambridge Univ. Press 1974, 1–9.

49. Stable Homotopy and Generalised Homology, Univ. of Chicago Press 1974.

50. Sub-Hopf-algebras of the Steenrod algebra, Proc. Cambridge Philos. Soc. 76 (1974), 45–52.

51. Geometric dimension of bundles over RP^n, Publ. Res. Inst. Math. Sci., Kyoto Univ. (1974), 1–17.

52. Idempotent functors in homotopy theory, in Manifolds — Tokyo 1973, Univ. of Tokyo Press 1975, 247–253.

53. (with A. Liulevicius) Buhstaber's work on two-valued formal groups, Topology 14 (1975), 291–296.

54. (with P. Hoffman) Operations on K-theory of torsion-free spaces, Math. Proc. Cambridge Philos. Soc. 79 (1976), 483–491.

55. (with Z. Mahmud) Maps between classifying spaces, Invent. Math. 35 (1976), 1–41.

56. Primitive elements in the K-theory of BSU, Quart. J. Math. Oxford Ser. (2) 27 (1976), 253–262.

57. The work of H. Cartan in its relation with homotopy theory, Astérisque 32–33 (1976), 29–41.

58. (with S. B. Priddy) Uniqueness of BSO, Math. Proc. Cambridge Philos. Soc. 80 (1976), 475– 509.

59. (with F. W. Clarke) Stable operations on complex K-theory, Illinois J. Math. 21 (1977), 826–829.

60. Maps between classifying spaces, Enseign. Math. (2) 24 (1978), 79–85. (Also in Lecture Notes in Math. 658, Springer 1978, 1–8.)

61. Infinite Loop Spaces, Ann. of Math. Stud. 90, Princeton Univ. Press 1978.

62. Maps between classifying spaces II, Invent. Math. 49 (1978), 1–65.

63. (with C. W. Wilkerson) Finite H-spaces and algebras over the Steenrod algebra, Ann. of Math. 111 (1980), 95–143; with a correction, ibid. 113 (1981), 621–622.

64. (with W. H. Lin, D. M. Davis and M. E. Mahowald) Calculation of Lin's Ext groups, Math. Proc. Cambridge Philos. Soc. 87 (1980), 459–469.

65. Graeme Segal's Burnside ring conjecture, in Topology Symposium, Siegen 1979, Lecture Notes in Math. 788, Springer 1980, 378–395.

66. Finite H-spaces and Lie groups, J. Pure Appl. Algebra 19 (1980), 1–8.

67. Spin(8), triality, F_4 and all that, in Superspace and Supergravity, ed. S. W. Hawking and M. Roček, Cambridge Univ. Press 1981, 435–445.

68. Graeme Segal's Burnside ring conjecture, Bull. Amer. Math. Soc. (NS) 6 (1982), 201–210.

(Also in Proc. Symp. Pure Math. 39, Part 1, Amer. Math. Soc. 1983, 77–86.)

69. Graeme Segal's Burnside ring conjecture, Contemp. Math. 12, Amer. Math. Soc. 1982, 9–18.

70. Maps from a surface to the projective plane, Bull. London Math. Soc. 14 (1982), 533–534.

71. (with Z. Mahmud) Maps between classifying spaces III, in Topological topics, London Math. Soc. Lecture Note Ser. 86, Cambridge Univ. Press 1983, 136–153.

72. Prerequisites (on equivariant stable homotopy) for Carlsson's lecture, in Algebraic Topology, Aarhus 1982, Lecture Notes in Math. 1051, Springer 1984, 483–532.

73. The fundamental representations of E_8, Contemp. Math. 37, Amer. Math. Soc. 1985, 1–10.

74. La conjecture de Segal, Sém. Bourbaki 1984–85, No. 645, Astérisque 133–134 (1986), 255–260.

75. Maxwell Herman Alexander Newman, Biographical Memoirs of Fellows of the Royal Society 31 (1985), 437–452.

76. (with J. H. Gunawardena and H. Miller) The Segal conjecture for elementary abelian p-groups, Topology 24 (1985), 435–460.

77. 2-Tori in E_8, Math. Ann. 278 (1987), 29–39.

78. (with J.-P. Haeberly, S. Jackowski and J. P. May) A generalisation of the Segal conjecture, Topology 27 (1988), 7–21.

79. (with J.-P. Haeberly, S. Jackowski and J. P. May) A generalisation of the Atiyah-Segal completion theorem, Topology 27 (1988), 1–6.

80. (with N. J. Kuhn) Atomic spaces and spectra, Proc. Edinburgh Math. Soc. (2) 32 (1989), 473–481.

81. (with Z. Wojtkowiak) Maps between p-completed classifying spaces, Proc. Roy. Soc. Edinburgh 112A (1989), 231–235.

82. Talk on Toda's work, in Homotopy theory and related topics, Lecture Notes in Math. 1418, Springer 1990, 7–14. 1990.

83. Two theorems of J. Lannes, Collected Works, Cambridge Univ. Press, to appear.

84. The work of M. J. Hopkins, Collected Works, Cambridge Univ. Press, to appear.

Twisted tensor products of DGA's and the Adams-Hilton model for the total space of a fibration

by

Kathryn P. Hess*

Abstract.

Let $F \longrightarrow E \longrightarrow B$ be a fibration. We show that an Adams-Hilton model for the total space E can be constructed in a straightforward manner from Adams-Hilton models for the base B and fiber F.

1. Introduction

In their now-classic paper [A-H], Adams and Hilton constructed an important new differential graded algebra (DGA) model for simply-connected spaces. An Adams-Hilton model for a simply-connected space X is a free DGA $\mathcal{A}(X)$, together with a morphism of DGAs

$$\theta_X : \mathcal{A}(X) \xrightarrow{\simeq} C_*\Omega X,$$

where the symbol $\xrightarrow{\simeq}$ denotes a map inducing an isomorphism on homology. An Adams-Hilton model is thus a tool for studying the Pontryagin algebra $H_*\Omega X$ of a topological space X. Furthermore, since Adams-Hilton models can be constructed to be relatively "small", they are quite useful for computational purposes. Explicit Adams-Hilton models are also reasonably easy to construct. If X is a simply-connected CW complex with a fixed cellular decomposition, then $\mathcal{A}(X)$ can be chosen to have generators corresponding to the cells in the decomposition and a differential encoding their attaching maps.

While much is known about how Adams-Hilton models can be computed in general, until recently the relationships among the Adams-Hilton models for the fiber, base space and total space of a fibration had not been thoroughly investigated. In his paper [A2], Anick studied the special case of a fibration over a sphere. The goal of this article is to construct an Adams-Hilton model over a field k for the total space of an arbitrary fibration from Adams-Hilton models for the base space and fiber.

As could be expected, an Adams-Hilton model for the total space of an arbitrary fibration can be obtained by "perturbing" in a well-defined manner an Adams-Hilton model for a product of the base and fiber. We will therefore first describe the Adams-Hilton model for a product of two spaces, as constructed in the main theorem of this article (Theorem 5.1).

* Supported by the NFR-Sweden during the course of this research

We will then explain how these simpler models should be perturbed in order to obtain models for the total space of an arbitrary fibration.

For any graded k-vector space V, let TV denote the free associative graded algebra generated by V. We can then state

Theorem 1. *Suppose that $\mathcal{A}\ (F) = (\mathrm{T}W, e)$ and $\mathcal{A}\ (B) = (\mathrm{T}V, d)$ are Adams-Hilton models over k for F and B, respectively. Then the Adams-Hilton model for $B \times F$, $\mathcal{A}\ (B \times F)$, can be chosen to be $(\mathrm{T}(V \oplus W \oplus s(W \otimes V)), D)$, where $Dv = dv$, $Dw = ew$, and*

$$Ds(w \otimes v) = w \otimes v + (-1)^{\deg w \deg v} v \otimes w + \alpha$$

where $\alpha \in [\mathrm{T}(V \oplus W) \amalg \mathrm{T}^+ s(W \otimes V)] \oplus [(\mathrm{T}V)_{<\deg v} \otimes \mathrm{T}W]$.

Note that the generators $s(W \otimes V)$ of $\mathcal{A}\ (B \times F)$ correspond to the product cells of $B \times F$. Furthermore, the form of the differential on $s(W \otimes V)$ is analogous to the description of the attaching maps of the product cells.

We now make the situation more complicated by considering an arbitrary fibration $F \longrightarrow E \longrightarrow B$.

Theorem 2. *Suppose that $\mathcal{A}\ (F) = (\mathrm{T}W, e)$ and $\mathcal{A}\ (B) = (\mathrm{T}V, d)$ are Adams-Hilton models over k for F and B, respectively. Then the Adams-Hilton model for E, $\mathcal{A}\ (E)$, can be chosen to be $\mathrm{T}(V \oplus W \oplus s(W \otimes V)), \bar{D})$, where $\bar{D}w = ew$, $\bar{D}v - dv \in (\mathrm{T}V)_{<\deg v - 1} \otimes \mathrm{T}^+ W$, and*

$$\bar{D}s(w \otimes v) = w \otimes v + (-1)^{\deg w \deg v} v \otimes w + \alpha$$

where $\alpha \in [\mathrm{T}(V \oplus W) \amalg \mathrm{T}^+ s(W \otimes V)] \oplus [(\mathrm{T}V)_{<\deg v} \otimes \mathrm{T}W]$.

In other words, the differential of $\mathcal{A}\ (E)$ is a perturbation of the differential of $\mathcal{A}\ (B \times F)$, and the underlying graded algebras are the same. This perturbation of the differential corresponds to tweaking the attaching maps in the product to obtain the CW structure for the total space of the fibration.

Note that the above theorem can be used as a tool in determining which are the possible total spaces E in a fibration with given base space B and fiber F. For example, suppose that $B = S^{n+1}$ and $F = S^{m+1}$, which have Adams-Hilton models $\mathcal{A}(B) = (\mathrm{T}\langle v_n \rangle, 0)$ and $\mathcal{A}(F) = (\mathrm{T}\langle w_m \rangle, 0)$. The conditions from the above theorem tell us that if

$$S^{m+1} \longrightarrow E \longrightarrow S^{n+1}$$

is a fibration then a possible Adams-Hilton model for E is

$$\mathcal{A}(E) = (\mathrm{T}\langle v, w, s(w \otimes v) \rangle, \bar{D}),$$

where $\bar{D}w = 0$ and $\bar{D}s(w \otimes v) = w \otimes v - (-1)^{mn} v \otimes w$. The possible choices $\bar{D}v$ are 0 and w^k, if $mk = n - 1$. If $n = 3$ and $m = 2$ and $\bar{D}v = 0$, then we obtain a model for $S^3 \times S^4$. If instead $\bar{D}v = w$, then $E = S^7$, and the fibration is the Hopf fibration.

Our construction of the Adams-Hilton model for the total space of a fibration is based on work of Brown [B], who proved that for any fibration $F \longrightarrow E \longrightarrow B$, there exists a weak equivalence of chain complexes of C_*E with a "twisted tensor product" $C_*B \otimes_{\Phi_B} C_*F$ of C_*B and C_*F, i.e., that there exists a morphism of chain complexes

$$\phi_E : C_*B \otimes_{\Phi_B} C_*F \xrightarrow{\simeq} C_*E.$$

Given a fibration $F \longrightarrow E \longrightarrow B$, we will show that is possible to define a product on $C_*\Omega B \otimes_{\Phi_{\Omega B}} C_*\Omega F$ so that $\phi_{\Omega E}$ is actually a morphism of DGA's. In fact, this product can be chosen so that $C_*\Omega B \otimes_{\Phi_{\Omega B}} C_*\Omega F$ has the structure of what we will call a "twisted tensor product of the DGA's $C_*\Omega B$ and $C_*\Omega F$", i.e., a perturbation of the usual tensor product of DGA's. This concept of a twisted tensor product of DGA's seems to be useful, and we devote some time to investigating it. In particular, we show how to construct a free model for a twisted tensor product of two DGA's, given free models for the component DGA's.

We then apply this knowledge about free models for twisted tensor products of DGA's to the case of the free models

$$(\mathrm{T}V, d) = \mathcal{A}(B) \xrightarrow{\simeq} C_*\Omega B$$
$$(\mathrm{T}W, e) = \mathcal{A}(F) \xrightarrow{\simeq} C_*\Omega F,$$

building from $\mathcal{A}(B)$ and $\mathcal{A}(F)$ a free model for the DGA $C_*\Omega B \otimes_{\Phi_{\Omega B}} C_*\Omega F$

$$(\mathrm{T}(V \oplus W \oplus s(W \otimes V)), D) \xrightarrow{\simeq} C_*\Omega B \otimes_{\Phi_{\Omega B}} C_*\Omega F.$$

Composing this with the equivalence $\phi_{\Omega E}$ yields a free model

$$(\mathrm{T}(V \oplus W \oplus s(W \otimes V)), D) \xrightarrow{\simeq} C_*\Omega E,$$

that is, an Adams-Hilton model for E.

The article is organized as follows. Section 2 contains the necessary algebraic background material, as well as an introduction to the concept of a twisted tensor product of DGA's. Free models for twisted tensor products of DGA's are constructed in Section 3. A technical theorem concerning conditions which guarantee the existence of a twisted tensor product of two DGA's is also presented in Section 3. Section 4 is devoted to an examination of Brown's equivalence. The unification of the algebraic results from Section 3 and the topological results of Section 4 is presented in Section 5, in the construction of an Adams-Hilton model for the total space of a fiber space from the Adams-Hilton models of the base space and the fiber.

Remark. In [L-S], Lambe and Stasheff proved a result similar to that presented in this article. They showed, via rather different methods from those applied here, that there is a weak equivalence

$$C_*\Omega E \xrightarrow{\simeq} \tilde{\Omega}(H_*B \otimes H_*F).$$

The symbol $\tilde{\Omega}$ denotes the tilde cobar construction of Stasheff [St], i.e., a certain perturbation of the usual cobar construction on the coalgebra $H_*B \otimes H_*F$.

2. Algebraic Preliminaries

The category in which we will be working throughout most of this paper is the category of connected associative chain algebras (DGA's) over R, a fixed commutative ring with identity. A DGA (A, d) is said to be free if $A = TV$, i.e., A is the tensor algebra on a free graded R-module V. A DGA (A, d) is called r-reduced if $A_0 = R$ and $A_k = 0$ for all $0 < k < r$. If A and B are two DGA's, then their sum in the category of DGA's will be denoted $A \amalg B$. Given two differential graded (DG) R-modules (M, d) and (N, e), we will write $(M \otimes N, d \otimes e)$ to denote their tensor product with differential given by $d \otimes e(m \otimes n) = dm \otimes n + (-1)^{\deg m} m \otimes en$.

A morphism of DG R-modules or DGA's will be called a quism if it induces an isomorphism on homology. If for two DGA's (A, d) and (B, e) there exists a series of quisms

$$(A, d) = (A_0, d_0) \xrightarrow{f_1} (A_1, d_1) \xleftarrow{f_2} \cdots \xrightarrow{f_n} (A_n, d_n) = (B, e),$$

then (A, d) and (B, e) are said to be weakly equivalent.

Because we are interested in extending a result of Brown which concerns the twisted tensor product of a DG coalgebra (see [B] for definition) with a DGA, we recall the following definition.

Definition. *Let (H, δ) be a DG coalgebra with coproduct Δ and let (A, d) be a DGA with augmentation ϵ. Let $C^*(H; A)$ be the DG R-module defined by $C^p(H; A) = \mathrm{Hom}_R(H_p, A)$. A twisting cochain is an element $\phi = \sum \phi_q \in C^*(H; A)$ such that*

(i.) $\phi_q \in C^q(H; A)$, $\phi_0 = 0$, $\phi_q(H_q) \subseteq A_{q-1}$; and

(ii.) $\epsilon\phi_1 = 0$ and $d\phi_q(h) = \phi_{q-1}\delta(h) - \sum_{k=1}^{q-1}(-1)^k \phi_k(h_i')\phi_{q-k}(h_i'')$, where $\Delta(h) = \sum_i h_i' \otimes h_i''$.

The twisted tensor product of H with an A-module (M, e) with respect to ϕ is the DG R-module $(H \otimes_\phi M, D_\phi)$ satisfying

(i.) as a graded R-module $H \otimes_\phi M \approx H \otimes M$; and

(ii.) $D_\phi(h\otimes m) = \delta h\otimes m + (-1)^{\deg h}\big(h\otimes em + \sum_i h'_i\otimes(\phi(h_i")\bullet m)\big)$, *where* $\Delta(h) = \sum_i h'_i \otimes h_i"$ *and* $\bullet$ *denotes the action of* A *on* M.

A straightforward calculation shows that $D_\phi^2 = 0$ and $\epsilon D_\phi = 0$.

For the purposes of this paper the most important example of a DG coalgebra will be C_*X, which will denote the non-degenerate singular cubical chains on a topological space X. For any $n \geq 0$ and $(n-1)$-connected pointed topological space X, $C_*^n X$ will denote the subcomplex of C_*X of simplices sending the $(n-1)$-skeleton of cubes to the basepoint of X. An example of a possible coassociative coproduct on C_*X is presented in [A1; Sec. 7]. Note that since we are using cubical instead of singular chains, the usual "front p-face tensor back $(n-p)$-face" Alexander-Whitney diagonal does not work.

We will now introduce an algebraic concept known as the "twisted tensor product" of two DGA's, which is analogous to Tanré's twisted product of two differential graded Lie algebras ([T]) and is a generalization of the semi-tensor product of a Hopf algebra with an algebra on which it acts ([S]).

Our study of twisted tensor products of DGA's is motivated by interest in Brown's natural equivalence of chain complexes (i.e., DG R-modules)

$$\phi_E : C_*B \otimes_{\Phi_B} C_*F \longrightarrow C_*E$$

where $F \longrightarrow E \longrightarrow B$ is a fibration and $C_*B \otimes_{\Phi_B} C_*F$ is a twisted product of the DG coalgebra C_*B with the $C_*\Omega B$-module C_*F. In section 5, we will loop the above fibration and examine the resulting natural equivalence, $\phi_{\Omega E}$. We will show that we can define a product on $C_*\Omega B \otimes_{\Phi_{\Omega B}} C_*\Omega F$ in such a way that it has the DGA structure of a twisted tensor product of $C_*\Omega B$ and $C_*\Omega F$.

The ultimate goal of Section 3 is to construct a free model for a twisted tensor productof two DGA's. This will enable us to obtain a free model for $C_*\Omega B \otimes_{\Phi_{\Omega B}} C_*\Omega F$. We will therefore have a weak equivalence between the Adams-Hilton model for E and the free model of a twisted tensor productof the Adams-Hilton models for B and F.

With motivation from Tanré's twisted product of differential graded Lie algebras, we now make the following

Definition. *A twisted tensor product of two DGA's* (A, d) *and* (B, e), *denoted* $(A \tilde{\otimes} B, D)$, *is a DGA which satisfies the following conditions:*

(i.) *as a graded R-module* $A \tilde{\otimes} B \approx A \otimes B$;

(ii.) *for all* $a \in A$ *and* $b \in B$,

$$(1\otimes b)\odot(a\otimes 1) - (-1)^{\deg a \deg b} a\otimes b \in A_{<\deg a}\otimes B_+$$

and $(a\otimes 1)\odot(1\otimes b) = a\otimes b$, *where* $\odot$ *denotes the product in* $A\tilde{\otimes} B$;

(iii.) The sequence

$$0 \to (B,e) \xrightarrow{\iota} (A \tilde{\otimes} B, D) \xrightarrow{\pi} (A,d) \to 0$$

is a sequence of DGA morphisms, exact at (A,d) and (B,e), where $\iota(b) = 1 \otimes b$ and $\pi(a \otimes b) = \epsilon(b)a$. Here, ϵ denotes the augmentation of B.

If there exists a morphism of graded algebras $\rho : A \longrightarrow A \tilde{\otimes} B$ which is a right inverse to π, the twisted tensor product is said to be simple.

It is important to remark at this point that a twisted tensor product of DGA's (A,d) and (B,e) is simply a perturbation of a certain type of the usual tensor product of DGA's $(A \otimes B, d \otimes e)$. Recall that the product in $(A \otimes B, d \otimes e)$ is defined so that both (A,d) and (B,e) are sub DGA's and $(1 \otimes b) \odot (a \otimes 1) = (-1)^{\deg a \deg b} a \otimes b$. Thus, for example, the product $(1 \otimes b) \odot (a \otimes 1)$ in a twisted tensor product has been perturbed by adding a term in $A_{<\deg a} \otimes B_+$.

3. Models for twisted tensor products of DGA's

We now direct our attention to the construction of free models for twisted tensor products of DGA's. A free model for a DGA (C, d_C) is a free DGA (TV, d) together with a quism $f : (TV, d) \longrightarrow (C, d_C)$.

The usual spectral sequence argument shows that if $f : A \longrightarrow A'$ and $g : B \longrightarrow B'$ are quisms of DGA's and $f \otimes g : A \tilde{\otimes} B \longrightarrow A' \tilde{\otimes} B'$ is a map of twisted tensor products, then $f \otimes g$ is also a quism. What is not immediately evident is that, given quisms f and g as above and a twisted tensor product $A' \tilde{\otimes} B'$, one can put the structure of a twisted tensor product on $A \otimes B$ so that it is weakly equivalent to $A' \tilde{\otimes} B'$ as DGA's. Under the conditions of the proposition below, however, it is possible to do so. Before proving the proposition, we present a lemma which will be useful in its demonstration.

Lemma 3.1. *Let (A, d_A) be a DGA, and let B be a graded R -module. Suppose that $f : A \longrightarrow B$ is a morphism of graded R -modules such that $d_A(\ker f) \subseteq \ker f$ and $\ker f$ is an ideal in A. Suppose further that f has a right inverse ι. Then it is possible to define a product on B, together with a differential d_B so that $f : (A, d_A) \longrightarrow (B, d_B)$ is a DGA morphism.*

Proof. Let $\mu_A : A \otimes A \longrightarrow A$ be the product on A. Define d_B to be $f d_A \iota$. Define $\mu_B : B \otimes B \longrightarrow B$ by $\mu_B = f \circ \mu_A \circ (\iota \otimes \iota)$. We need to show that f commutes with both the differential and the product, $d_B^2 = 0$, μ_B is associative, and d_B is a derivation with respect to μ_B.

To see that $f d_A = d_B f$, note first that $\iota f(a) - a \in \ker f$ for all $a \in A$. Then, since $d(\ker f) \in \ker f$, $f d_A(\iota f(a) - a) = 0$. Thus, $f d_A(a) = f d_A(\iota f(a)) = d_B f(a)$.

We can prove that $f \circ \mu_A = \mu_B \circ f \otimes f$ by noticing that since ker$f$ is an ideal in A it follows that

$$f\mu_A(a \otimes a) = f\mu_A(\iota f(a) \otimes a) = f\mu_A \circ (\iota \otimes \iota) \circ (f \otimes f)(a \otimes a) = \mu_B \circ (f \otimes f)(a \otimes a).$$

The remainder of the necessary identities can be proved similarly to the above two and are left to the reader. **QED**

Proposition 3.2. *Suppose that*

$$f : (\mathrm{T}V, d_V) \rightarrow (A, d_A)$$

and

$$g : (\mathrm{T}W, d_W) \rightarrow (B, d_B)$$

are free r-reduced models for the DGA's (A, d_A) and (B, d_B). If $(A \tilde{\otimes} B, D)$ is a twisted tensor product of DGA's, then there exists a twisted tensor product $(\mathrm{T}V \tilde{\otimes} \mathrm{T}W, D')$, together with a DGA (C, d) and two surjective quisms of DGA's

$$h_0 : (C, d) \xrightarrow{\simeq} (A \tilde{\otimes} B, D)$$

and

$$h_1 : (C, d) \xrightarrow{\simeq} (\mathrm{T}V \tilde{\otimes} \mathrm{T}W, D').$$

Proof. The differential D of the twisted tensor product is a perturbation of $d_A \otimes d_B$, i.e., there is a degree -1 R-module map $\sigma : A \longrightarrow A \otimes B$ such that $\sigma(A_n) \subset A_{<n-1} \otimes B$ and $D(a \otimes b) = (d_A \otimes d_B)(a \otimes b) + \sigma(a)b$. We would like to define an analogous perturbation $\sigma' : \mathrm{T}V \longrightarrow \mathrm{T}V \otimes \mathrm{T}W$ with $f \otimes g \circ \sigma' = \sigma \circ f \otimes g$ and $d_V \otimes d_W \circ \sigma' + \sigma' \circ d_V \otimes d_W + (\sigma')^2 = 0$. Then, if D' were defined to be the degree -1 map given by

$$D'(x \otimes y) = (d_V \otimes d_W)(x \otimes y) + \sigma'(x)y,$$

the conditions satisfied by σ' would guarantee that $f \otimes g \circ D' = D \circ f \otimes g$ and $(D')^2 = 0$, respectively. Unfortunately, if f and g are not surjective, it is not certain that such a σ' can be defined. We can work around this problem, however, by means of the following construction.

Let V' and W' be graded R-modules, isomorphic as graded R-modules to $A \oplus s^{-1}A$ and $B \oplus s^{-1}B$, respectively. Define differentials $d_{V'}$ and $d_{W'}$ on V' and W' by choosing $d_{V'}(a) = s^{-1}a$, $d_{W'}(b) = s^{-1}b$, $d_{V'}(s^{-1}a) = 0$ and $d_{W'}(s^{-1}b) = 0$. Then the DGA's $(\mathrm{T}V', d_{V'})$ and $(\mathrm{T}W', d_{W'})$ are acyclic.

Let $\delta = d_V \amalg d_{V'}$ and $\delta' = d_W \amalg d_{W'}$ denote the differentials on $\mathrm{T}(V \oplus V')$ and $\mathrm{T}(W \oplus W')$, respectively. Define $\bar{f} : \mathrm{T}(V \oplus V') \longrightarrow A$ to be the DGA morphism extending f on $\mathrm{T}V$ and 1_A on $\mathrm{T}V'$; define $\bar{g} : \mathrm{T}(W \oplus W') \longrightarrow B$ analogously. Thus $\bar{f}$ and $\bar{g}$ are surjective quisms of DGA's.

We now wish to define a differential $\bar{D}$ on $T(V \oplus V') \otimes T(W \oplus W')$ so that

$$h : (T(V \oplus V') \otimes T(W \oplus W'), \bar{D}) \longrightarrow (A \tilde{\otimes} B, D)$$

is a morphism of DG R-modules and so that $\bar{D}$ is of the correct form to be a differential on a twisted tensor productof the DGA's $T(V \oplus V')$ and $T(W \oplus W')$. Let K denote the R-submodule $[(TV \amalg T^+V') \otimes T(W \oplus W') + T(V \oplus V') \otimes (TW \amalg T^+W')]$. We wish furthermore to define $\bar{D}$ so that

$$\bar{D}K \subseteq K.$$

This last condition will enable us to use Lemma 3.1 to construct the necessary differential on $TV \otimes TW$.

The differential $\bar{D}$ will be defined via an inductive construction of a graded R-module morphism, homogeneous of degree -1,

$$\bar{\sigma} : T(V \oplus V') \longrightarrow T(V \oplus V') \otimes T^+(W \oplus W'),$$

satisfying the following two conditions.

$$\bar{f} \otimes \bar{g} \circ \bar{\sigma} = \sigma \circ \bar{f} \tag{1}$$

and

$$\delta \otimes \delta' \circ \bar{\sigma} + \bar{\sigma} \circ \delta + \bar{\sigma}^2 = 0 \tag{2}$$

We will require also that

$$\bar{\sigma}(TV \amalg T^+V') \subseteq K. \tag{3}$$

The existence of such a $\bar{\sigma}$ is clearly equivalent to the existence of a $\bar{D}$ as described above.

If $v \in T(V \oplus V')_r$, then it is clear that we can choose $\bar{\sigma}(v) = 0$. Suppose $\bar{\sigma}$ has been constructed on $T(V \oplus V')_{<N}$ so that conditions (1), (2), and (3) have been satisfied.

Define $\bar{D}$ on $(T(V \oplus V')_{<N}) \otimes T(W \oplus W')$ by

$$\bar{D}(x \otimes y) = \delta x \otimes y + (-1)^{\deg x}(x \otimes \delta' y + \bar{\sigma}(x)y).$$

Then, $\bar{f} \otimes \bar{g} \circ \bar{D} = D \circ \bar{f} \otimes \bar{g}$ in $(T(V \oplus V')_{<N}) \otimes T(W \oplus W')$. Condition (3) implies that $[(TV_{<N} \amalg T^+V'_{<N}) \otimes T(W \oplus W') + T(V \oplus V')_{<N} \otimes (TW \amalg T^+W')]$ is a subDG R-module of $((T(V \oplus V')_{<N}) \otimes T(W \oplus W'), \bar{D})$. Furthermore, both

$$(\ker(\bar{f} \otimes \bar{g}) \cap (T(V \oplus V')_{<N-1}) \otimes T(W \oplus W'), \bar{D})$$

and

$$(\ker(\bar{f} \otimes \bar{g}) \cap K \cap (T(V \oplus V')_{<N-1}) \otimes T(W \oplus W'), \bar{D})$$

are acyclic, by the usual spectral sequence argument.

Choose $x \in \mathrm{T}((V \oplus V')_N)$. Since the restriction of $\bar{f} \otimes \bar{g}$ to K is surjective, there exists $z \in K$ such that $\bar{f} \otimes \bar{g}(z) = \sigma \circ \bar{f}(x)$. Then since $D^2 = 0$

$$\bar{f} \otimes \bar{g} \circ (\bar{D}z + \bar{\sigma}(\delta x)) = D\sigma \circ \bar{f}(x) + \sigma(d_A \bar{f}(x)) = 0.$$

Thus $\bar{D}z + \bar{\sigma}(\delta x)$ is a cycle of degree $N-2$ in $\ker(\bar{f} \otimes \bar{g})$, or, if $x \in \mathrm{T}V \amalg \mathrm{T}^+V'$, in $\ker(\bar{f} \otimes \bar{g}) \cap K$. Therefore, there exists $w \in \ker(\bar{f} \otimes \bar{g})$, or, when $x \in \mathrm{T}V \amalg \mathrm{T}^+V'$, in $\ker(\bar{f} \otimes \bar{g}) \cap K$, of degree $N-1$, such that $\bar{D}w = \bar{D}z + \bar{\sigma}(\delta x)$.

Set $\bar{\sigma}(x) = w - z$. Then it is clear that conditions (1), (2), and (3) hold. Thus, by induction, the required $\bar{D}$ can be constructed.

The morphism $\bar{f} \otimes \bar{g} : (\mathrm{T}(V \oplus V') \otimes \mathrm{T}(W \oplus W'), \bar{D}) \longrightarrow (A \tilde{\otimes} B, D)$ now fits the hypotheses of Theorem 3.1. In fact, since the restriction of $\bar{f} \otimes \bar{g}$ to K is surjective, for every $a, a' \in \mathrm{T}(V \oplus V')$ and $b, b' \in \mathrm{T}(W \oplus W')$, there exists

$$x \in (\mathrm{T}(V \oplus V'))_{<\deg a + \deg a'} \otimes \mathrm{T}(W \oplus W') \cap K$$

such that

$$(\bar{f}(a) \otimes \bar{g}(b)) \odot (\bar{f}(a') \otimes \bar{g}(b')) - (-1)^{\deg a' \deg b} \bar{f}(aa') \otimes \bar{g}(bb') = (\bar{f} \otimes \bar{g})(x).$$

Here, $\odot$ denotes the product in $A \tilde{\otimes} B$. Furthermore, the standard spectral sequence argument shows that K itself is acyclic with respect to the differential $\bar{D}$, as $\mathrm{T}V \amalg \mathrm{T}^+V'$ and $\mathrm{T}W \amalg \mathrm{T}^+W'$ are acyclic with respect to the δ and δ', respectively.

It is thus clear that in applying Lemma 3.1 to this situation, the product μ on $\mathrm{T}(V \oplus V') \tilde{\otimes} \mathrm{T}(W \oplus W')$ can be chosen so that K is an ideal of $\mathrm{T}(V \oplus V') \tilde{\otimes} \mathrm{T}(W \oplus W')$. Now let

$$\pi : \mathrm{T}(V \oplus V') \tilde{\otimes} \mathrm{T}(W \oplus W') \longrightarrow \mathrm{T}V \otimes \mathrm{T}W$$

be the graded R-module morphism with kernel K. The morphism π fits the hypotheses of Lemma 3.1, so there exist a product $\mu' = \pi\mu \circ (\iota \otimes \iota)$ and a differential $D' = \pi \bar{D} \iota$ on $\mathrm{T}V \otimes \mathrm{T}W$ such that π is a DGA morphism. Further, since K is acyclic, π is in fact a DGA quism. Note finally that the fact that $\mathrm{T}(V \oplus V') \tilde{\otimes} \mathrm{T}(W \oplus W')$ is a twisted tensor productof DGA's implies that $(\mathrm{T}V \otimes \mathrm{T}W, D')$ satisfied the conditions to be a twisted tensor product of $\mathrm{T}V$ and $\mathrm{T}W$ with respect to the product μ'. Write $\mathrm{T}V \tilde{\otimes} \mathrm{T}W$ for this twisted tensor product.

The DGA $\mathrm{T}V \tilde{\otimes} \mathrm{T}W$ is thus the required twisted tensor product, which is weakly equivalent to $A \tilde{\otimes} B$. **QED**

Our attention is now focused on the construction of a free model for a twisted tensor products of two free DGA's. Note that sX denotes the suspension of a graded R-module X, i.e. $(sX)_n = X_{n-1}$.

Theorem 3.3. *Suppose that* $(\mathrm{T}V \tilde{\otimes} \mathrm{T}W, D)$ *is a twisted tensor product of free DGA's* $(\mathrm{T}V, d)$ *and* $(\mathrm{T}W, e)$. *Let* $\mathcal{M}$ *denote the free algebra* $\mathrm{T}(V \oplus W \oplus s(W \otimes V))$. *Let*

$$\pi : \mathcal{M} \longrightarrow \mathrm{T}V \tilde{\otimes} \mathrm{T}W$$

be the algebra morphism extending $\pi(v) = v$, $\pi(w) = w$, *and* $\pi(s(w \otimes v)) = 0$. *There exists a differential* $\bar{D}$ *on* $\mathcal{M}$ *such that* π *is a quism, i.e., such that* $(\mathcal{M}, \bar{D})$ *is a free model for* $(\mathrm{T}V \tilde{\otimes} \mathrm{T}W, D)$. *In particular* $\bar{D}$ *can be chosen so that* $\bar{D}v = Dv$, $\bar{D}w = ew$ *and* $\bar{D}s(w \otimes v) - w \otimes v + (-1)^{\deg w \deg v} v \otimes w \in [\mathrm{T}(V \oplus W) \amalg \mathrm{T}^+ s(W \otimes V)] \oplus [(\mathrm{T}V)_{<\deg v} \otimes \mathrm{T}W]$.

Proof. The problem of defining a differential such that π is a quism is equivalent to that of constructing the differential so that $\ker\pi$ is acyclic. It is thus important to know what the elements of $\ker\pi$ are.

Note that π will also be used to denote the restriction of π to the subcomplex $\mathrm{T}V \otimes \mathrm{T}W$.

We begin by constructing an element of $\ker\pi$ for each $w \otimes v \in W \otimes V$ and then show that these elements, together with $s(W \otimes V)$, generate $\ker\pi$ as a subalgebra of $\mathcal{M}$. The first step is to define a map of DG R-modules $\rho : \mathrm{T}V \otimes \mathrm{T}W \longrightarrow \mathrm{T}V \otimes \mathrm{T}W$ by

$$\rho(a \otimes b) = \pi(a \otimes b) - a \otimes b.$$

Recall that for any $a \in \mathrm{T}V$, $b \in \mathrm{T}W$, $\pi(a \otimes b) = a \otimes b + \sum_{\deg a_i < \deg a} a_i \otimes b_i$. Thus $\rho(a \otimes b) \in (\mathrm{T}V)_{<\deg a} \otimes \mathrm{T}W$. Since the multiplication in $\mathrm{T}V \tilde{\otimes} \mathrm{T}W$ commutes with the differential, the map ρ does as well.

Next, let $\chi : \mathrm{T}V \otimes \mathrm{T}W \longrightarrow \mathrm{T}V \otimes \mathrm{T}W$ be the DG R-module morphism given by $\chi(a \otimes b) = \sum_{n \geq 0} (-1)^n \rho^n(a \otimes b)$, where ρ^0 is defined to be the identity. Note that the sum will always be finite, as the degree of the TV-component of $\rho(a \otimes b)$ is always less than $\deg a$.

It is easy to see that $\pi\chi$ is the identity on $\mathrm{T}V \otimes \mathrm{T}W$. It follows therefore that $\pi(x - \chi\pi(x)) = 0$ for all $x \in \mathrm{T}V \amalg \mathrm{T}W$. In particular $w \otimes v - \chi\pi(w \otimes v) \in \ker\pi$ for all $w \in W$ and $v \in V$.

Note that if $x, y \in \mathrm{T}V \otimes \mathrm{T}W \subset \mathcal{M}$ and $\pi(x) = \pi(y)$, then $x = y$. Thus $\chi\pi(x)$ is in fact the unique element of $\mathrm{T}V \otimes \mathrm{T}W$ such that $\pi\chi\pi(x) = \pi(x)$, for any $x \in \mathrm{T}V \amalg \mathrm{T}W$.

Let $\Phi(x) = x - \chi\pi(x)$ for all $x \in \mathrm{T}V \amalg \mathrm{T}W$. Then as a further consequence of the fact that $\pi\chi = 1_{\mathrm{T}V \otimes \mathrm{T}W}$, we have the identities, for $u \in V \oplus W$, $x \in \mathrm{T}V \amalg \mathrm{T}W$

$$\begin{aligned} \Phi(u \otimes x) &= u \otimes \Phi(x) + \Phi(u \otimes \chi\pi(x)) \\ \Phi(x \otimes u) &= \Phi(x) \otimes u + \Phi(\chi\pi(x) \otimes u). \end{aligned} \tag{1}$$

Note that, in particular, $\Phi(x \otimes w) = \Phi(x) \otimes w$ and $\Phi(v \otimes x) = v \otimes \Phi(x)$ for $w \in W$ and $v \in V$.

Suppose that $\pi(x) = \pi(y)$ for some $x, y \in \mathrm{T}V \amalg \mathrm{T}W$. Then $\chi\pi(x) = \chi\pi(y)$, so that the element $x - y$ in $\ker\pi$ can be written as

$$(x - \chi\pi(x)) - (y - \chi\pi(y)) = \Phi(x) - \Phi(y).$$

Using the identities (1), it is easy to see that

$$\Phi(x) \in \mathrm{T}V \amalg \mathrm{T}W \amalg R\langle \Phi(w \otimes v) : w \otimes v \in W \rangle.$$

Thus $\ker\pi = \mathcal{M} \amalg R\langle s(w \otimes v), \Phi(w \otimes v) : w \otimes v \in W \otimes V \rangle$.

Now, in order to construct the necessary differential $\bar{D}$, we begin by defining a degree +1 morphism of graded R -modules

$$S : (\mathrm{T}V \amalg \mathrm{T}W) \amalg (W \otimes V) \longrightarrow \mathcal{M}$$

which extends $S(w \otimes v) = s(w \otimes v)$. Looking to the identities (1) for motivation, we extend S by

$$\begin{aligned} S(v \otimes x) &= (-1)^{\deg v} v \otimes S(x) \\ S(w \otimes x) &= (-1)^{\deg w} w \otimes S(x) \\ S(x \otimes v) &= S(x) \otimes v + S(\chi\pi(x) \otimes v) \\ S(x \otimes w) &= S(x) \otimes w \end{aligned}$$

for $x \in (\mathrm{T}V \amalg \mathrm{T}W) \amalg (W \otimes V)$ and $v \in V$, $w \in W$. It follows easily that for any $\alpha \in \mathrm{T}V \amalg \mathrm{T}W$ and $\beta \in \mathrm{T}W$

$$S(\alpha \otimes x) = (-1)^{\deg \alpha} \alpha \otimes S(x)$$

and

$$S(x \otimes \beta) = S(x) \otimes \beta.$$

We can now define a deg -1 derivation on $\mathcal{M}$ by setting $\bar{D}v = Dv$, $\bar{D}w = Dw$ and $\bar{D}s(w \otimes v) = \Phi(w \otimes v) - S\bar{D}(w \otimes v)$ and extending to a derivation. Then $\bar{D}^2 v = 0$ and $\bar{D}^2 w = 0$ for all $v \in V$ and $w \in W$. It remains to show that $\bar{D}^2 s(w \otimes v) = 0$, which will be an immediate consequence of the following claim.

Claim: Let $\mathcal{B}$ denote the algebra $\mathrm{T}V \amalg \mathrm{T}W \amalg (W \otimes V)$. Then

$$\bar{D}S(x) = \Phi(x) - S\bar{D}(x)$$

for all $x \in \mathcal{B}$.

Proof of claim: The claim holds by definition in the lowest possible degree, since the only x of lowest degree are equal to $w \otimes v$ for some $v \in V$ and $w \in W$.

Suppose the claim is true up to degree n. Choose $x \in \mathcal{B}_n$.

Case 1: $x = v \otimes y$, $v \in V$, $y \in \mathcal{B}_{n-1}$.

$$\begin{aligned}\bar{D}S(x) =&\bar{D}(-1)^{\deg v} v \otimes S(y)\\ =&(-1)^{\deg v}\bar{D}v \otimes S(y) + v \otimes \bar{D}S(y)\\ =&(-1)^{\deg v}\bar{D}v \otimes S(y) + v \otimes \Phi(y) - v \otimes S\bar{D}(y)\\ =&\Phi(v \otimes y) - S\bar{D}(v \otimes y)\end{aligned}$$

Case 2: $x = y \otimes w$, $w \in W$, $y \in \mathcal{B}_{n-1}$.

The proof of this case is quite similar to that of case 1 and is left to the reader.

Case 3: $x = y \otimes v$, $v \in V$, $y \in \mathcal{B}_{n-1}$.

$$\begin{aligned}&\bar{D}S(x)\\ =&\bar{D}[S(y) \otimes v + S(\chi\pi(y) \otimes v)]\\ =&\bar{D}S(y) \otimes v + (-1)^{\deg y+1} S(y) \otimes \bar{D}v + \Phi(\chi\pi(y) \otimes v) - S\bar{D}(\chi\pi(y) \otimes v)\\ =&\Phi(y) \otimes v - S\bar{D}(y) \otimes v - (-1)^{\deg y} S(y) \otimes \bar{D}(v)\\ &+ \Phi(\chi\pi(y) \otimes v) - S(\chi\pi(\bar{D}y) \otimes v) - (-1)^{\deg y} S(\chi\pi(y) \otimes \bar{D}v)\\ =&\Phi(y \otimes v) - S\bar{D}(y \otimes v)\end{aligned}$$

Therefore, by induction, the claim holds. **QED**

It now follows immediately that $\bar{D}^2 s(w \otimes v) = 0$ as

$$\begin{aligned}\bar{D}^2 s(w \otimes v) =&\bar{D}[\Phi(w \otimes v) - S\bar{D}(w \otimes v)]\\ =&\bar{D}\Phi(w \otimes v) - \Phi\bar{D}(w \otimes v) + S\bar{D}^2 s(w \otimes v)\\ =&0.\end{aligned}$$

Using the language of Lambe and Stasheff ([L-S]), we can thus say that we have strong deformation retract data (SDR-data):

$$(TV \tilde{\otimes} TW, D) \underset{\pi}{\overset{\chi}{\rightleftharpoons}} (\mathcal{M}, \bar{D}); S$$

since $\pi\chi = 1_{TV \tilde{\otimes} TW}$ and $\chi\pi = 1_{\mathcal{M}} + S\bar{D} + \bar{D}S$. Since $\pi\chi$ and $\chi\pi$ are both quisms, π is a quism as well. **QED**

Combining Proposition 3.2 and Theorem 3.3, we now know how to construct a free model for any twisted tensor product of DGA's, given free models for the component DGA's.

Corollary 3.4. *Suppose that*

$$f : (\mathrm{T}V, d_V) \longrightarrow (A, d_A)$$

and

$$g : (\mathrm{T}W, d_W) \longrightarrow (B, d_B)$$

are free r-reduced models for the DGA's (A, d_A) and (B, d_B). Let $\mathcal{M}$ denote the free algebra $\mathrm{T}(V \oplus W \oplus s(W \otimes V))$. If $(A \tilde{\otimes} B, D)$ is a twisted tensor product of DGA's, then there exist a differential $\bar{D}$ on $\mathcal{M}$, as in Theorem 3.3, so that $(\mathcal{M}, \bar{D})$ is a DGA, and a quism of DGA's

$$(\mathcal{M}, \bar{D}) \xrightarrow{\simeq} A \tilde{\otimes} B,$$

i.e. $(\mathcal{M}, \bar{D})$ is a free model for $A \tilde{\otimes} B$.

We conclude this section by presenting a theorem giving conditions under which the existence of a twisted tensor product of two DGA's is guaranteed. This theorem will be a key tool in proving results later in the paper.

Theorem 3.5. *Let k be a field. Let (A, d_A), (B, d_B), and (C, d_C) be r-reduced DGA's ($r \geq 1$), with products $\mu_A : A \otimes A \longrightarrow A$, $\mu_B : B \otimes B \longrightarrow B$, and $\mu_C : C \otimes C \longrightarrow C$. Let $(A \tilde{\otimes} B, D)$ denote a DG k-vector space such that as a k-vector space $A \tilde{\otimes} B \approx A \otimes B$ and the sequence*

$$0 \to (B, e) \xrightarrow{\iota} (A \tilde{\otimes} B, D) \xrightarrow{\pi} (A, d) \to 0$$

is a sequence of morphisms of DG k-vector spaces. Suppose that $\phi : A \tilde{\otimes} B \to C$ is a DG k-vector space quism which satisfies

(i.) $\mu_C(\phi(a \otimes b) \otimes \phi(1 \otimes b')) = \phi(a \otimes \mu_B(b \otimes b'))$ *for all $a \in A$ and $b, b' \in B$;*

(ii.) for every $a, a' \in A$ and $b \in B$,

$$\mu_C(\phi(a \otimes b) \otimes \phi(a' \otimes 1)) = (-1)^{\deg a' \deg b} \phi[\mu_A(a \otimes a') \otimes b] + \lambda$$

where $\lambda \in \phi(A_{<\deg a + \deg a'} \otimes B_+)$; and

(iii.) the subDG k-vector space $(A \tilde{\otimes} B_+) \cap \ker \phi$ is acyclic.

Then there exists a product $\mu : (A \tilde{\otimes} B) \otimes (A \tilde{\otimes} B) \longrightarrow A \tilde{\otimes} B$ such that $A \tilde{\otimes} B$ is a twisted tensor product of A and B with respect to μ, and ϕ is a morphism of DGA's.

This theorem is proved by induction on degree in the complex $(A \tilde{\otimes} B) \otimes (A \tilde{\otimes} B)$. The proof proceeds in a fairly straightforward manner, though care must be taken to ensure the linearity and associativity of μ simultaneously. This is done by carrying out the following two-step procedure.

First, assuming that μ has been defined on $[(A\tilde{\otimes}B)\otimes(A\tilde{\otimes}B)]_{<N}$ so that it satisfies

(a) $\phi\mu = \mu_C \circ (\phi\otimes\phi)$ and $D\mu = \mu(D\otimes D)$;

(b) for all $a\otimes b\otimes a'\otimes b' \in [(A\tilde{\otimes}B)\otimes(A\tilde{\otimes}B)]_{<N}$,

$$\mu(a\otimes b\otimes a'\otimes 1) - (-1)^{\deg a' \deg b}\mu_A(a\otimes a')\otimes b \in A_{<\deg a + \deg a'}\otimes B_+$$

and $\mu(a\otimes b\otimes a'\otimes b') = [\mu(a\otimes b\otimes a'\otimes 1)]b'$; and

(c) if $x, y, z \in A\tilde{\otimes}B$ and $\deg x + \deg y + \deg z < N$, then $\mu(\mu(x\otimes y)\otimes z) = \mu(x\otimes\mu(y\otimes z))$,

we write

$$(A\tilde{\otimes}B)_{<N} = \mu[((A\tilde{\otimes}B)\otimes(A\tilde{\otimes}B))_{<N}]\oplus W.$$

Construct μ linearly on $[(A\otimes B)\otimes W]_N$ so that a), b), and c) are satisfied. Define then

$$\mu(x\otimes(y\otimes z)) =: \mu(\mu(x\otimes y)\otimes z)$$

for all $x, y \in A\otimes B$ and $z\in W$ such that $\deg x + \deg y + \deg z = N$.

4. An examination of Brown's equivalence

Let $q : E \longrightarrow B$ be a map of pointed topological spaces. Define

$$U_q = \{(\ell, x) \in PB\times E : \epsilon(\ell) = q(x)\},$$

where $PB = \{\ell : [0,r]\to B : \ell \text{continuous}\}$, the (unbased) Moore paths on B. Here, $\epsilon(\ell) = \ell(r)$, the endpoint of the path ℓ. Suppose $\lambda : U_q \longrightarrow E$ is a map. We say that λ is a lifting function if $q\lambda(\ell, x) = \epsilon(\ell)$ for all $(\ell, x)\in U_q$. Use the notation ℓ_b to denote the path of duration 0 at the point b in B. A lifting function λ is transitive if

(i.) $\lambda(\ell_b, x) = x$ for all $b\in B$ and $x\in E$ such that $q(x) = b$; and

(ii.) $\lambda(\ell_0 * \ell_1, x) = \lambda(\ell_0, \lambda(\ell_1, x))$ when the path composition $\ell_0 * \ell_1$ is defined and when $(\ell_1, x)\in U_q$.

We call λ weakly transitive if (i.) holds for $b = *$, the basepoint of B, and if condition (ii.) holds when $\ell_1(0) = \epsilon(\ell_0) = *$. If λ is a weakly transitive lifting function, the 4-tuple (E, B, q, λ) is said to be a weakly transitive fiber space, or simply a fiber space. Note that a lifting function determines an associative left action of ΩB on F, where $F = q^{-1}(*)$, and therefore an associative left action $\lambda_\# : C_*\Omega B\otimes C_*F \longrightarrow C_*F$.

A pair (f, g) of maps, $f : B \longrightarrow B'$ and $g : E \longrightarrow E'$, is a map of the fiber spaces (E, B, q, λ) and (E', B', q', λ') if the following two diagrams both commute.

$$\begin{array}{ccc} E & \xrightarrow{g} & E' \\ \downarrow{\scriptstyle q} & & \downarrow{\scriptstyle q'} \\ B & \xrightarrow{f} & B' \end{array}$$

$$\begin{array}{ccc} U_q & \xrightarrow{Pf \times g} & U_{q'} \\ \downarrow{\scriptstyle \lambda} & & \downarrow{\scriptstyle \lambda'} \\ E & \xrightarrow{g} & E' \end{array}$$

Denote the category of weakly transitive fiber spaces by $\mathcal{FS}$.

Remark. Any (Hurewicz) fibration is homotopic to a fibration which also has the structure of a transitive fiber space. [B]

For any fiber space (E, B, q, λ), let $\tilde{C}_* E$ be the subcomplex of $C_* E$ consisting of simplices which send the vertices of the cube to F.

Theorem 4.1. *([B]) Assume that we have fixed a natural choice of twisting cochains Φ_B, one for each topological space B. Let $\mathbf{DGM}(R)$ denote the category of DG R-modules, and let $T_1, T_2 : \mathcal{FS} \longrightarrow \mathbf{DGM}(R)$ denote the functors defined by $T_1(E, B, q, \lambda) = C_*^1 B \otimes_{\Phi_B} C_* F$ and $T_2(E, B, q, \lambda) = \tilde{C}_* E$. Then there is a natural equivalence $\phi : T_1 \longrightarrow T_2$.*

The theorem is proved by the method of acyclic models. We do not present the entire proof, for which the reader is refered to [B], but do recall the models in $\mathcal{FS}$ defined in [B], as they will be useful in choosing the Φ_B naturally.

Let $X_{n,m}$ be the fiber space $(E_{n,m}, B_{n,m}, q_{n,m}, \lambda_{n,m})$, where

$$E_{n,m} = P^* \bar{I}^n \times I^m$$

and

$$B_{n,m} = \bar{I}^n.$$

For any n-cube I^n, $\bar{I}^n$ denotes the cube with all vertices identified to the basepoint. For a pointed space Y, P^*Y denotes the Moore paths on Y ending at the basepoint. Define

$$q_{n,m} : P^* \bar{I}^n \times I^m \longrightarrow \bar{I}^n$$

by $q_{n,m}(\ell, x) = \ell(0)$, so that $F_{n,m} = \Omega \bar{I}^n \times \bar{I}^m$. Define

$$\lambda_{n,m} : U_{q_{n,m}} \longrightarrow E_{n,m}$$

by $\lambda_{n,m}(\ell,(\omega,x))=(\ell*\omega,x)$.

Let X_n be the fiber space (E_n,B_n,q_n,λ_n) with

$$E_n=\{(\ell,x)\in P\bar{I}^n\times I^n:\epsilon(\ell)=\bar{x}\}$$

and

$$B_n=\bar{I}^n.$$

Define $q_n:E_n\longrightarrow B_n$ by $q_n(\ell,x)=\ell(0)$, so that

$$F_n=\{(\ell,x)\in P\bar{I}^n\times I^n:\epsilon(\ell)=\bar{x},\ell(0)=*\}.$$

Define $\lambda_n:U_{q_n}\longrightarrow E_n$ by $\lambda_n(\ell,(\omega,x))=(\ell*\omega,x)$.

The set of models in $\mathcal{FS}$ used by Brown [B] is

$$\mathcal{M}=\{X_n,X_{n,m}:n,m\geq 0\}.$$

The goal of this section is to show that we have a certain amount of control over the weak equivalence

$$C^1_*B\otimes_{\Phi_B}C_*F\longrightarrow\tilde{C}_*E,$$

which will aid us in proving the existence of an associative product on $C^1_*\Omega B\otimes_{\Phi_{\Omega B}}C_*\Omega F$. One can choose the twisting cochains Φ_B in a manner somewhat different from Brown [B], so that the topological situation we are considering fits the conditions of the algebraic theorems in Section 3.

We first define $\beta:I\longrightarrow\Omega I$ to be the loop on ΩI such that for any $t\in I$, $\beta(t)$ is the loop given by

$$\beta(t)(s)=\begin{cases}\begin{cases}4s, & \text{if } 0\leq s\leq\frac{t}{2}\\ 4(t-s), & \text{if } \frac{t}{2}\leq s\leq t\end{cases} & \text{if } 0\leq t\leq\frac{1}{2}\\ \begin{cases}4(1-s), & \text{if } 0\leq s\leq\frac{1-t}{2}\\ 4(1-t-s), & \text{if } \frac{1-t}{2}\leq s\leq 1-t.\end{cases} & \text{if } \frac{1}{2}\leq t\leq 1\end{cases}$$

In other words, for $t\leq\frac{1}{2}$, $\beta(t)$ is the loop of duration t from 0 to $2t$ and back. For $t\geq\frac{1}{2}$, $\beta(t)$ is the loop of duration $1-t$ from 0 to $2(1-t)$ and back. Thus, in particular, $\beta(0)=\beta(1)=\ell_*$.

Now let β^m be the map $\beta^m:I^m\longrightarrow\Omega I^m$ defined by

$$\beta^m(t_1,...,t_m)(s)=(\beta(t_1),0,...,0)*\cdots*(0,...,0,\beta(t_m))(s).$$

Note that $\beta^m(v)=\ell_*$ for any vertex v of I^m.

For all $n,m\geq 0$, let $\xi_n:I^n\to\bar{I}^n=B_n=B_{n,m}$ be the identification map, and let $\eta_{n,m}:I^m\to\Omega\bar{I}^n\times I^m=F_{n,m}$ be defined by $\eta_{n,m}(x)=(\ell_*,x)$. The set of simplices $\{\xi_n\otimes\eta_{n,m}:n,m\geq 0\}$ is a basis for the functor T_1 of Theorem 4.1 ([B]).

Theorem 4.2. *There is a set of twisting cochains $\Phi_B \in C^*(C^1_*B; C_*\Omega B)$ one for each topological space B, such that*

(i.) if $U \in C^1_{\leq 1}B$, then $\Phi_B(U) = 0$;

(ii.) if $f : B \longrightarrow B'$, then $(\Omega f)_\# \circ \Phi_B = \Phi_{B'} \circ f_\#$;

(iii.) $\phi_{n,0}(\xi_n \otimes \eta_{n,0})$ can be chosen to be $\xi_n \circ \beta^n$ for all n; and

(iv.) $\phi_{n,m}(\xi_n \otimes \eta_{n,m})$ can be chosen to be $(\phi_{n,0}(\xi_n \otimes \eta_{n,0}), \eta_{0,m})$.

Proof. Because every simplex $U \in C^1_*B$ is represented by a map $U : I^n \longrightarrow B$ which factors through $\bar{I^n}$, for some n, it will suffice to define $\Phi_{B_{n,0}}(\xi_n)$ for every n. Then $\Phi_B(U) = (\Omega\bar{U})_\# \circ \Phi_{B_{n,0}}(\xi_n)$, where $\bar{U} : \bar{I^n} \longrightarrow B$ is the map induced by U. Start the proof by defining $\Phi_{B_{1,0}}(\xi_1) = 0$, so that condition (i.) is fulfilled.

Let Δ denote the coproduct on C_*B. For any $U \in C^1_*B$, write $\Delta(U) = \sum_{i\in I_1} U'_i \otimes U''_i$ and, in general,

$$(1^{\otimes n-1} \otimes \Delta)\cdots(1 \otimes \Delta)\Delta(U) = \sum_{i\in I_n} U_i^{(1)} \otimes \cdots \otimes U_i^{(n+1)}.$$

Recall that Δ is coassociative. If $U \in C^1_*B$, let $\partial_U : C_*B_{n-1,0} \longrightarrow C_*B$ be such that $\partial_U(\xi_{n-1}) = DU$, where D is the differential on C_*B. It is easy to see how to define such a ∂_U.

Define $\Phi_{B_{n,0}}(\xi_n)$ recursively by

$$\Phi_{B_{n,0}}(\xi_n) = (-1)^{n+1}\partial_{\xi_n} \circ \xi_{n-1} \circ \beta^{n-1} - \sum_{i\in I_1}((\xi_n)'_i \circ \beta^{\deg(\xi_n)'_i}) * \Phi_{B_{n,0}}((\xi_n)''_i).$$

It is then a matter of complicated, though straightforward, computation to show that (iii) and (iv) hold and that

$$D(\Phi_B)_q = (\Phi_B)_{q-1}D - \sum_{k=1}^{q-1}(-1)^k(\Phi_B)_k \cup (\Phi_B)_{q-k},$$

where $(\Phi_B)_j$ denotes the component of Φ_B in $C^j(C_*B; C_*\Omega B)$. **QED**

5. The Adams-Hilton model for the total space of a fibration

Since the ultimate goal of this paper, and this section in particular, is to understand the Adams-Hilton model of the total space of a fibration in terms of the Adams-Hilton models of the base space and the fiber, it is important to review some of the most vital properties of the Adams-Hilton model.

If X is a simply-connected CW complex, an Adams-Hilton model for X is a free DGA $\mathcal{A}(X)$, unique only up to homotopy, together with a quism $\theta_X : \mathcal{A}(X) \xrightarrow{\simeq} C_*^1 \Omega X$. The Adams-Hilton model possesses the following properties, among others:

(i.) For any map $f : X \longrightarrow Y$ between simply-connected CW complexes and any choice of models $(\mathcal{A}(X), \theta_X)$ and $(\mathcal{A}(Y), \theta_Y)$, there exists a map of DGA's
$$\mathcal{A}(f) : \mathcal{A}(X) \longrightarrow \mathcal{A}(Y)$$
such that $\theta_Y \circ \mathcal{A}(f)$ is homotopic to $C_*^1 \Omega f \circ \theta_X$.

(ii.) If $X = * \cup \bigcup_{\alpha \in I} e_\alpha$ is a cell decomposition of X, $\deg e_\alpha \geq 2$, then $\mathcal{A}(X)$ may be taken to be $TR\langle b_\alpha : \alpha \in I \rangle$, where $\deg b_\alpha = \deg e_\alpha - 1$.

(iii.) If f and g are homotopic, then $\mathcal{A}(f)$ is homotopic to $\mathcal{A}(g)$. Furthermore if f is a homotopy equivalence, then $\mathcal{A}(f)$ is as well.

(iv.) If $X_0 \subseteq X$ is a subcomplex and $\mathcal{A}(X_0)$ is any model of X_0, then there exists a model of X, $\mathcal{A}(X)$, which is an extension of $\mathcal{A}(X_0)$. Also, if $f : X \longrightarrow Y$ is a map, $f|_{X_0} = f_0$, and $\mathcal{A}(f_0)$ is a model for f_0, then there exists a model $\mathcal{A}(f)$ of f extending $\mathcal{A}(f_0)$.

The above represents merely a small selection of the major properties of the Adams-Hilton model. For a more in-depth description, as well as a discussion of homotopy in the category of DGA's, the reader is refered to [A1].

Throughout the remainder of this section (E, B, q, λ) will denote a fibration of pointed spaces, which also has the structure of a (weakly transitive) fiber space. As was remarked in Section 4, this is not a restrictive hypothesis, since every fibration is homotopic to a fibration with a transitive lifting function.

The goal of this section is to prove

Theorem 5.1. *Let (E, B, q, λ) be a fiber space, and let $F = q^{-1}(*)$. Suppose that $\mathcal{A}(F) \cong (TW, e)$ and $\mathcal{A}(B) \cong (TV, d)$ are Adams-Hilton models for F and B. Then the Adams-Hilton model for E, $\mathcal{A}(E)$, is weakly equivalent to a twisted tensor product of $\mathcal{A}(B)$ and $\mathcal{A}(F)$ and can be chosen to be $(TW \amalg TV \amalg Ts(W \otimes V), D)$, where $Dv = dv$, $Dw = ew$ and*
$$\bar{D}s(w \otimes v) = w \otimes v + (-1)^{\deg w \deg v} v \otimes w + \alpha$$

where $\alpha \in [\mathrm{T}(V \oplus W) \amalg \mathrm{T}^+ s(W \otimes V)] \oplus [(\mathrm{T}V)_{<\deg v} \otimes \mathrm{T}W]$. *Furthermore, given a map f between two fiber spaces, together with choices of Adams-Hilton models for the fiber and base spaces, there is a DGA morphism v such that the following diagram commutes.*

$$\begin{array}{ccc} \mathrm{T}W \amalg \mathrm{T}V \amalg \mathrm{T}s(W \oplus V) & \xrightarrow{v} & \mathrm{T}W' \amalg \mathrm{T}V' \amalg \mathrm{T}s(W' \oplus V') \\ \downarrow \simeq & & \downarrow \simeq \\ C_*^1 \Omega E & \xrightarrow{C_*^1 \Omega f} & C_*^1 \Omega E' \end{array}$$

The first step in the proof of Theorem 5.1 is to show that a product can be defined on $C_*^1\Omega B \otimes_{\Phi_{\Omega B}} C_*^1 \Omega F$ making it into a twisted tensor product of the DGA's $C_*^1\Omega B$ and $C_*^1\Omega F$ and such that the weak equivalence $C_*^1\Omega B \otimes_{\Phi_{\Omega B}} C_*^1\Omega F \longrightarrow C_*^1\Omega E$ is a map of DGA's. This is the content of the next theorem.

Remark. It is clear from Theorem 4.2 that $C_*^1\Omega B \otimes C_*^1\Omega F$ is closed under the differential defined in terms of the twisting cochain $\Phi_{\Omega B}$. Let ι denote the inclusion

$$C_*^1\Omega B \otimes_{\Phi_{\Omega B}} C_*^1\Omega F \hookrightarrow C_*^1\Omega B \otimes_{\Phi_{\Omega B}} C_*\Omega F.$$

Also obvious as a result of Theorem 5.2 is that $\mathrm{Im}(\phi \circ \iota) \subseteq C_*^1\Omega E$. For the remainer of this section, ϕ' will denote the composition $\phi \circ \iota$.

Before continuing further, it is necessary to define an appropriate lifting function for the looped fibration. It will be easier for us to work with a fiber space which is homotopy equivalent to the looped fibration, instead of the looped fibration itself. Since $q\lambda(\ell, x) = \epsilon(\ell)$, the image of the restriction of λ to $\Omega B \times F$ is contained in F. Thus $\lambda(-,*)$ defines a basepoint preserving map from ΩB to F. By taking the cone on $\lambda(-,*)$, we can assume that ΩB is a subspace of F and the action of ΩB on F is an extension of the loop multiplication on ΩB.

The sequence

$$\Omega F \longrightarrow \Omega E \xrightarrow{\Omega q} \Omega B$$

is then homotopy equivalent to the fiber space

$$\Omega F \longrightarrow \bar{E} \xrightarrow{\bar{q}} \Omega B$$

where $\bar{E} = \{(b, \ell) \in \Omega B \times P^*F : \ell(0) = b\}$ and $\bar{q}(b,\ell) = b$. The map $\bar{q}$ is clearly an H-map, where the product on $\bar{E}$ is given by $(b,\ell)\cdot(b',\ell') = (b * b', \ell'')$, where $\ell''(t) = \lambda(b, \ell'(-)) * \ell(t)$. Define $\bar{\lambda} : U_{\bar{q}} \longrightarrow \bar{E}$ by $\bar{\lambda}(\omega, (b,\ell)) = (\omega(0), \omega * \ell)$. Note that we will often write just ℓ instead of $(\ell(0), \ell)$ for elements of $\bar{E}$.

Theorem 5.2. *Let (E, B, q, λ) be a fiber space. There exists an associative product*

$$\nu : (C^1_*\Omega B \otimes_{\Phi_{\Omega B}} C^1_*\Omega F) \otimes (C^1_*\Omega B \otimes_{\Phi_{\Omega B}} C^1_*\Omega F) \longrightarrow C^1_*\Omega B \otimes_{\Phi_{\Omega B}} C^1_*\Omega F$$

such that $C^1_*\Omega B \otimes_{\Phi_{\Omega B}} C^1_*\Omega F$ is a twisted tensor product of the DGA's $C^1_*\Omega B$ and $C^1_*\Omega F$ with respect to the product ν. Furthermore the diagram

$$\begin{array}{ccc} (C^1_*\Omega B \otimes_{\Phi_{\Omega B}} C^1_*\Omega F) \otimes (C^1_*\Omega B \otimes_{\Phi_{\Omega B}} C^1_*\Omega F) & \xrightarrow{\nu} & C^1_*\Omega B \otimes_{\Phi_{\Omega B}} C^1_*\Omega F \\ \downarrow \phi' \otimes \phi' & & \downarrow \phi' \\ C^1_*\bar{E} \otimes C^1_*\bar{E} & \xrightarrow{C_*\mu \circ \tau} & C^1_*\bar{E} \end{array}$$

commmutes, where $\mu : \bar{E} \times \bar{E} \longrightarrow \bar{E}$ is the multiplication as defined above, i.e., $C_*\mu \circ \tau$ is the usual product on $C_*\bar{E}$, and τ is the Eilenberg-Zilber equivalence.

In order to be able to construct such a product ν, it is clearly necessary to have $\mathrm{Im}(C_*\mu \circ \tau \circ (\phi' \otimes \phi')) \subseteq \mathrm{Im}\phi'$. This is a consequence of Theorem 4.2, as will be shown in the next lemma.

Recall that $\tau : C_*\bar{E} \otimes C_*\bar{E} \longrightarrow C_*(\bar{E} \times \bar{E})$ can be chosen so that $\tau(S \otimes T) = (S, T)$ for simplices $S, T \in C_*\bar{E}$.

Lemma 5.3. *Under the hypotheses of the previous theorem and with the above choice of τ,*

$$\mathrm{Im}(C_*\mu \circ \tau \circ (\phi' \otimes \phi')) \subseteq \mathrm{Im}\phi'(1 \otimes C^1_*\Omega F).$$

Proof. Choose simplices $S, U \in C^1_*\Omega B$ and $T, V \in C^1_*\Omega F$. Suppose that $\deg S = m$ and $\deg U = n$. Then, according to Theorem 4.2,

$$\begin{aligned} &C_*\mu \circ \tau \circ (\phi' \otimes \phi')(S \otimes T \otimes U \otimes V) \\ &= C_*\mu \circ ((S \circ \phi_{m,0}(\xi_m \otimes \eta_{m,0}) * T), (U \circ \phi_{n,0}(\xi_n \otimes \eta_{n,0}) * V)) \\ &= (S \circ \xi_m \circ \beta^m) * T * (U \circ \xi_n \circ \beta^n) * V. \end{aligned}$$

This is a composition of loops on F, as $\xi_k \circ \beta^k$ is a loop for each k. Thus

$$C_*\mu \circ \tau \circ (\phi' \otimes \phi')(S \otimes T \otimes U \otimes V) = \phi'(1 \otimes (S \circ \xi_m \circ \beta^m) * T * (U \circ \xi_n \circ \beta^n) * V).$$

QED

Before proceeding to the proof of Theorem 5.2, we wish to prove two further lemmas which will be useful in the construction of ν. In fact, once we have the results in the following two lemmas, the proof of Theorem 5.2 will be reduced to a straightforward application of Theorem 3.5.

Lemma 5.4. *Let S and U be simplices in $C^1_*\Omega B$ and T and V be simplices in $C_*\Omega F$. Suppose that $\deg S = m$, $\deg T = p$ and $\deg U = n$. Then there exists $W \in C_*\Omega F$ such that*

$$\phi'(\lambda(S,U))*T*V-(-1)^{np}\phi'(S)*T*\phi'(U)*V-D\phi'(1\otimes W) \in \phi'(1\otimes C_*\Omega F).$$

Proof. The proof of this lemma is similar to the standard proof that the group of homotopy classes of maps from a given space into the loops on an H-space is abelian.

We start by defining a second action, in addition to the previously defined $\bar{\lambda}$, namely

$$\lambda' : P^*\Omega B \times \Omega F \longrightarrow \bar{E}.$$

For $\ell \in PZ$, Z a topological space, let $\mathrm{d}(\ell)$ denote the duration of the path ℓ, that is $\ell : [0,r] \longrightarrow Z$. Let ℓ_*^r denote the path of duration r which sits at the basepoint. Define λ' by

$$\lambda'(\ell,\omega)(t) = \begin{cases} \begin{cases} \lambda(\ell(t),\omega(t)), & \text{if } t \leq \mathrm{d}(\omega) \\ \ell(t), & \text{if } \mathrm{d}(\omega) \leq t \leq \mathrm{d}(\ell) \end{cases} & \text{if } \mathrm{d}(\omega) \leq \mathrm{d}(\ell) \\ \begin{cases} \lambda(\ell(t),\omega(t)), & \text{if } t \leq \mathrm{d}(\ell) \\ \lambda(\epsilon(\ell),\omega(t)), & \text{if } \mathrm{d}(\ell) \leq t \leq \mathrm{d}(\omega) \end{cases} & \text{if } \mathrm{d}(\ell) \leq \mathrm{d}(\omega) \end{cases}$$

for $\ell \in P^*\Omega B$ and $\omega \in \Omega F$.

Note that for $\ell_0, \ell_1 \in \Omega(\Omega B)$ and $\omega_0, \omega_1 \in \Omega F$

$$\lambda'(\ell_0,\omega_0)*\lambda'(\ell_1,\omega_1) = \begin{cases} \lambda'(\ell_0 * \ell_1, \omega_0 * \ell_*^{\mathrm{d}(\ell_0)-d(\omega_0)} * \omega_1), & \mathrm{d}(\ell_0) \geq \mathrm{d}(\omega_0) \\ \\ \lambda'(\ell_0 * \ell_*^{\mathrm{d}(\omega_0)-\mathrm{d}(\ell_0)} * \ell_1, \omega_0 * \omega_1), & \mathrm{d}(\omega_0) \geq \mathrm{d}(\ell_0). \end{cases}$$

Next suppose that $V \in C^1_q\Omega F$. Let $N = m+n+p+q$. For any point $x = (x_1, ..., x_N)$ in I^N and any $1 \leq i \leq j \leq N$, let x_i^j denote the subsequence $(x_i, ..., x_j)$. Define $\tau_i : I^N \longrightarrow I^N$ by $\tau_i(x_1, ..., x_N) = (x_1, ..., x_{i+1}, x_i, ..., x_N)$. Then τ_i is homotopic to the identity, via a homotopy $G_i : I^{N+1} \longrightarrow I^N$.

Define $H_i : I^{N+1} \longrightarrow \bar{E}$, $i = 1, 2$, by

$$H_1(t,x) = [(S \circ \xi_m \circ \beta^m) * \lambda'(\ell_*^{(1-t)\mathrm{d}(T(x))} * U \circ \xi_n \circ \beta^n, T) * V](x)$$

$$H_2(t,x) = [(S \circ \xi_m \circ \beta^m) * \lambda'(U \circ \xi_n \circ \beta^n, (\ell_*^{t\mathrm{d}[U\circ\xi_n\circ\beta^n(x)]} * T)) * V](x)$$

where $S \circ \xi_m \circ \beta^m$ is evaluated on x_1^m, $U \circ \xi_n \circ \beta^n$ on x_{m+1}^{m+n}, T on x_{m+n+1}^{m+n+p}, and V on x_{N-q+1}^N. Then

$$D(H_1 + H_2 - \sum_{i=m+1}^{m+n} \sum_{k=0}^{p-1} \phi'(S) * T * \phi'(U) * V \circ \tau_i^k \circ G_i)$$
$$= \phi'(\lambda(S,U)) * T * V - (-1)^{np}\phi'(S) * T * \phi'(U) * V + \tilde{H}$$

where $\tilde{H}(t,x) = (H_1 + H_2 - \sum_{i=m+1}^{m+n} \sum_{k=0}^{p-1} \phi'(S) * T * \phi'(U) * V \circ \tau_i^k \circ G_i)(t, \partial(x))$ for $x \in I^{N-1}$, $t \in I$. Here, $\partial : I^{N-1} \longrightarrow I^N$ is the alternating sum of the face maps.

Thus, since the images of the H_i and of $\phi'(S) * T * \phi'(U) * V \circ \tau_i^k \circ G_i$ lie in $1 \otimes \Omega F$ and the restriction of ϕ' to $1 \otimes C_*^1 \Omega F$ is simply the inclusion, the lemma is true. **QED**

Lemma 5.5. *Let D denote the differential in $C_*^1 \Omega B \otimes_{\Phi_{\Omega B}} C_*^1 \Omega F$. If*

$$\Psi \in [C^1_{>k+1} \Omega B \otimes_{\Phi_{\Omega B}} C^1_{<N-k} \Omega F]_{N+1}$$

and

$$D\Psi \in [C^1_{\leq k} \Omega B \otimes C^1_{\geq N-k} \Omega F]_N,$$

then there exists $\Psi' \in [C^1_{\leq k+1} \Omega B \otimes C^1_{\geq N-k} \Omega F]_{N+1}$ which satisfies: (1) $D\Psi' = D\Psi$ and (2) $\phi'(\Psi') = \phi'(\Psi)$.

The proof, a somewhat delicate induction on k, is omitted here, due to its technicality.

As an immediate consequence of Lemma 5.5, we have the following

Corollary 5.6. $\ker \phi' \cap [C_*^1 \Omega B \otimes C_+^1 \Omega F]$ *is acyclic.*

We are now ready to prove Theorem 5.2.

Proof of Theorem 5.2 Consider the diagram

$$\begin{array}{ccc} (C_*^1 \Omega B \otimes_{\Phi_{\Omega B}} C_*^1 \Omega F) \otimes (C_*^1 \Omega B \otimes_{\Phi_{\Omega B}} C_*^1 \Omega F) & & C_*^1 \Omega B \otimes_{\Phi_{\Omega B}} C_*^1 \Omega F \\ \downarrow{\scriptstyle \phi' \otimes \phi'} & & \downarrow{\scriptstyle \phi'} \\ C_*^1 \bar{E} \otimes C_*^1 \bar{E} & \xrightarrow{C_* \mu \circ \tau} & C_*^1 E. \end{array}$$

We claim that the situation of Theorem 5.2 exactly fits the hypotheses of Theorem 3.5, with $r = 1$. The differential on $C_*^1 \Omega B \otimes_{\Phi_{\Omega B}} C_*^1 \Omega F$ as defined in terms of the twisting cochain $\Phi_{\Omega B}$ certainly has the required form. Since the restriction of ϕ' to $C_*^1 \Omega F$ is an inclusion of DGA's, condition (i.) is also satisfied. Condition (ii.) is just Lemma 5.4, while Corollary 5.6 says that condition (iii.) is fulfilled as well. Thus Theorem 3.5 assures us of the existence of the product

$$\nu : (C_*^1 \Omega B \otimes_{\Phi_{\Omega B}} C_*^1 \Omega F) \otimes (C_*^1 \Omega B \otimes_{\Phi_{\Omega B}} C_*^1 \Omega F) \longrightarrow C_*^1 \Omega B \otimes_{\Phi_{\Omega B}} C_*^1 \Omega F.$$

QED

Theorem 5.1, the result motivating this article, is now within our reach.

Proof of Theorem 5.1. The proof is quite straightforward at this point, involving applying results of Section 3 to Theorem 5.2.

Suppose that $\theta_B : \mathcal{A}(B) \longrightarrow C_*^1(\Omega B)$ and $\theta_F : \mathcal{A}(F) \longrightarrow C_*^1(\Omega F)$ are Adams-Hilton models. Let TV and TW be the underlying algebras for $\mathcal{A}(B)$ and $\mathcal{A}(F)$, respectively. Corollary 3.4 assures us of the existence of a free model for $C_*^1(\Omega B) \tilde{\otimes} C_*^1(\Omega F)$ of the form

$$(TW \amalg TV \amalg Ts(W \oplus V), \bar{D}) \xrightarrow{\rho} C_*^1(\Omega B) \tilde{\otimes} C_*^1(\Omega F).$$

Let $h : \bar{E} \longrightarrow \Omega E$ be the homotopy equivalence discussed at the beginning of the section. Then the composition

$$(TW \amalg TV \amalg Ts(W \oplus V), \bar{D}) \xrightarrow{h \circ \phi' \circ \rho} C_*^1 \Omega E \qquad (\star)$$

is a quism and therefore is an acceptable Adams-Hilton model for E.

Given a map between two fibrations and choices of models for the base and fiber spaces, constructing

$$v : TW \amalg TV \amalg Ts(W \oplus V) \longrightarrow TW' \amalg TV' \amalg Ts(W' \oplus V')$$

is a simple matter of applying the lifting lemma for DGA's. [A1]
QED

References.

[A1] D. Anick, *Hopf algebras up to homotopy*, JAMS **2** (1989) 417-453

[A2] D. Anick, *The Adams-Hilton model for a fibration over a sphere*, (1990) preprint

[A-H] J.F. Adams and P.J. Hilton, *On the chain algebra of a loop space*, Comm. Math. Helv. **30** (1955) 305-330

[B] E. Brown, *Twisted tensor products, I*, Ann. Math. (2) **69** (1959) 223-246

[L-S] L. Lambe and J.D. Stasheff, *Applications of perturbation theoryto iterated fibrations*, Manuscripta Math. **58** (1987) 363-376

[S] L. Smith, *Split extensions of Hopf algebras and semi-tensor products*, Math. Scand. **26** (1970) 17-41

[St] J.D. Stasheff, *Homotopy associativity of H-spaces II*, Trans AMS **108** (1963) 293-312

[T] D. Tanré, *Homotopie rationelle: Modèles de Chen, Quillen, Sullivan* LNM **1025** (1983)

Département de mathématiques
Ecole Polytechnique Fédérale de Lausanne
CH-1015 Lausanne Switzerland

HOCHSCHILD HOMOLOGY, CYCLIC HOMOLOGY, AND THE COBAR CONSTRUCTION

J.D.S. JONES AND J. MCCLEARY

University of Warwick and Vassar College

To the memory of J.F. Adams

Introduction. The free loop space appears in two important computations in cyclic homology. Let X be a simply-connected space with based loop space ΩX and free loop space $\mathcal{L}X = \mathrm{map}(S^1, X)$. Using the Alexander-Whitney product the singular cochains $C^*(X)$ becomes a differential graded algebra and using the product induced by loop multiplication the singular chains $C_*(\Omega X)$ also becomes a differential graded algebra; then

$$\mathrm{HH}_*(C^*(X)) \cong \mathrm{H}^*(\mathcal{L}X), \qquad \mathrm{HC}_*^-(C^*(X)) \cong \mathrm{H}_{\mathrm{T}}^*(\mathcal{L}X)$$
$$\mathrm{HH}^*(C_*(\Omega X)) \cong \mathrm{H}(\mathcal{L}X), \qquad \mathrm{HC}^*(C_*(\Omega X)) \cong \mathrm{H}_{\mathrm{T}}^*(\mathcal{L}X).$$

The functors which appear in these isomorphisms are as follows:

(1) HH_* and HH^* are Hochschild homology and cohomology,
(2) HC_*^- is negative cyclic homology,
(3) HC^* is cyclic cohomology,
(4) $\mathrm{H}_{\mathrm{T}}^*$ is (Borel) equivariant cohomology for the circle group T.

The Hochschild and cyclic groups are described in more detail in §2 and §5. It is to be understood that the circle, T, acts on $\mathcal{L}X$ by rotating loops. Proofs of the isomorphisms involving $C^*(X)$ are given in [J]. Those involving $C_*(\Omega X)$ are proved in [G]; there are also proofs in [B-F] and [J].

The problem which motivated this paper is to understand the relation between these two, apparently different, computations which lead to the same answer. The key to relating them is a theorem of Adams [A]. Using the Alexander-Whitney coproduct the singular chains $C_*(X)$ becomes a differential graded co-algebra whose dual is the differential graded algebra $C^*(X)$. We can form the cobar construction $F(C_*(X))$ on the co-algebra $C_*(X)$ and Adams proves that there is an equivalence, of differential graded algebras,

$$C_*(\Omega X) \simeq F(C_*(X)).$$

Therefore it follows from the isomorphisms above that

$$\mathrm{HH}_*(C^*(X)) \cong \mathrm{HH}^*(F(C_*(X))), \qquad \mathrm{HC}_*^-(C^*X) \cong \mathrm{HC}^*(F(C_*(X))).$$

This leads very naturally to our main observation.

Theorem. *Suppose (C, ∂) is a simply-connected, counital, coaugmented, differential graded coalgebra and let (Γ, d) be the dual differential graded algebra. Then there are natural isomorphisms*

$$\mathrm{HH}^*(F(C)) \cong \mathrm{HH}_*(\Gamma), \qquad \mathrm{HC}^*(F(C)) \cong \mathrm{HC}_*^-(\Gamma).$$

Throughout we use homology grading, thus differentials have degree -1 and the algebra Γ is graded in negative degrees. Our proof is very straightforward. The key point is that the cobar construction is free as a graded algebra so, from [L-Q] and [V], we get special (explicit) chain complexes which compute Hochschild and cyclic homology. When we examine these special chain complexes in the case of the cobar construction $F(C)$ we see that they are precisely, that is at the chain level, the duals of the standard Hochschild complex for computing the Hochschild homology of Γ and Connes's b-B complex for computing cyclic homology. The result in Hochschild homology is also proved in [H-V] by rather different methods.

The above theorem, while not explicitly stated in the literature, can be derived from known results by direct methods. However we feel the result is of some interest for two main reasons:

(1) It illuminates the relation between the two, seemingly different, appearances of the free loop space in cyclic theory by making the general underlying algebraic fact explicit.
(2) This general algebraic fact may be of some use in the theory of cyclic homology since it shows that the algebras Γ and $F(C)$ are, in some sense, dual.

Thus it is worth drawing attention to this result and giving a proof.

The computation of the cyclic theory of $C_*(\Omega X)$ and its relation with the free loop space is very important in applications of cyclic theory to Waldhausen K-theory. It is shown in [G-J-P] that if X is a smooth manifold and $\mathcal{A}^*(X)$ is the algebra of differential forms on X, then the computation of the cyclic theory of $\mathcal{A}^*(X)$ and its relation with the free loop space is related to index theory. It would be very interesting to establish some direct link between the invariants of Waldhausen K-theory derived from cyclic theory and index theory and, in view of the above computations, it is very tempting to think that this conjectural link has something to do with free loop spaces. This is one of the main reasons for trying to understand the relation between the two computations.

We would like to thank Ezra Getzler for suggesting that the above result can be interpreted in terms of the bivariant cyclic theory of [J-K]; we carry this out in §8.

§1 Coalgebras and the cobar construction. Let (C, ∂) denote a co-unital, supplemented, differential graded coalgebra over a field k. We assume that $C_0 = k$ and that $C_n = 0$ if $n < 0$. In this case both the co-unit and the supplementation give isomorphisms of C_0 with k. Let JC denote the cokernel of the supplementation and let $\bar{C}$ be the kernel of the co-unit. This gives differential graded morphisms

$$\bar{C} \xrightarrow{i} C \xrightarrow{\pi} JC$$

and differential graded isomorphisms

$$\bar{C} \cong \bigoplus_{i>0} C_i \cong JC.$$

Define the *reduced coproduct* as the composite

$$\bar{\Delta} : JC \cong \bar{C} \xrightarrow{i} C \xrightarrow{\Delta} C \otimes C \xrightarrow{\pi \otimes \pi} JC \otimes JC.$$

If $c \in C_n$ we write

$$\Delta(c) = \sum_{i=0}^{n} \Delta_{i,n-i}(c) \in \bigoplus_{i=0}^{n} C_i \otimes C_{n-i}$$

and we use a similar notation for the components of $\bar{\Delta}$.

The *cobar construction* on C is defined as follows. We adopt Adams's conventions and denote the cobar construction by $(F(C), \partial_F)$. The desuspension $s^{-1}V$ of a graded vector space V is defined by $(s^{-1}V)_n = V_{n+1}$. As an algebra

$$F(C) = T(s^{-1}JC),$$

is the tensor algebra on the graded vector space $s^{-1}JC$. We use the notation

$$\langle c_1 | \cdots | c_n \rangle = s^{-1}c_1 \otimes \cdots \otimes s^{-1}c_n \in F(C), \quad c_i \in JC$$

to denote elements in the cobar construction. Note that

$$\deg_{F(C)} \langle c_1 | \cdots | c_n \rangle = \deg_C c_1 + \cdots \deg_C c_n - n.$$

Let $i : s^{-1}JC \to F(C)$ be the inclusion. The differential ∂_F is the derivation on $F(C)$ determined by the map

$$\partial_F \circ i : s^{-1}JC \to F(C),$$

defined by

$$\partial_F \circ i\langle c \rangle = \langle \partial c \rangle + \sum_i (-1)^{i-1} \langle \bar{\Delta}_{i,n-i}(c) \rangle$$

for $c \in JC_n$. With this structure $(F(C), \partial_F)$ becomes a differential graded algebra.

§2 Hochschild homology. Let (A, d) denote a unital, augmented, differential graded algebra over a field k. We assume that $A_0 = k$ and that either $A_n = 0$ for $n < 0$ or $A_n = 0$ for $n > 0$. If C is a coalgebra, as in §1, and in addition is simply connected, that is $C_1 = 0$, then $F(C)$ satisfies the first hypothesis and the dual algebra, negatively graded, satisfies the second. The *normalised bar construction* on A, denoted $(\bar{B}(A), d_B)$, is defined in an exactly dual manner to the cobar construction on a coalgebra, described in §1, using the augmentation ideal IA in place of JC. Thus $\bar{B}(A)$ is the free coalgebra generated by $sI(A)$. We use the notation

$$[a_1|\cdots|a_n] = sa_1 \otimes \cdots \otimes sa_n \in \bar{B}(A), \quad a_i \in I(A).$$

Let $p : \bar{B}(A) \to sI(A)$ be the projection. The differential d_B is the coderivation determined by the map

$$p \circ d_B : \bar{B}(A) \to sI(A),$$

defined by

$$\begin{aligned} p \circ d_B(1) &= 0, \\ p \circ d_B[a] &= [da], \\ p \circ d_B[a_1|a_2] &= (-1)^{|a_1|+1}[a_1 a_2], \\ p \circ d_B[a_1|\cdots|a_n] &= 0, \qquad \text{for } n \geq 2. \end{aligned}$$

Now define

$$\theta : A \otimes \bar{B}(A) \to A \otimes \bar{B}(A)$$

of degree -1, by the formula

$$\theta(a \otimes [u_1|\ldots|u_n]) = (-1)^{|a|} au_1 \otimes [u_2|\ldots|u_n] - (-1)^{\varepsilon} u_n a \otimes [u_1|\ldots|u_{n-1}]$$

where the sign is given by

$$\varepsilon = (|u_n| + 1)(|a| + |u_1| + \cdots + |u_{n-1}| + n + 1).$$

Define the total differential d on $A \otimes \bar{B}(A)$ by

$$d(a \otimes [u_1|\cdots|u_n]) = -da \otimes [u_1|\cdots|u_n] + (-1)^{|a|} a \otimes d_B[u_1|\cdots|u_n].$$

Then it is a standard computation to check that

$$d\theta + \theta d + \theta \circ \theta = 0.$$

The *normalised Hochschild complex* of A is defined as follows. Let b be the differential $A \otimes \bar{B}(A)$ given by

$$b = d + \theta.$$

Then $b^2 = 0$ and this operator b is the *Hochschild boundary operator*. The complex $(A \otimes \bar{B}(A), b)$ will be denoted by $\mathcal{C}(A)$ and is refered to as the *Hochschild complex* of A. The *Hochschild homology* of A is given by

$$\mathrm{HH}_*(A) = \mathrm{H}_*(\mathcal{C}(A))$$

and the *Hochschild cohomology* of A, is the cohomology of the dual complex

$$\mathrm{HH}^*(A) = \mathrm{H}^*(\mathcal{C}^*(A)).$$

In order to have all differentials of degree -1 we grade the complex $\mathcal{C}^*(A)$ negatively, that is, $\mathcal{C}^{-n}(A) = \mathrm{Hom}(C_n(A), k)$. With this grading convention $\mathrm{HH}^n(A) = 0$ if $n > 0$.

§3 The Hochschild homology of a free algebra. Now suppose that the algebra A, as in §2, is free as an algebra over k, that is $A = T(V)$ is the tensor algebra on a graded vector space V. Then a much smaller complex can be used to compute $\mathrm{HH}_*(A)$. This complex is introduced in [L-Q] for algebras over k and it is extended to differential graded algebras in [B-V] and [V].

Define maps

$$\begin{aligned} j &: A \otimes A \to A \otimes V \\ \delta' &: A \otimes V \to A \otimes V \\ \tau &: A \otimes V \to A \\ \delta &: A \otimes V \to A \otimes V \end{aligned}$$

by the following formulas:

$$\begin{aligned} j(a \otimes 1) &= 0 \\ j(a \otimes v_1 \cdots v_p) &= \sum_{i=1}^{p} (-1)^{\mu_i} v_{i+1} \cdots v_p a v_1 \cdots v_{i-1} \otimes v_i \\ \delta'(a \otimes v) &= da \otimes v + (-1)^{|a|} j(a \otimes dv) \\ \tau(a \otimes v) &= av - (-1)^{|a||v|} va \\ \delta &= \delta' + (-1)^{\text{total degree}} \tau \end{aligned}$$

Here $a \in A$, $v, v_i \in V$ and the sign in the definition of j is given by

$$\mu_i = (|v_{i+1}| + \cdots + |v_p|)(|a| + |v_1| + \cdots + |v_i|).$$

In case there might be any ambiguity, the terms in the definition of j which occur when $i = 1$ and $i = p$ are

$$(-1)^{(|v_2|+\cdots+|v_p|)(|a|+|v_1|)} v_2 \cdots v_p a \otimes v_1, \qquad a v_1 \cdots v_{p-1} \otimes v_p$$

The following formulas can be checked by direct computations:

(1) $jd = \delta' j$,
(2) $(\delta')^2 = 0$,
(3) $\delta^2 = 0$,
(4) $\tau j(a \otimes u) = [a_1, u] = (-1)^{|a|}\theta(a \otimes [u])$, where θ is as in §2.

The following two formulas are very useful when computing with j. If $a_1, a_2 \in A$ and $v \in V$, then

$$j(a_1 \otimes a_2 v) = a_1 a_2 \otimes v + (-1)^{|v|(|a_1|+|a_2|)} j(v a_1 \otimes a_2)$$
$$j(a_1 \otimes v a_2) = j(a_1 v \otimes a_2) + (-1)^{|a_2|(|a_1|+|v|)} a_2 a_1 \otimes v.$$

Now consider the graded vector space $W = k \oplus sV$ where k has degree zero and sV is the suspension of the graded vector space V, that is, $(sV)_n = V_{n-1}$. We use the obvious notation $a \otimes 1$, $a \otimes sv$ for elements in $A \otimes W$. In view of the above formulas we define a map

$$\bar{\jmath} : A \otimes \bar{B}(A) \to A \otimes W$$

as follows:

$$\bar{\jmath}(a \otimes 1) = a$$
$$\bar{\jmath}(a \otimes [u]) = (-1)^{|a|}(1 \otimes s) j(a \otimes u), \qquad \text{if } u \in IA$$
$$\bar{\jmath}(a \otimes [u_1|\cdots|u_n]) = 0, \qquad \text{if } n \geq 2.$$

By construction we get a chain map

$$\bar{\jmath} : (\mathcal{C}(A), b) \to (A \otimes W, \delta).$$

Theorem [V]. *The chain map $\bar{\jmath}$ induces an isomorphism*

$$\mathrm{HH}_*(A) \cong \mathrm{H}_*(A \otimes W, \delta)$$

Proof. We may consider $A \otimes \bar{B}(A)$ as the total complex $C_{pq} = (A \otimes IA^{\otimes p})_q$ with differentials given by the components of $d + \theta$. Similarly $A \otimes W$ is the total complex of a double complex with $\bar{C}_{0q} = A_q$ and $\bar{C}_{1q} = (A \otimes V)_q$ and differentials δ' and τ. Filter these double complexes by

$$\mathcal{F}_n C = \bigoplus_{q \leq n} C_{*q}, \quad \mathcal{F}_n \bar{C} = \bigoplus_{q \leq n} \bar{C}_{*q}.$$

The mapping $\bar{\jmath}$ preserves the filtration and induces a mapping of the spectral sequences associated with the filtrations with

$$E_1 \cong \mathrm{HH}_*(A, 0) \to \bar{E}_1 \cong \mathrm{H}_*(A \otimes W, \tau).$$

Here $\mathrm{HH}_*(A, 0)$ means the Hochschild homology of A equipped with the zero differential. According to [L-Q] and [B-V], $\bar{\jmath}$ induces an isomorphism on E_1-terms and the standard argument now completes the proof.

§4 The duality in Hochschild homology. Now let (C, ∂) denote a differential graded coalgebra, satisfying the hypotheses of §1 and in addition we assume that C is simply connected. Let (Γ, d) be the dual differential graded algebra. In order to work with differentials all of degree -1 we adopt the grading convention $\Gamma^{-n} = \mathrm{Hom}(C_n, k)$ and therefore grade Γ in non-positive dimensions. With this grading convention the assumptions on the coalgebra C imply that $\mathrm{HH}_n(\Gamma) = 0$ for $n > 0$.

Since $F(C)$ is a free algebra we may apply the construction of §3 to compute Hochschild homology. As graded vector spaces, the chain complex of the previous section can be identified with $F(C) \otimes C$. Using the formula of §1 for the differential ∂_F we obtain the following formula for the total differential $\delta = \delta' + (-1)^{\text{total degree}}\tau$ on $F(C) \otimes C$:

$$\delta = \partial + \varphi$$

where ∂ is the total differential in $F(C) \otimes C$ given by

$$\partial(a \otimes u) = \partial_F(a) \otimes u + (-1)^{|a|} a \otimes \partial(u) + (-1)^{|a|+|u|}\tau(a \otimes u),$$

and

$$\varphi(a \otimes c) = (-1)^{|a|}\bar{\jmath}(a \otimes \langle \bar{\Delta}(c) \rangle).$$

Let us compare this with the Hochschild complex $\mathcal{C}(\Gamma)$ of Γ. As vector spaces $F(C) \otimes C$ and $\Gamma \otimes \bar{B}(\Gamma)$ are exactly dual in that

$$(\Gamma \otimes \bar{B}(\Gamma))^{-n} = \mathrm{Hom}((F(C) \otimes C)_n, k).$$

By definition the operator d is the dual of ∂. Expressing θ and φ in terms of $\bar{\Delta}$, $(1 \otimes s)$, and the twist map $T(a \otimes b) = (-1)^{|a||b|} b \otimes a$ shows that θ is the dual of φ. This allows us to conclude the following duality theorem in Hochschild homology.

Theorem A. *Let (C, ∂) denote a simply connected, counital, coaugmented, differential graded coalgebra and let (A, d) be the dual algebra. Then there is a natural isomorphism*

$$\mathrm{HH}^*(F(C)) \cong \mathrm{HH}_*(\Gamma).$$

Notice that with our grading conventions both $\mathrm{HH}^*(F(C))$ and $\mathrm{HH}_*(\Gamma)$ are non-positively graded. This theorem is proved in [H-V] by rather different methods.

§5 Cyclic theory. The extra ingredient in the definition of cyclic theory is *Connes's B-operator.* In general this is the operator

$$B : A \otimes \bar{B}(A) \to A \otimes \bar{B}(A)$$

defined by the formula

$$B(a \otimes [u_1|\cdots|u_n]) = \sum_{i=0}^{n} (-1)^{(\varepsilon_{i-1}+1)(\varepsilon_n - \varepsilon_{i-1})} 1 \otimes [u_i|\cdots|u_n|a|u_1|\cdots|u_{i-1}]$$

where $\varepsilon_i = |a| + |u_1| + \cdots + |u_i| - i$ (and $\varepsilon_{-1} = 1$). A standard computation shows that $B^2 = bB + Bb = 0$.

By a *mixed complex*, in the sense of Kassel [K], we simply mean a graded vector space C equipped with two operators b, B where b has degree -1 and B has degree 1 which satisfy the identities

$$b^2 = B^2 = bB + Bb = 0.$$

We will assume that the mixed complex is either bounded below or else bounded above. The theory becomes unmanageable if C has an infinite number of non-zero terms in both the positive and negative directions. We denote the mixed complex $(A \otimes \bar{B}(A), b, B)$ by $\mathcal{C}(A)$. We assume that A satisfies one of the boundedness hypotheses in §2 in order to ensure that $\mathcal{C}(A)$ is either bounded above or below. Now we define the *negative cyclic homology* and the *cyclic cohomology* of a mixed complex (C, b, B).

Let $k[u]$ be a polynomial algebra on a single indeterminate u of degree -2. Form the *completed tensor product* $C \hat{\otimes} k[u]$ where the completion means we take the direct product of the homogeneous terms rather than the direct sum. This completion is important in the case where the mixed complex is not bounded above since in this case there may be an infinite number of terms with fixed total degree. If the mixed complex is bounded above then there are only a finite number of terms with fixed degree and the completion is irrelevant. Then $b + uB$ defines a differential, of degree -1 on $C\hat{\otimes}k[u]$. We denote the resulting complex by $\mathcal{B}^-(C)$ and then

$$\mathrm{HC}_*^-(C) = \mathrm{H}_*(\mathcal{B}^-(C), b + uB).$$

We define $\mathrm{HC}_*^-(A)$ by

$$\mathrm{HC}_*^-(A) = \mathrm{HC}_*^-(\mathcal{C}(A)).$$

To compute the cyclic cohomology of (C, b, B) let C^* be the dual of C and let b^* and B^* be the dual operators. As usual we grade C^* negatively. Now form the complex

$$\mathcal{B}^*(C) = (C^*\hat{\otimes}k[u], b^* + uB^*)$$

and then

$$\mathrm{HC}^*(C) = \mathrm{H}^*(\mathcal{B}^*(C), b^* + uB^*),$$

and

$$\mathrm{HC}^*(A) = \mathrm{HC}^*(\mathcal{C}(A)).$$

§6 The B-operator for free algebras. Now suppose that (A, d) is a differential graded algebra with $A = T(V)$ a free algebra over k satisfying the hypotheses of §3. To extend the duality to cyclic theory we need to find the operator on the complex $(A \otimes W, \delta)$ which corresponds, under the equivalence $\bar{\jmath}$ of §3 to Connes's B-operator. Once more there is an explicit formula for this operator which can be found in [L-Q] in the case where the differential in A is zero and in [B-V] and [V] in the general case.

Define

$$\beta : A \to A \otimes V$$

by the formula

$$\beta(v_1 \cdots v_p) = \sum_{i=1}^{p} (-1)^{\mu_i} v_{i+1} \cdots v_p v_1 \cdots v_{i-1} \otimes v_i.$$

The sign is given by

$$\mu_i = (|v_{i+1}| + \cdots |v_p|)(|v_1| + \cdots + |v_i|)$$

The terms which occur when $i = 1$ and $i = p$ are

$$(-1)^{|v_1|(|v_2|+\cdots+|v_p|)} v_2 \cdots v_p \otimes v_1, \qquad v_1 \cdots v_{p-1} \otimes v_p.$$

We extend β to $A \otimes W$ by defining it to be β on $A \subset A \otimes W$ and to be zero on $A \otimes sV$. Then a direct computation shows that

$$\delta\beta = \beta\delta$$

and so, with these operators, $A \otimes W$ becomes a mixed complex which we will refer to simply as $A \otimes W$. Another computation shows that

$$\bar{\jmath} : (\mathcal{C}(A), b, B) \to (A \otimes W, \delta, \beta)$$

is a map of mixed complexes.

Theorem [V]. *The induced map*

$$\mathrm{HC}^-_*(A) = \mathrm{HC}^-_*(\mathcal{C}(A)) \to \mathrm{HC}^-_*(A \otimes W)$$

is an isomorphism.

This follows from Theorem 2 using the standard result that if a map of mixed complexes induces an isomorphism in homology with respect to b, then it induces an isomorphism in HC^-_*.

§7 The duality in cyclic theory. Now let (C, ∂) denote a differential graded coalgebra, which satisfies the hypotheses of §1 and is simply connected. Let (Γ, d) denote the dual differential graded algebra; we continue to use the negative grading convention for Γ as in §4. We identify the underlying vector space of the mixed complex used in §6 with $F(C) \otimes C$. In view of the duality between the graded vector spaces

$$A \otimes \bar{B}(A), \qquad F(C) \otimes C$$

we now look for the relation between the operators B on $A \otimes \bar{B}(A)$ and β on $F(C) \otimes C$.

Lemma. $\beta^* = B$

Proof. We express B (without signs) as the following composite.

$$\begin{aligned} A \otimes (sIA)^{\otimes n-1} &\xrightarrow{s\otimes 1} (sIA)^{\otimes n} \\ &\xrightarrow{\Delta} \bigoplus_{k=1}^{n-1} (sIA)^{\otimes k} \otimes (sIA)^{n-k} \\ &\xrightarrow{T} \bigoplus_{k=1}^{n-1} (sIA)^{\otimes n-k} \otimes (sIA)^{k} \\ &\xrightarrow{\Delta^*} (sIA)^{\otimes n}. \end{aligned}$$

Likewise we can express β as the composite

$$\begin{aligned} (s^{-1}JC)^{\otimes n} &\xrightarrow{\Delta} \bigoplus_{k=1}^{n-1} (s^{-1}JC)^{\otimes k} \otimes (s^{-1}JC)^{n-k} \\ &\xrightarrow{T} \bigoplus_{k=1}^{n-1} (s^{-1}JC)^{\otimes n-k} \otimes (s^{-1}JC)^{k} \\ &\xrightarrow{\Delta^*} (s^{-1}JC)^{\otimes n} \\ &\xrightarrow{1\otimes s} (s^{-1}JC)^{\otimes n-1} \otimes JC. \end{aligned}$$

The mapping T introduces the usual sign $T(a \otimes b) = (-1)^{|a||b|} b \otimes a$. From these composites and the self-duality of T and s, it is clear that $B = \beta^*$.

Theorem B. *Let (C, ∂) denote a simply connected, counital, coaugmented differential graded coalgebra and let (Γ, d) be the dual algebra. Then there is a natural isomorphism*

$$\mathrm{HC}^n(F(C)) \cong \mathrm{HC}^-_{-n}(\Gamma).$$

§8 Bivariant cyclic theory. Theorem B has a very nice interpretation, suggested to us by Ezra Getzler, in the bivariant cyclic theory of [J-K]. Recall the basic definitions of bivariant cyclic theory. Let Λ be the exterior algebra over k on one generator ε of degree 1. Then a mixed complex C is nothing more than a differential graded module over Λ where the differential defines the b operator and multiplication by ε defines the B operator. Given two differential graded Λ-modules M and N define

$$\mathrm{HC}^*(M, N) = \mathrm{Ext}^*_\Lambda(M, N)$$

and then, for any two differential graded algebras A and B,

$$\mathrm{HC}^*(A, B) = \mathrm{Ext}^*_\Lambda(\mathcal{C}(A), \mathcal{C}(B)).$$

It is shown in [J-K] that

$$\mathrm{HC}^*(M, k) = \mathrm{HC}^*(M), \qquad \mathrm{HC}^*(k, N) = \mathrm{HC}^-_*(N).$$

Theorem C. *Let (C, ∂) denote a simply connected, counital, coaugmented differential graded coalgebra and let (Γ, d) be the dual differential graded algebra. Let A and B be two further algebras. Then there are natural isomorphisms*

$$\mathrm{HC}^*(A \otimes F(C), B) \cong \mathrm{HC}^*(A, \Gamma \otimes B).$$

Proof. We have exhibited a map of differential graded Λ modules

$$\mathcal{C}(\Gamma) \to \mathcal{C}(F(C))^*$$

which is an isomorphism in homology with respect to the internal differential. From the basic properties of Ext it follows that

$$\mathrm{Ext}^*_\Lambda(M, \mathcal{C}(\Gamma) \otimes N) \cong \mathrm{Ext}^*_\Lambda(M \otimes \mathcal{C}(F(C)), N).$$

This shows that

$$\mathrm{HC}^*(\mathcal{C}(A) \otimes \mathcal{C}(F(C)), \mathcal{C}(B)) \cong \mathrm{HC}^*(\mathcal{C}(A), \mathcal{C}(\Gamma) \otimes \mathcal{C}(B)).$$

However the Künneth theorem of [H-J] and [K] (compare [J-K, §3]) shows that

$$\mathrm{HC}^*(\mathcal{C}(A) \otimes \mathcal{C}(F(C)), \mathcal{C}(B)) \cong \mathrm{HC}^*(\mathcal{C}(A \otimes F(C)), \mathcal{C}(B)) = \mathrm{HC}^*(A \otimes F(C), B)$$

and also

$$\mathrm{HC}^*(\mathcal{C}(A), \mathcal{C}(\Gamma) \otimes \mathcal{C}(B)) \cong \mathrm{HC}^*(\mathcal{C}(A), \mathcal{C}(\Gamma \otimes B)) = \mathrm{HC}^*(A, \Gamma \otimes B).$$

§9 Loop spaces. Recall the following theorem, due to Adams, relating the cobar construction and the based loop space ΩX of a pointed, simply connected, topological space X.

Theorem [A]. *Let X be a simply connected space and let $C_*(X;k)$ be the coalgebra of cochains on X with coefficients in k. There is a homomorphism of differential graded algebras*

$$\Phi : F(C_*(X;k)) \to C_*(\Omega X;k)$$

which induces an isomorphism on homology

$$\mathrm{H}_*(F(C_*(X;k)) \cong \mathrm{H}_*(\Omega X;k).$$

An immediate consequence of Theorem A above is the fact, for a differential graded coalgebra of the type considered, $\mathrm{HH}_*(F(C_*),\partial_F)$ pairs perfectly with $\mathrm{HH}_*(C_{-*},\delta_*)$. When we apply Adams's theorem, we obtain a very direct proof of the following well-known theorem:

Corollary D. *For X a simply-connected space,*

$$\mathrm{HH}_*(C_*(\Omega X;k),\partial) \cong \mathrm{H}_*(\mathcal{L}X;k),$$

where $\mathcal{L}X$ denotes the free loop space of X, $\mathrm{map}(S^1,X)$.

Proof. Consider the pullback diagram (see [S1] and [D])

$$\begin{array}{ccc} \mathcal{L}X & \longrightarrow & X^I \\ \downarrow & & \downarrow{\scriptstyle (\mathrm{eval}_0,\mathrm{eval}_1)} \\ X & \xrightarrow{\mathrm{diag}} & X \times X. \end{array}$$

By a theorem of Eilenberg and Moore [E-M], we have

$$\mathrm{HH}_*(C^*(X;k),\delta) \cong \mathrm{Tor}_{C^*(X)\otimes C^*(X)}(C^*(X),C^*(X)) \cong \mathrm{H}^*(\mathcal{L}X;k).$$

But, by Theorem A, $\mathrm{HH}_*(C^*(X;k),\delta)$ is dual to $\mathrm{HH}_*(F(C_*(X;k)),\partial_F)$, which is isomorphic to $\mathrm{HH}_*(C_*(\Omega X;k),\partial)$ by Adams's theorem.

This is a considerably simpler proof of a theorem of Goodwillie [G], Jones [J], and Burghelea-Fiedorowicz [B-F] in the case where X is simply connected. Notice how this proof replaces the theory of simplicial sets and spaces in the usual proof, by the Eilenberg-Moore theorem and requires simple connectivity for the application of the Eilenberg-Moore theorem.

REFERENCES

[A] J.F. Adams, *On the cobar construction*, Proc. Nat. Acad. Sci. USA **42** (1956), 409–412.

[B-F] D. Burghelea and Z. Fiedorowicz, *Cyclic homology and the algebraic K-theory of spaces II*, Topology **25** (1986), 303–317.

[B-V] D. Burghelea and M. Vigué-Poirrier, *Cyclic homology of commutative algebras I*, Springer LNM **1318** (1988), 51–72.

[C] A. Connes, *Non-commutative differential geometry*, Publ. Math. IHES **62** (1985), 41–144.

[D] Dinh Lanh Dang, *The homology of free loop spaces*, Princeton thesis (1978).

[E-M] S. Eilenberg and J.C. Moore, *Homology and fibrations: Coalgebras, cotensor product, and its derived functors*, Comm. Math. Helv. **40** (1966), 199–236.

[G] T.G. Goodwillie, *Cyclic homology, derivations, and the free loop space*, Topology **24** (1985), 187–215.

[G-J-P] E. Getzler, J.D.S. Jones, and S. Petrack, *Differential forms on loop spaces and the cyclic bar complex*, Topology **30** (1991), 339–372.

[H-V] S. Halperin and M. Vigué-Poirrier, *The homology of a free loop space*, Pac. J. Math. **147** (1991), 311–324.

[H-J] C.E. Hood and J.D.S. Jones, *Some algebraic properties of cyclic homology groups*, K-theory **1** (1987), 361–384.

[J] J.D.S. Jones, *Cyclic homology and equivariant homology*, Inv. Math. **87** (1987), 403–423.

[J-K] J.D.S Jones and C. Kassel, *Bivariant cyclic theory*, K-theory **3** (1989), 339–365.

[K] C. Kassel, *Cyclic homology, comodules, and mixed complexes*, J. of Algebra **107** (1987), 195–216.

[L-Q] J.-L. Loday and D. G. Quillen, *Cyclic homology and the Lie algebra homology of matrices*, Comment. Math. Helv. **59** (1984), 565–591.

[S1] L. Smith, *On the characteristic zero cohomology of the free loop space*, American J. Math. **103** (1981), 887–910.

[S2] L. Smith, *The Eilenberg-Moore spectral sequence and the mod 2 cohomology of certain free loop spaces*, Ill. J. Math. **38** (1984), 516–522.

[V] M. Vigue-Poirrier, *Homologie de Hochschild et homologie cyclique des algèbres différentielles graduées*, Publ. IRMA, Lille # VIII **17** (1989).

MATHEMATICS INSTITUTE, UNIVERSITY OF WARWICK, COVENTRY CV4 7AL, ENGLAND

DEPARTMENT OF MATHEMATICS, VASSAR COLLEGE, POUGHKEEPSIE, NEW YORK 12601, USA

Hermitian A_∞ Rings and Their K-theory

Z. Fiedorowicz, R. Schwänzl and R. Vogt

This paper is meant to be a guide through a series of papers to come, which contain the detailed and often very technical conditions and proofs. § 1 contains the geometrical motivation of our work. In § 2 we illustrate in detail the problems one encounters when one tries to put a Hermitian structure on the A_∞ ring $\Omega^\infty S^\infty(\Omega M_+)$, the fundamental example for geometric applications. In § 3 we introduce the notion of Hermitian K-theory of Hermitian A_∞ rings. We give an indication of the coherence machinery" involved and discuss the problems we encountered before coming up with a reasonable definition. Here, in one particular example, we demonstrate in some more detail our general approach. In § 4 we give a short report on properties of Hermitian K-theory and geometric applications. We limit ourselves to occasional comments about the new techniques necessary for proving these properties and refer the reader to our forthcoming papers in this subject.

§ 1 An outline of the general program

In Burghelea-Fiedorowicz [BF] it was shown that a suitable definition of Hermitian K-theory provides a unifying framework for understanding the homotopy type of automorphism groups of topological manifolds in the stable range. Traditionally the approach has been to split this into two problems. Given a compact topological manifold M one denotes by Homeo(M), resp. $H(M)$, the space of homeomorphisms, resp. self-homotopy equi- valences, of M fixing the boundary. One then notes that the quotient $H(M)/\mathrm{Homeo}(M)$ has an involution and that away from 2 there results stably a splitting

$$H(M)/\mathrm{Homeo}(M) \simeq H(M)/\widetilde{\mathrm{Homeo}}(M) \times \widetilde{\mathrm{Homeo}}(M)/\mathrm{Homeo}(M)$$

where $\widetilde{\mathrm{Homeo}}(M)$ is the space of block homeomorphisms of M. The homotopy type of $H(M)/\widetilde{\mathrm{Homeo}}(M)$ can be analyzed in terms of surgery theory.

The other factor $\widetilde{\text{Homeo}}(M)/\text{Homeo}(M)$ can be analyzed using concordance theory: The classifying space of stable concordances of M, denoted $\mathcal{B}\mathcal{C}(M)$, carries a natural involution which splits it (away from 2) into two pieces, one of which is $\widetilde{\text{Homeo}}(M)/\text{Homeo}(M)$.

Waldhausen showed that $\mathcal{C}(M)$ can be analyzed in terms of what he called the algebraic K-theory of M, denoted $A(M)$. He constructed a fibration sequence

$$\Omega^\infty(M \wedge A(*)) \to A(M) \to Wh^A(M)$$

and showed that in the stable range $\Omega_0^2 Wh^A(M)$ has the homotopy type of $\mathcal{C}(M)$.

In [BF] Burghelea and Fiedorowicz rationally obtain a fibration sequence

$$\Omega^\infty(M \wedge {}_\varepsilon L(*)) \to {}_\varepsilon L(M) \to Wh^{\varepsilon L}(M)$$

where ${}_\varepsilon L(X)$ is the Hermitian algebraic K-theory of the simplicial Hermitian ring $\mathbb{Z}[\Omega X]$, obtained using the Kan loop group as model for ΩX, and $\Omega_0 Wh^{\varepsilon L}(M)$ has the homotopy type of $H(M)/\text{Homeo}(M)$ in the stable range, i.e. up to dimension k with $\dim M \geq \max(2k+7, 3k+4)$. Thus, in principle, Hermitian algebraic K-theory of topological spaces gives direct access to the homotopy type of $H(M)/\text{Homeo}(M)$.

However, [BF] only provided a suitable definition of Hermitian algebraic K-theory rationally. There ${}_\varepsilon L(X)$ was defined as the Hermitian algebraic K-theory of the simplicial ring $\mathbb{Z}[\Omega X]$. As was first noted by Waldhausen, in order to obtain global results, one must replace $\mathbb{Z}[\Omega X]$ by the A_∞ ring $\Omega^\infty S^\infty(\Omega X_+)$. We have thus developed a Hermitian algebraic K-theory of A_∞ rings which extends these results to $\frac{1}{2}$ local results, in particular, we obtain access to the $\frac{1}{2}$ local homotopy type of $H(M)/\text{Homeo}(M)$.

§ 2 Hermitian structures on A_∞ rings

In order to define Hermitian K-theory of an A_∞ ring we need to formalize two additional operations in the A_∞ context: An involution and the operation of negation. These operations are needed to codify the fundamental notion of Hermitian K-theory: Taking $\pm$ the conjugate transpose of a matrix.

For the simplicial ring $\mathbb{Z}[\Omega M]$ the appropriate involution is given by the formula $\sum m_i g_i \to \sum m_i w(g_i) g_i^{-1}$, where $w : \Omega M \to \mathbb{Z}/2 = \{\pm 1\}$ is the loop of the first Stiefel-Whitney class $M \to B\mathbb{Z}/2$ of the tangent bundle of M.

Up to homotopy, the appropriate involution on $\Omega^\infty S^\infty(\Omega M_+)$ is equally simple to describe. Take the composite

$$\Omega M_+ \xrightarrow{(\Omega\xi,\eta)} F_+ \wedge \Omega^\infty S^\infty(\Omega M_+) \hookrightarrow \Omega^\infty S^\infty \wedge \Omega^\infty S^\infty(\Omega M_+) \xrightarrow{\mu} \Omega^\infty S^\infty(\Omega M)_+$$

and extend by the universal property of $\Omega^\infty S^\infty(\ )$. Here $\xi : M \to BF$ is the classifying map of the tangent bundle regarded as a stable spherical fibration, η is given by taking the inverse of a loop and including in $\Omega^\infty S^\infty(\Omega M_+)$, and μ denotes smash product. Of course the difficulty is that there is no canonical choice of smash product, so we have to use the language of operads and we have to describe the basic relations $\overline{x+y} = \bar{x} + \bar{y}, \overline{xy} = \bar{y}\bar{x}$, and $\bar{\bar{x}} = x$ in these terms.

There is one particularly serious difficulty which arises when one attempts to codify the two relations $\overline{xy} = \bar{y}\bar{x}$ and $\bar{\bar{x}} = x$ simultaneously. In the group ring $\mathbb{Z}[\Omega X]$ this involves the following manipulations

$$\overline{g_1 g_2} = w(g_1 g_2)(g_1 g_2)^{-1} = w(g_1)w(g_2)g_2^{-1}g_1^{-1} = w(g_2)g_2^{-1}w(g_1)g_1^{-1} = \overline{g_2}\,\overline{g_1}$$

$$\bar{\bar{g}} = \overline{w(g)g^{-1}} = w(g)w(g^{-1})(g^{-1})^{-1} = w(g)w(g)^{-1}g = g.$$

We see that these manipulations involve commuting factors past each other in a product and cancellation of multiplicative inverses. However, Schwänzl and Vogt [SV2] have shown that it is impossible to simultaneously codify up to coherent homotopy the notion of a commutative H-space together with a homotopy inverse unless the underlying space is a product of Eilenberg-MacLane spaces. Fortunately for us, it turns out that the Hermitian relations only require restricted forms of commutativity and cancellation and thus can be codified up to higher coherence homotopies.

For much the same reasons, we also have to be careful with the operation of negation. We only codify up to higher coherence homotopies the multiplicative relations

$$\textbf{(2.1)} \qquad -(-x) = x, \quad (-x)y = x(-y) = -(xy), \quad (-x)(-y) = xy,$$

not the additive relation $x + (-x) = 0$.

Our starting point is to note that while there is no obvious way to describe the Hermitian structure on $\Omega^\infty S^\infty(G_+)$ where G is a topological group, there are closely related A_∞ rings which have Hermitian structures that are easy to describe. Namely A_∞ rings of the form $\Omega^\infty S^\infty((\mathbb{Z}/2 \times G)_+)$ can be endowed with involutions arising from involutions on $\mathbb{Z}/2 \times G$ of the form

$$(\lambda, g) \to (\lambda w(g), g^{-1})$$

where $w : G \to \mathbb{Z}/2$ is a homomorphism. Also multiplication by -1 on $\mathbb{Z}/2$ induces an operation on $\Omega^\infty S^\infty((\mathbb{Z}/2 \times G)_+)$ which formally satisfies relations (2.1) in a sense which can be precisely described. A closer investigation of this example leads to the following definition.

Definition 2.2 A Hermitian operad pair $(\mathcal{C}, \mathcal{G}, \tau)$ is an A_∞ operad pair $(\mathcal{C}, \mathcal{G})$ together with involutions τ defined on each space $\mathcal{G}(n)$ of the multiplicative operad satisfying
(1) $1 \cdot \tau = 1$ where $1 \in \mathcal{G}(1)$ is the unit.
(2) $\gamma(g \cdot \tau; g_1, g_2, ..., g_k) = \gamma(g; g_k \cdot \tau, g_{k-1} \cdot \tau, ..., g_1 \cdot \tau)\tau$,where γ is the structure map of $\mathcal{G}$.
(3) $\rho(g \cdot \tau; a_1, a_2, ..., a_n) = \rho(g; a_n, a_{n-1}, ..., a_1) \cdot \tau\langle j_1, ..., j_n\rangle$ where ρ denotes the action of $\mathcal{G}$ on $\mathcal{C}$, $a_i \in \mathcal{C}(j_i)$ and $\tau\langle j_1, ..., j_n\rangle \in \Sigma_{j_1 j_2 ... j_n}$ is the permutation which changes the order on $j_1 \times j_2 \times ... \times j_n$ from lexicographic to reverse lexicographic.

A *Hermitian* $(\mathcal{C}, \mathcal{G}, \tau)$ space X is a $(\mathcal{C}, \mathcal{G})$ space in the sense of [M1] together with two involutions $\tau, \varepsilon : X \to X$ representing the antiinvolution and multiplication by -1 respectively, both being $\mathcal{C}$-homomorphisms and both commuting with the $\mathcal{G}$-action and the involution on $\mathcal{G}$ in the appropriate way. (Alternatively, we could incorporate τ and ε into $\mathcal{C}$ and $\mathcal{G}$ to obtain an operad pair $(\mathcal{C}', \mathcal{G}')$, and a Hermitian $(\mathcal{C}, \mathcal{G}, \tau)$-space is simply a $(\mathcal{C}', \mathcal{G}')$-space [FSV1].) Of course the example to keep in mind is $\Omega^\infty S^\infty((\mathbb{Z}/2 \times G)_+)$ with involution τ and ε as described above. The appropriate operad pair is the Steiner operad pair $(\mathcal{A}, \mathcal{L})$, with $\mathcal{L}$ the linear isometries operad, regarded as an A_∞ operad [St1]. The involution τ on $\mathcal{L}(n)$ is given by the action of the totally order reversing permutation with respect to the E_∞ structure of $\mathcal{L}$ (which is otherwise neglected).

$\Omega^\infty S^\infty((X \times \mathbb{Z}/2 \times G)_+)$ with X any A_∞ space is a slightly more general example of an $(\mathcal{A}, \mathcal{L}, \tau)$ space.

The construction of a Hermitian structure on $\Omega^\infty S^\infty(G_+)$ is quite involved and constitutes the main body of our first paper [FSV1]. We replace $\Omega^\infty S^\infty(G_+)$ by the two-sided bar resolution

$$B_*(\Omega^\infty S^\infty((\ )_+), F \times -, SF \times \mathbb{Z}/2 \times G)$$

whose space of n-simplices is

$$\Omega^\infty S^\infty((F^n \times SF \times \mathbb{Z}/2 \times G)_+)$$

and thus carries an A_∞ Hermitian structure as described above. One should note however that the bar resolution here is simplicial only in the sense of ūp to coherent homotopy". The iterated face and degeneracy operations are described in terms of operad actions. We also implicitly use the splitting $F \simeq SF \times \mathbb{Z}/2$ as infinite loop spaces, which requires localization away from 2.

The bar resolution described above produces an object which is a Hermitian A_∞ ring space in a much more complicated sense than Def. 2.2. However the rectification machinery of May [M2] can be adapted to simplify the Hermitian structure of this object. We obtain the following theorem.

Theorem 2.3 Let G be a topological group and $\xi : BG \to BF$ a map. Then there is a Hermitian $(\mathcal{A}, \mathcal{L}, \tau)$ space $X(G)$ which is equivalent as an A_∞ ring space to the Bousfield-Kan localization $\mathbb{Z}\left[\frac{1}{2}\right] \Omega^\infty S^\infty(G_+)$. Under this equivalence $\varepsilon : X(G) \to X(G)$ corresponds up to homotopy to the negative of the identity map and $\tau : X(G) \to X(G)$ to the map $\Omega^\infty S^\infty(G_+) \to \Omega^\infty S^\infty(G_+)$ given by $g \to \Omega\xi(g) \cdot g^{-1}$.

We should point out that the Bousfield-Kan localization only localizes each path component, leaving π_0 untouched. By the arguments of [BF], this only affects Hermitian K-theory at the prime 2, which is poorly understood even for discrete rings.

§ 3 The construction of Hermitian K-theory

Given a Hermitian A_∞ ring as described above, we in [FSV2] develop its Hermitian K-theory along the same lines as [BF] but using the techniques of [SV1]. Recall that the Hermitian K-theory ${}_\delta L(R)$ ($\delta = \pm 1$) of a simplicial ring R was defined to be the group completion of the two-sided bar construction

$$\coprod_n B({}_\delta\mathrm{Symm}_n(R), \widetilde{GL_n}(R), *)$$

where ${}_\delta\mathrm{Symm}_n(R)$ is the space of δ-symmetric matrices

$${}_\delta\mathrm{Symm}_n(R) = \{A \in \widetilde{GL_n}(R) \mid A^* = \delta A\}$$

and $\widetilde{GL_n}(R)$ is the pullback

$$\begin{array}{ccc} \widetilde{GL}_n(R) & \longrightarrow & M_n(R) \\ \downarrow & & \downarrow \\ GL_n(\pi_0 R) & \longrightarrow & \pi_0 M_n(R) = M_n(\pi_0 R) \end{array}$$

$M_n(R)$ is the space of all $(n \times n)$-matrices and $\widetilde{GL}_n(R)$ acts on ${}_\delta\mathrm{Symm}_n(R)$ by $(A, \lambda) \to \lambda^* A \lambda$.

The first problem we must consider is what δ-symmetric might mean in the A_∞ ring context. To begin with note that we can alternatively describe ${}_\delta\mathrm{Symm}_n(R)$ as the fixed point space of the involution $A \to \delta A^*$ on $\widetilde{GL}_n(R)$. We can formulate this involution in the A_∞ ring context but only up to coherent homotopy. Thus we have no control on the actual fixed point space (which might very well be empty). The obvious remedy for this is to use the homotopy fixed point space $F_{\mathbb{Z}/2}(E\mathbb{Z}/2, \widetilde{GL}_n(R))$ instead. If R has $\frac{1}{2}$ local path components, this is easily seen to have the same $\frac{1}{2}$ local homotopy type as ${}_\delta\mathrm{Symm}_n(R)$. It is even more convenient to note that in that case

$$\coprod_n B({}_\delta\mathrm{Symm}_n(R), \widetilde{GL}_n(R), *)$$

has the same $\frac{1}{2}$ local homotopy type as

$$F_{\mathbb{Z}/2}(E\mathbb{Z}/2, \coprod_n B(\widetilde{GL}_n(R), \widetilde{GL}_n(R), *))$$

where $\mathbb{Z}/2$ acts trivially on the second slot of the bar construction.

The actual construction begins by codifying the formal operations of a semiring with Hermitian structure (i.e. an involution and a unary operation satisfying the multiplicative properties of negation) as a theory $\Theta_{r,H}$. This is a category whose objects $[m]$ are indexed by the nonnegative integers. The morphisms from $[m]$ to $[n]$ are n-tuples of polynomials of the form

(3.1) $\quad p(x_1, x_2, ..., x_m, \bar{x}_1, ..., \bar{x}_m) + \varepsilon q(x_1, x_2, ..., x_m, \bar{x}_1, \bar{x}_2, ..., \bar{x}_m)$

where p and q are polynomials in $2m$ noncommuting variables $x_1, x_2, ..., x_m$, $\bar{x}_1, \bar{x}_2, ..., \bar{x}_m$ with nonnegative integer coefficients. Composition is defined by substitution and reduction to normal form using (noncommutative) semiring relations and the Hermitian relations

$$\overline{u+v} = \bar{u} + \bar{v}, \quad \overline{uv} = \bar{v}\bar{u}, \quad \bar{\bar{u}} = u, \quad \varepsilon^2 = 1, \quad \bar{\varepsilon} = \varepsilon, \quad \varepsilon u = u\varepsilon \quad \forall u.$$

By construction a topological semiring with Hermitian structure is the same thing as a product preserving functor $R : \Theta_{r,H} \to Top$. Given such a functor, $R[1]$ (which we call the *underlying space* of R) has the structure of a topological semiring with Hermitian structure.

A *Hermitian A_∞ ring theory* is an algebraic theory Θ with topologized sets of morphisms, which augments over $\Theta_{r,H}$. As in the non-Hermitian A_∞ ring case the augmentation is a homotopy equivalence over a distinguished subset of morphisms in $\Theta_{r,H}$, which we call simple. A Θ-space, i.e. a continuous product preserving functor $\Theta \to Top$, is called a *Hermitian A_∞ ring.*

Extending a construction of Steiner [St2] to the Hermitian ring case, one can associate to each Hermitian operad pair $(\mathcal{C}, \mathcal{G}, \tau)$ a Hermitian A_∞ ring theory $\Theta(\mathcal{C}, \mathcal{G}, \tau)$ such that each $(\mathcal{C}, \mathcal{G}, \tau)$-space X gives rise to a $\Theta(\mathcal{C}, \mathcal{G}, \tau)$-space with underlying space X, so that X is a Hermitian A_∞ ring.

Hence the space $X(G)$ of Theorem 2.3 determines a Hermitian A_∞ ring

$$X(G) : \Theta(\mathcal{A}, \mathcal{L}, \tau) \to Top$$

where we abusively denote the functor also by $X(G)$ or even more simply by X. Starting from this functor we construct an infinite loop space ${}_\varepsilon L(X)$ which we will call the Hermitian K-theory of X.

To illustrate the concepts behind our construction, we begin by showing how to construct the two-sided bar construction

$$B({}_\delta\mathrm{Symm}_n(X), \widetilde{GL_n}(X), *).$$

We first construct a functor

$$\Psi_n : \Delta^{op} \times \mathbb{Z}/2 \to \Theta_{r,H}$$

so that for any Hermitian topological ring R the composite

$$\Delta^{op} \times \mathbb{Z}/2 \xrightarrow{\Psi_n} \Theta_{r,H} \xrightarrow{R} Top$$

codifies the two-sided bar construction

$$B(M_n(R), M_n(R), *)$$

with $M_n(R)$ in the second slot acting on $M_n(R)$ in the first slot by $(A, B) \to B^*AB$, and with $\mathbb{Z}/2$-action given by $A \to \delta A^*$ on $M_n(R)$ in the first slot. The functor Ψ_n is given by

$$\Psi_n([m], *) = [(m+1)n^2]$$

on objects. Think of $[(m+1)n^2]$ as the $(m+1)n^2$ entries of an $(m+1)$-tuple $(Y_0, Y_1, ..., Y_m)$ of matrices $Y_j \in M_n(R)$, the ordering given by

$$Y_j = \begin{pmatrix} y_{jn^2+1} & , & y_{jn^2+2} & ,\dots, & y_{jn^2+n} \\ \vdots & & \vdots & & \vdots \\ y_{jn^2+(n-1)n+1} & , & y_{jn^2+(n-1)n+2} & ,\dots, & y_{(j+1)n^2} \end{pmatrix}$$

The value of Ψ_n on the face maps $\partial_i : [m] \to [m-1]$

$$\Psi_n(\partial_i, 1) : [(m+1)n^2] \to [mn^2]$$

is the mn^2-tuple of polynomials (recall (3.1); we again use the matrix notation)

$$\begin{array}{ll} (X_1^* \cdot X_0 \cdot X_1, X_2, ..., X_m) & \text{if } i = 0 \\ (X_0, ..., X_i \cdot X_{i+1}, ..., X_m) & \text{if } 0 < i < m \\ (X_0, ..., X_{m-1}) & \text{if } i = m \end{array}$$

where " $\cdot$ " is matrix multiplication. On degeneracies,

$$\Psi_n(s_i, 1) : [(m+1)n^2] \to [(m+2)n^2]$$

is the $(m+2)n^2$-tuple

$$(X_0, X_i, ..., X_i, I_n, X_{i+1}, ..., X_m) \quad 0 \le i \le m$$

with I_n standing for the $n \times n$ identity matrix. For the non-trivial element τ of $\mathbb{Z}/2$, $\Psi_n(id, \tau) : [(m+1)n^2] \to [(m+1)n^2]$ is the tuple

$$(\delta X_0^*, X_1, ..., X_m).$$

Now from the pullback diagram, where Φ is the augmentation

$$(3.2) \qquad \begin{array}{ccc} \mathcal{P} & \xrightarrow{\tilde{\Psi}_n} & \Theta(\mathcal{A}, \mathcal{L}, \iota) \\ \Big\downarrow \tilde{\Phi} & & \Big\downarrow \Phi \\ \Delta^{op} \times \mathbb{Z}/2 & \xrightarrow{\Psi_n} & \Theta_{r,H} \end{array}$$

We note that the image of Ψ_n contains only simple morphims, which implies that $\tilde{\Phi}$ is an equivalence on the space of morphisms.

Now consider the composite

$$\mathcal{P} \xrightarrow{\tilde{\Psi}_n} \Theta(\mathcal{A}, \mathcal{L}, \iota) \xrightarrow{X} Top.$$

The value of $X \circ \tilde{\Psi}_n$ on objects is

$$X \circ \tilde{\Psi}_n([m], *) = M_n(X)^{m+1}$$

where $M_n(X)$ is the space of $n \times n$ matrices with values in X. Let $\widetilde{GL_n}(X)$ again denote the subspace of $M_n(X)$ consisting of homotopy invertible matrices. It is easy to see that for any morphism $f : ([m], *) \to ([p], *)$ in $\mathcal{P}$ we have

$$X \circ \tilde{\Psi}_n(f)(\widetilde{GL_n}(X)^{m+1}) \subseteq \widetilde{GL_n}(X)^{p+1}.$$

Thus we obtain an induced subfunctor

$$F : \mathcal{P} \to Top$$

with $F([m], *) = \widetilde{GL_n}(X)^{m+1}$.

We can then apply the Segal push-down [S; App. B] along the functor $\tilde{\Phi}$ to obtain a functor $T = \tilde{\Phi}_* F : \Delta^{op} \times \mathbb{Z}/2 \to Top$. Since $\tilde{\Phi}$ is an equivalence on morphisms, $T([m], *)$ has the same homotopy type as $\widetilde{GL_n}(X)^{m+1}$. We note that $T([m], *)$ has a $\mathbb{Z}/2$-action and so we define

$$B({}_\delta\mathrm{Symm}_n(X), \widetilde{GL_n}(X), *) = |m \to F_{\mathbb{Z}/2}(E\mathbb{Z}/2, T([m], *))|.$$

By the remarks at the beginning of this section, it is not hard to show that if X is a Hermitian topological ring such that each path component of X is $\frac{1}{2}$ local, then this new definition of $B({}_\delta\mathrm{Symm}_n(X), \widetilde{GL_n}(X), *)$ agrees up to homotopy with the standard one.

Of course, this construction is insufficient by itself to define Hermitian algebraic K-theory of Hermitian A_∞ rings: We must also codify the direct sum operation in this framework. We obtain a Segal Γ-space

$$U : \Gamma^{op} \to Top$$

with $U[1] \simeq \coprod_{n=0}^{\infty} B({}_\delta\mathrm{Symm}_n(X), \widetilde{GL_n}(X), *)$. We define the Hermitian algebraic K-theory ${}_\delta L(X)$ to be the group completion of U.

§ 4 Properties of Hermitian Algebraic K-theory and Geometric Applications

The first important property of ${}_\epsilon L(X)$, which follows more or less directly from the construction (using e.g. general properties of the bar construction) is the following consistency result [FSV2].

Proposition 4.1 If the A_∞ Hermitian ring R is the geometric realization of a simplicial Hermitian ring R, then Hermitian K-theory ${}_\epsilon L(R)$ of § 3, is equivalent away from 2 with the Hermitian K-theory ${}_\epsilon L(R_*)$ of [BF].

Another result, proved along similar lines is homotopy invariance.

Proposition 4.2 Let $f : X \to Y$ be a homomorphism of Hermitian A_∞ ring spaces, satisfying
(i) $f_* : \pi_0 X \xrightarrow{\cong} \pi_0 Y$
(ii) f is n-connected modulo a Serre class $\mathcal{C}$.
Then the induced map ${}_\epsilon L(f) : {}_\epsilon L(X) \to {}_\epsilon L(Y)$ is $(n+1)$-connected modulo $\mathcal{C}$.

This result can be generalized to homotopy homomorphisms of Hermitian A_∞ rings. It is also worth pointing out that rationally any Hermitian A_∞ ring is equivalent to the geometric realization of a simplicial Hermitian A_∞ ring. Thus by combining Proposition 4.1 and 4.2 we conclude that for any Hermitian A_∞ ring X, the rationalization of ${}_\epsilon L(X)$ can be described as the Hermitian K-theory of a simplicial Hermitian ring as defined in [BF].

The next result concerns the relation between the Hermitian algebraic K-theory ${}_\epsilon L(X)$ and the ordinary algebraic K-theory $K(X)$ (with the Hermitian structure on the A_∞ ring X neglected) as defined in [M1], [Stb] and later developed in [St2] and [SV1].

Proposition 4.3 (a) There are involutions σ and τ on the infinite loop spaces ${}_\epsilon L(X)$ and $K(X)$ respectively inducing $\frac{1}{2}$ local splittings

$$\begin{aligned} {}_\epsilon L(X) &\simeq {}_\epsilon L^+(X) \times {}_\epsilon L^-(X) \\ K(X) &\simeq K^+(X) \times K^-(X) \end{aligned}$$

(b) There are infinite loop maps

$$\begin{aligned} F : {}_\epsilon L(X) &\to K(X) \qquad \text{(forgetful map)} \\ H : K(X) &\to {}_\epsilon L(X) \qquad \text{(hyperbolic map)} \end{aligned}$$

which decompose away from 2 into composites

$$\begin{aligned} {}_\epsilon L(X) \xrightarrow{proj} {}_\epsilon L^+(X) \xrightarrow{f} K^+(X) \hookrightarrow K(X) \\ K(X) \xrightarrow{proj} K^+(X) \xrightarrow{h} {}_\epsilon L^+(X) \hookrightarrow {}_\epsilon L(X) \end{aligned}$$

where f and h are mutually inverse $\frac{1}{2}$ local equivalences.

We indicate some of the flavor of the proof. The main idea is to imbed the diagram used to construct Hermitian K-theory into larger diagrams, perform the Segal pushdown and then use the fact that up to equivalence taking sub-diagrams commutes with the Segal pushdown. Thus we begin by constructing a pullback diagram for the forgetful map of the form

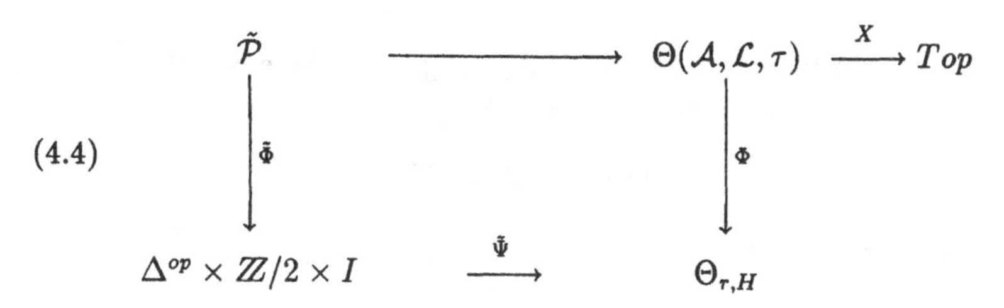

where I denotes the category $0 \to 1$, and $\tilde{\Psi}$ is a functor whose restriction to $0 \subseteq I$ is the functor in diagram (3.2) and whose restriction to $1 \subseteq I$ factors through the functor

$$\Delta^{op} \to \Theta_{\tau,H}$$

used in [SV1] to construct ordinary algebraic K-theory of X. Again we restrict the functor of (4.4) to invertible components, take the Segal pushdown and take the homotopy fixed point sets with respect to the $\mathbb{Z}/2$-actions, and geometric realization with respect to Δ^{op}. Extending this construction so that it also codifies direct sum operations we obtain a functor

$$\Gamma^{op} \times I \to Top$$

specifying a map of Γ-spaces. By using the restriction property of Segal pushdown we can identify the domain of this map with ${}_\varepsilon L(X)$ and the target with $K(X)$, thus constructing

$$F: {}_\varepsilon L(X) \to K(X).$$

The construction of the hyperbolic map

$$H: K(X) \to {}_\varepsilon L(X)$$

uses the same idea together with the observation that $K(X) \simeq {}_\varepsilon L(X \times X^{op})$ (also used in [BF]). This observation is also used to construct the involution τ on $K(X)$.

The construction of the involution σ on ${}_\varepsilon L(X)$ is the only point where our line of argument departs significantly from that in [BF]. For a close analysis of that argument shows that it uses the additive relation $1+(-1) = 0$, which, as mentioned in § 1, is not available in the A_∞ world. Instead we show directly that $H \cdot F - 1$ defines a homotopy involution on ${}_\epsilon L(X)$. (This raises some extra difficulties when considering the effect of pairings, but these difficulties can be overcome.)

Next we define Gersten-Wagoner type spectra [FSV3].

Proposition 4.5 If X is a Hermitian A_∞ ring, then there is a Hermitian A_∞ ring sX together with an infinite loop map

$$ {}_\epsilon L(X) \to \Omega_\epsilon L(sX) $$

which induces a $\frac{1}{2}$ local equivalence on basepoint components.

The construction of sX (suspension of X) is given in [FSSV] which proves the corresponding result for $K(X)$. The proof of Proposition 4.5 is an adaptation of the arguments used there to the Hermitian case.

Using Proposition 4.5, we define a nonconnective Hermitian K-theory spectrum ${}_\epsilon \underline{\underline{L}}(X) = \{\Omega_\epsilon L(s^{k+1}X)\}_{k=0}^{\infty}$. Then we introduce a notion of pairing of Hermitian A_∞ rings $R \times S \to T$ and show that these give rise to pairings of infinite loop spaces [FSV3]

$$ {}_\epsilon L(R) \times {}_\delta L(S) \to {}_{\epsilon\delta} L(T). $$

In particular, we construct pairings of the form

$$ s^k Q(\Omega X_+) \times s^l Q(\Omega Y_+) \to s^{k+l} Q(\Omega(X \times Y)_+) $$

inducing pairings of spectra

$$ \text{(4.6)} \qquad {}_\epsilon \underline{\underline{L}}(Q(\Omega X_+)) \times {}_\delta \underline{\underline{L}}(Q(\Omega Y_+)) \to {}_{\epsilon\delta} \underline{\underline{L}}(Q(\Omega(X \times Y)_+)). $$

These pairings are associative, commutative and unital up to homotopy. Moreover their behavior with respect to the splitting

$$ {}_\epsilon \underline{\underline{L}}(R) \simeq {}_\epsilon \underline{\underline{L}}^+(R) \times {}_\epsilon \underline{\underline{L}}^-(R) $$

induced by the involution $\sigma = H \circ F - 1$ is such that (4.6) induces pairings

$$ {}_\epsilon \underline{\underline{L}}^+(Q(\Omega X_+)) \times {}_\delta \underline{\underline{L}}^+(Q(\Omega Y_+)) \to {}_{\epsilon\delta} \underline{\underline{L}}^+(Q(\Omega(X \times Y)_+) $$
$$ {}_\epsilon \underline{\underline{L}}^-(Q(\Omega X_+)) \times {}_\delta \underline{\underline{L}}^-(Q(\Omega Y_+)) \to {}_{\epsilon\delta} \underline{\underline{L}}^-(Q(\Omega(X \times Y)_+). $$

In particular, we get a pairing

$$(4.7) \qquad {}_{-1}\underline{\underline{L}}^-(Q(S^0)) \times {}_{\epsilon}\underline{\underline{L}}^-(Q(\Omega X_+)) \to {}_{-\epsilon}\underline{\underline{L}}^-(Q(\Omega X_+))$$

from which we in [FSV4] obtain:

Proposition 4.8 (Karoubi Periodicity) There is a $\frac{1}{2}$ local equivalence

$${}_{\epsilon}\underline{\underline{L}}^-(Q(\Omega X_+) \simeq \Omega^2{}_{-\epsilon}\underline{\underline{L}}^-(Q(\Omega X_+))$$

Proof. Sketch. Karoubi constructs elements $u \in \pi_2({}_{-1}\underline{\underline{L}}^-(\mathbb{Z}))$ and $v \in \pi_{-2}({}_{-1}\underline{\underline{L}}^-(\mathbb{Z}))$ whose product is 32 times the unit element of $\pi_0({}_{+1}L^-(\mathbb{Z}))$. Since

$$QS^0 \to \mathbb{Z}$$

is 2-connected away from 2, and an equivalence on π_0, it follows from Proposition 4.2 and 4.3 that the induced map

$${}_{-1}\underline{\underline{L}}(QS^0) \to_{-1} \underline{\underline{L}}(\mathbb{Z})$$

is 3-connected. Hence we obtain corresponding elements $u \in \pi_2({}_{-1}\underline{\underline{L}}^-(QS^0))$ and $v \in \pi_{-2}({}_{-1}\underline{\underline{L}}^-(QS^0))$, which we can use with the pairing (4.7) to construct the $\frac{1}{2}$ local equivalences.

We obtain as a corollary

Corollary 4.9 The map $Q(\Omega X_+) \to \mathbb{Z}[\pi_1 X]$ induces a $\frac{1}{2}$ local equivalence

$$\epsilon\underline{\underline{L}}^-(Q(\Omega X_+)) \to {}_{\epsilon}\underline{\underline{L}}^-(\mathbb{Z}[\pi_1 X]) = {}_{\epsilon}\mathbb{L}(\pi_1 X),$$

where $\mathbb{L}$ stands for Quinn's surgery spectrum.

Proof. This follows from periodicity and homotopy invariance.

Finally we use pairings to construct an assembly map

$$\Omega^\infty(X \wedge {}_{\epsilon}\underline{\underline{L}}(Q(S^0))) \to {}_{\epsilon}\underline{\underline{L}}(Q(\Omega X_+))$$

and we define the Hermitian Whitehead space $Wh^{\epsilon\underline{\underline{L}}}(X)$ to be the cofiber of this map in the category of infinite loop spaces.

We now obtain the following theorem

Theorem 4.10 Let M be an even dimensional manifold. Then there is a map

$$H(M)/\mathrm{Homeo}(M) \to \Omega Wh^{\cdot\underline{\underline{L}}}(M)$$

which is a $\frac{1}{2}$ local equivalence in the stable range on basepoint components.

Proof.Sketch. In the stable range we have the $\frac{1}{2}$ local splitting

$$H(M)/\mathrm{Homeo}(M) \simeq H(M)/\widetilde{\mathrm{Homeo}}(M) \times \widetilde{\mathrm{Homeo}}(M)/\mathrm{Homeo}(M)$$

of [BF]. The splittings of Proposition 4.3 induce splittings

$$\begin{aligned} Wh^{\cdot\underline{\underline{L}}}(M) &\simeq Wh^{\cdot\underline{\underline{L}}}(M)^+ \times Wh^{\cdot\underline{\underline{L}}}(M)^- \\ Wh^{\underline{\underline{K}}}(M) &\simeq Wh^{\underline{\underline{K}}}(M)^+ \times Wh^{\underline{\underline{K}}}(M)^- \end{aligned}$$

and a $\frac{1}{2}$ local equivalence

$$Wh^{\cdot\underline{\underline{L}}}(M)^+ \simeq Wh^{\underline{\underline{K}}}(M)^+.$$

By results of Waldhausen [W], we have a $\frac{1}{2}$ local equivalence

$$\widetilde{\mathrm{Homeo}}(M)/\mathrm{Homeo}(M) \to \Omega Wh^{\underline{\underline{K}}}(M)^+$$

on basepoint components in the stable range. By the surgery exact sequence and Corollary 4.9 we get a $\frac{1}{2}$ local equivalence

$$H(M)/\widetilde{\mathrm{Homeo}}(M) \to \Omega Wh^{\cdot\underline{\underline{L}}}(M)^-.$$

References

[BF] D. Burghelea and Z. Fiedorowicz, Hermitian algebraic K-theory of simplicial rings and topological spaces, J. Math. Pures Appl. 64 (1985), 175-235

[FSSV] Z. Fiedorowicz, R. Schwänzl, R. Steiner and R. Vogt, Non-connective delooping of K-theory of an A_∞ ring space, Math.Z. 203 (1990), 43-57

[FSV1] Z. Fiedorowicz, R. Schwänzl, and R.M. Vogt, Hermitean structures on A_∞ ring spaces, to appear

[FSV2] Z. Fiedorowicz, R. Schwänzl, and R.M. Vogt, Hermitean K-theory of A_∞ rings. Homotopy invariance and the hyperbolic map, to appear

[FSV3] Z. Fiedorowicz, R. Schwänzl, and R.M. Vogt, Non-connective delooping of Hermitean K-theory of A_∞ rings and pairings, to appear

[FSV4] Z. Fiedorowicz, R. Schwänzl, and R.M. Vogt, Karoubi periodicity of Hermitian K-theory of A_∞ rings and geometric applications, to appear

[M1] J.P. May, A_∞ ring spaces and algebraic K-theory, Springer Lecture Notes in Math. 658 (1978), 240-315

[M2] J.P. May, Multiplicative infinite loop space theory, J. Pure Appl. Algebra 26 (1982), 1-69

[SV1] R. Schwänzl and R.M. Vogt, Matrices over homotopy ring spaces and algebraic K-theory, OSM, P73 (1984), Universität Osnabrück

[SV2] R. Schwänzl and R.M. Vogt, E_∞ monoids with coherent homotopy inverses are abelian groups, Topology 28 (1989), 481-484

[S] G. Segal, Categories and cohomology theories, Topology 13 (1974), 293-312

[Stb] M. Steinberger, On the equivalence of the two definitions of the algebraic K-theory of a topological space, Springer Lecture Notes in Math. 763 (1979), 317-331

[St1] R. Steiner, A canonical operad pair, Math. Proc. Camb. Phil. Soc. 86 (1979), 443-449

[St2] R. Steiner, Infinite loop structures on the algebraic K-theory of spaces, Math. Proc. Camb. Phil. Soc. 90 (1981), 85-111

[W] F. Waldhausen, Algebraic K-theory of topological spaces I, Proc. Symp. Pure Math. 32 (1978), 35-60

A splitting result for the second homology group of the general linear group

Dominique Arlettaz

1. Introduction

Let A be a ring with identity, $GL(A) = \varinjlim GL_m(A)$ the infinite general linear group over A (considered as a discrete group), $E(A)$ its subgroup generated by elementary matrices, and $BGL(A)^+$ and $BE(A)^+$ the infinite loop spaces obtained by performing the plus construction on the classifying spaces of $GL(A)$ and $E(A)$ respectively. Recall that these spaces, which have the same homology as the corresponding groups, provide Quillen's definition of the algebraic K-theory of the ring A : $K_nA := \pi_nBGL(A)^+$ for all $n \geq 1$, and $K_nA \cong \pi_nBE(A)^+$ for $n \geq 2$ since $BE(A)^+$ is the universal cover of $BGL(A)^+$ (cf. [L], Proposition 1.1.4).

The purpose of this note is to show that K_2A is always a direct summand of $H_2(GL(A);\mathbb{Z})$, in a natural way. More precisely, we first prove the following

Theorem 1. *For any ring A, the 2-dimensional Hurewicz homomorphism $h_2 : K_2A \to H_2(BGL(A)^+;\mathbb{Z}) \cong H_2(GL(A);\mathbb{Z})$ is split injective and*

$$H_2(GL(A);\mathbb{Z}) \cong K_2A \oplus \Lambda^2(K_1A),$$

where $\Lambda^2(-)$ denotes the second exterior power.

This is quite obvious if A is a commutative ring with $SK_1A = 0$ (cf. Remark 2.1), but the assertion is true in general, without any hypothesis on the ring A (in particular, without any finiteness condition and without the assumption that A is commutative). Let us mention that this theorem has also been established by K. Dennis [D] (and applied in [Hu]), but never published. The proof presented here is very short and uses only a topological argument : the assertion follows directly from the fact that $BGL(A)^+$ is a double loop space (cf. Section 2). Apparently, this simple and useful splitting result is not well-known.

Observe that the Serre spectral sequence of the fibration

$$BE(A)^+ \longrightarrow BGL(A)^+ \longrightarrow K(K_1A, 1)$$

produces the exact sequence

$$H_3(K_1A; \mathbb{Z}) \longrightarrow H_2(E(A); \mathbb{Z}) \longrightarrow H_2(GL(A); \mathbb{Z}) \longrightarrow H_2(K_1A; \mathbb{Z}) \longrightarrow 0.$$

The Hurewicz map $K_2A \to H_2(E(A); \mathbb{Z})$ is an isomorphism since $BE(A)^+$ is simply connected, and we may replace $H_2(K_1A; \mathbb{Z})$ by $\Lambda^2(K_1A)$ because $H_2(G; \mathbb{Z}) \cong \Lambda^2(G)$ for any abelian group G (cf. [B], p.123, Theorem 6.4). Thus, we get the exact sequence

$$H_3(K_1A; \mathbb{Z}) \longrightarrow K_2A \xrightarrow{h_2} H_2(GL(A); \mathbb{Z}) \longrightarrow \Lambda^2(K_1A) \longrightarrow 0.$$

Notice that H. Hopf has already explained in [Ho] that the cokernel of h_2 is always the second homology group of the fundamental group. Theorem 1 claims actually the existence of a homomorphism

$$\theta_A : H_2(GL(A); \mathbb{Z}) \longrightarrow K_2A$$

associated with any ring A, such that the composition $\theta_A h_2$ is the identity of K_2A. We then prove that this splitting is natural :

Theorem 2. *Let $\phi : A \to B$ be a ring homomorphism, then the diagram*

$$\begin{array}{ccc} H_2(GL(A); \mathbb{Z}) & \xrightarrow{\theta_A} & K_2A \\ \downarrow \phi_* & & \downarrow \phi_\sharp \\ H_2(GL(B); \mathbb{Z}) & \xrightarrow{\theta_B} & K_2B \end{array}$$

is commutative, where ϕ_* and $\phi_\sharp$ denote the homomorphisms induced by ϕ.

In order to generalize these results, it is possible to define integers $\bar{R}_n$ for $n \geq 3$ (which are independent of A) and to construct a map

$\theta_{A,n} : H_n(GL(A);\mathbb{Z}) \to K_nA$ for any ring A and any integer $n \geq 3$ with the following property : the composition of the Hurewicz homomorphism $h_n : K_nA \to H_n(GL(A);\mathbb{Z})$ with $\theta_{A,n}$ is multiplication of K_nA by a divisor of $\bar{R}_n$ (cf. [A2], Theorem 2.1 and Remark 2.3). Our paper [A2] contains a general discussion of the Hurewicz homomorphism in algebraic K-theory.

We shall prove Theorem 1 in Section 2 and Theorem 2 in Section 3. Throughout the paper, if the coefficients of homology (or homotopy) groups are not written explicitely, integral coefficients must be understood.

2. The second Postnikov section of a double loop space

Remark 2.1. If A is a commutative ring such that $SK_1A = 0$ (i.e., $E(A) = SL(A)$ and $K_1A \cong A^\times$), the fibration of infinite loop spaces $BSL(A)^+ \to BGL(A)^+ \to K(A^\times, 1)$, considered in the introduction, has a splitting $K(A^\times,1) \to BGL(A)^+$ which is induced by the inclusion $A^\times = GL_1(A) \hookrightarrow GL(A)$; consequently, $BGL(A)^+$ is a product :

$$BGL(A)^+ \simeq BSL(A)^+ \times K(A^\times, 1).$$

In this case, we obtain the statement of Theorem 1 directly by taking the second homology group: $H_2GL(A) \cong H_2SL(A) \oplus H_2A^\times \cong K_2A \oplus \Lambda^2(A^\times)$.

The goal of this section is to prove Theorem 1 in the general case. If X is a connected simple CW-complex, let us call $\alpha_i : X \to X[i]$ its i-th Postnikov section for $i \geq 1$ (i.e., $\pi_kX[i] = 0$ for $k > i$ and $(\alpha_i)_* : \pi_kX \to \pi_kX[i]$ is an isomorphism for $k \leq i$), and denote the Postnikov k-invariants of X by $k^{n+1}(X) \in H^{n+1}(X[n-1]; \pi_nX)$ for $n \geq 2$. The basic point of our argument is the next result which is proved in [A1].

Proposition 2.2. *If X is a connected double loop space, then its first k-invariant $k^3(X) \in H^3(K(\pi_1X, 1); \pi_2X)$ is trivial.*

This is of course equivalent to the existence of a homotopy equivalence

$$\lambda_X : X[2] \xrightarrow{\simeq} K(\pi_1X, 1) \times K(\pi_2X, 2).$$

(Let us mention that this homotopy equivalence is in general not unique.) This enables us to deduce the following immediate consequence.

Corollary 2.3. *If X is a connected double loop space, then the Hurewicz homomorphism $h_2 : \pi_2 X \to H_2 X$ is split injective and*

$$H_2X \cong \pi_2 X \oplus \Lambda^2(\pi_1 X).$$

Proof of Theorem 1. In fact, it is now sufficient to apply Corollary 2.3 to the infinite loop space $X = BGL(A)^+$. But let us explain more precisely the definition of the splitting

$$\theta_A : H_2GL(A) \longrightarrow K_2A.$$

Let ζ_A be the composition

$$BGL(A)^+ \xrightarrow{\alpha_2} BGL(A)^+[2] \underset{\simeq}{\xrightarrow{\lambda_{BGL(A)^+}}} K(K_1A,1) \times K(K_2A,2) \xrightarrow{p_A} K(K_2A,2) \underset{\simeq}{\xrightarrow{\eta_A}} K(K_2A,2),$$

where p_A denotes the projection onto the second factor (observe that $p_A \lambda_A \alpha_2$ induces an isomorphism on π_2), and η_A the homotopy equivalence which has the property that the homomorphism induced by ζ_A on $\pi_2 : \pi_2 BGL(A)^+ \to \pi_2 K(K_2A,2)$ is exactly the identity of K_2A. We then define the splitting θ_A as follows:

$$\theta_A : H_2GL(A) \cong H_2BGL(A)^+ \xrightarrow{(\zeta_A)_*} H_2K(K_2A,2) \xrightarrow{h_2^{-1}} K_2A,$$

where the first homomorphism is the homomorphism induced by the map ζ_A on the second homology group and the second the inverse of the Hurewicz isomorphism for the Eilenberg-MacLane space $K(K_2A,2)$.

The commutativity of the diagram

$$\begin{array}{ccc} K_2A = \pi_2BGL(A)^+ & \xrightarrow[\text{identity}]{(\zeta_A)_*} & \pi_2K(K_2A,2) = K_2A \\ \downarrow h_2 & & \uparrow h_2^{-1} \\ H_2BGL(A)^+ & \xrightarrow{(\zeta_A)_*} & H_2K(K_2A,2) \end{array}$$

implies immediately that the composition $K_2A \xrightarrow{h_2} H_2BGL(A)^+ \xrightarrow{\theta_A} K_2A$ is the identity since $\theta_A = h_2^{-1}(\zeta_A)_*$.

Remark 2.4. The map ζ_A is in general not unique, however $\theta_A = h_2^{-1}(\zeta_A)_* : H_2BGL(A)^+ \cong H_2K(K_1A,1) \oplus H_2K(K_2A,2) \to K_2A$ is uniquely characterized: since ζ_A is an infinite loop map, it induces a (Pontryagin) ring homomorphism $(\zeta_A)_* : H_*BGL(A)^+ \to H_*K(K_2A,2)$ which is clearly trivial in dimension 1, and $(\zeta_A)_*$ (consequently also θ_A) must then vanish on $H_2K(K_1A,1)$ because any element of $H_2K(K_1A,1)$ is a sum of products of two 1-dimensional homology classes (cf. [B], p.123, Theorem 6.4); on the other hand, it follows directly from the construction of θ_A that it is uniquely defined on $H_2K(K_2A,2)$.

Notice in particular that, if A is commutative, θ_A is not the obvious composition $H_2GL(A) \to \Lambda^2(K_1A) \to K_2A$, where the first arrow is the homomorphism induced on H_2 by the surjection $GL(A) \to GL(A)/E(A) = K_1A$ and the second the product operation.

3. The naturality of the splitting

Remark 3.1. The construction introduced in the proof of Theorem 1 produces actually a map $\zeta_X : X \to K(\pi_2X, 2)$ for all double loop spaces X, but this map is in general not natural.

Theorem 2 asserts the naturality of the splitting given by Theorem 1, in other words it claims that θ is a natural transformation from the functor $H_2GL(-)$ to the functor $K_2(-)$, defined on the category of rings.

Proof of Theorem 2. Consider a ring homomorphism $\phi : A \to B$ and the infinite loop map $f_\phi : BGL(A)^+ \to BGL(B)^+$ induced by ϕ. We must establish the commutativity of the diagram

$$\begin{array}{ccc} H_2K(K_1A,1) \oplus H_2K(K_2A,2) \cong H_2BGL(A)^+ & \xrightarrow{\theta_A} & K_2A \\ \downarrow \phi_* & & \downarrow \phi_\sharp \\ H_2K(K_1B,1) \oplus H_2K(K_2B,2) \cong H_2BGL(B)^+ & \xrightarrow{\theta_B} & K_2B \end{array}$$

where ϕ_* and $\phi_\sharp$ denote the homomorphisms induced by f_ϕ on H_2 and on π_2 respectively. It is easy to check the equality $\phi_\sharp \theta_A = \theta_B \phi_*$ on $H_2K(K_2A, 2)$, because of the fact that the composition $\theta_A h_2$, respectively $\theta_B h_2$, is the identity. On the other hand, it follows from the definition of θ_A (cf. Remark 2.4) that θ_A vanishes on $H_2K(K_1A, 1)$. Therefore, in order to complete the proof, it is sufficient to show that $\theta_B \phi_*$ is trivial on $H_2K(K_1A, 1)$.

But this is again a consequence of the argument given in Remark 2.4: since the infinite loop map $\zeta_B f_\phi$ induces a (Pontryagin) ring homomorphism $(\zeta_B)_* \phi_* : H_*BGL(A)^+ \to H_*K(K_2B, 2)$ which is clearly trivial on 1-dimensional classes, in dimension 2 $(\zeta_B)_* \phi_*$ is then zero on $H_2K(K_1A, 1)$ because any element of $H_2K(K_1A, 1)$ is a sum of products of two 1-dimensional classes. Consequently, $\theta_B \phi_* = h_2^{-1} (\zeta_B)_* \phi_*$ is also trivial on $H_2K(K_1A, 1)$.

References

[A1] D. Arlettaz : The first k-invariant of a double loop space is trivial, *Arch. Math.* 54 (1990), 84–92.

[A2] D. Arlettaz : The Hurewicz homomorphism in algebraic K-theory, to appear in *J. Pure Appl. Algebra*.

[B] K. Brown : Cohomology of Groups, *Graduate Texts in Math.* 87 (Springer 1982).

[D] K. Dennis : In search of new "homology" functors having a close relationship to K-theory, *preprint* (1976).

[Ho] H. Hopf : Fundamentalgruppe und zweite Bettische Gruppe, *Comment. Math. Helv.* 14 (1941–1942), 257–309.

[Hu] J. Huebschmann : The first k-invariant, Quillen's space BG^+ and the construction of Kan and Thurston, *Comment. Math. Helv.* 55 (1980), 314–318.

[L] J.-L. Loday : K-théorie algébrique et représentations de groupes, *Ann. Sci. Ecole Norm. Sup.* 9 (1976), 309–377.

Dominique Arlettaz
Institut de mathématiques
Université de Lausanne
CH–1015 Lausanne, Switzerland

Low dimensional spinor representations, Adams maps and geometric dimension

K.Y. Lam and D. Randall

§1. Introduction.

The use of low dimensional spinor representations as a tool to estimate the geometric dimension of vector bundles over real projective spaces first appeared in Gitler and Lam [7], and then more systematically, in Adams [1], where spin vector bundles of geometric dimension ≤ 5 were treated up to stable equivalence. In this note, we shall show that by a more delicate study of spinor representations results similar to Adams' can be obtained for vector bundles of geometric dimension 6 and 7. One of the novel features is to consider the Lie group $\text{spin}^c(n)$ and its representation Δ_n^c, which is similar to the classical spinor representations. This is done in §2. The main theorems, listed below, are proved in §3. Some of the theorems are best possible, as shown in §4 (with technical details concerning Adams maps deferred to §5).

This work is supported by an NSERC grant of Canada.

In all that follows, ξ_n denotes the canonical Hopf line bundle over the real projective space RP^n, and $k\xi_n$ denotes the k-fold Whitney sum of ξ_n with itself. Similarly, k or $k\epsilon$ denotes a k dimensional trivial bundle over any space, and $\epsilon_{\mathbf{C}} = \epsilon \otimes_R \mathbf{C}$ is the trivial complex line bundle. As is classical, the geometric dimension of $k\xi_n$ is the smallest positive integer d such that there exists a d-dimensional vector bundle η over RP^n stably equivalent to $k\xi_n$. We shall also assume $n \geq 12$ unless otherwise stated, because for $n \leq 11$, the geometric dimension of vector bundles over RP^n are all known.

Theorem 1. *Suppose that $4k\xi_n$ has geometric dimension ≤ 6. Then*

(i) $16k \equiv 0(\text{mod}\, 2^{[n/2]})$ *if k is even and*

(ii) $4k \equiv 4(\text{mod}\, 2^{[n/2]})$ *if k is odd.*

Here $[n/2]$ denotes the largest integer not exceeding $n/2$.

Theorem 2. *Suppose that $4k\xi_n$ has geometric dimension ≤ 7. Then*

(i) $32k \equiv 0 (\bmod 2^{\phi(n)})$ *if k is even and*
(ii) $32(k-1) \equiv 0 (\bmod 2^{\phi(n)})$ *if k is odd.*

Here $\phi(n)$ denotes the number of integers in the interval $[1, n]$ congruent to $0, 1, 2$ or $4 (\bmod 8)$.

Theorem 3. *Suppose that $(4k+2)\xi_n$ has geometric dimension ≤ 7. Then*

(i) $32k \equiv 0 (\bmod 2^{[n/2]})$ *if k is even and*
(ii) $32(k-1) \equiv 0 (\bmod 2^{[n/2]})$ *if k is odd.*

§2. Spinor representations.

The spinor groups spin($2r$) and spin($2r+1$) have well-known complex representations $\Delta^{\pm}_{2r}$: spin$(2r) \to U(2^{r-1})$ and Δ_{2r+1}: spin$(2r+1) \to U(2^r)$ respectively. With respect to tensor products and exterior powers over C, they have the following properties as recorded, for example, in Hüsmoller [8,p.188] or Bott [4,p.63].

$$\Delta_{2r+1} \otimes \Delta_{2r+1} = (\lambda_r + \lambda_{r-1} + \cdots + \lambda_0)\rho_{2r+1} \, .$$

(#)

$$\lambda_2(\Delta^+_{2r}) = \lambda_2(\Delta^-_{2r}) = \sum \lambda_i(\rho_{2r}), \; i \equiv r+2 (\bmod 4) \text{ and } 0 \leq i \leq r-1 \, ,$$

$$\lambda_2(\Delta_{2r+1}) = \sum \lambda_i(\rho_{2r+1}), \; i \equiv r+2 \text{ or } r+3 (\bmod 4) \text{ and } 1 \leq i \leq r-1 \, ;$$

where ρ_r denotes the complexification of the double covering representation spin$(r) \to SO(r)$. It is also customary to write Δ_{2r} for $\Delta^+_{2r} + \Delta^-_{2r}$.

Consider the Lie group spin$^c(r) = $ spin$(r) \times_T$spin(2), where T acts on spin$(r) \times$ spin(2) by the involution $(a, b) \to (-a, -b)$, with $-a$ meaning the image of a under the nontrivial deck transformation of the double cover spin$(r) \to SO(r)$. This group is obviously a double covering of $SO(r) \times SO(2)$ while spin$(r) \times$spin(2) is a fourfold covering.

The representation of spin$(r) \times$ spin(2) by unitary matrices defined by

$$(a, b) \to \Delta_r(a) \cdot \Delta^+_2(b) \, ,$$

where the dot means multiplication of a unitary matrix $\Delta_r(a)$ by a complex number $\Delta_2^+(b)$, factors through the T action and gives a representation of spin$^c(r)$ by unitary matrices, which we shall denote by Δ_r^c. In this note we shall be interested in $\Delta_{2r+1}^c : \mathrm{spin}^c(2r+1) \to U(2^r)$, and we have the following properties.

PROPOSITION (2.1). *Under tensor product over* C *one has*

$$\Delta_{2r+1}^c \otimes \Delta_{2r+1}^c = \pi^* \left((\lambda_r + \lambda_{r-1} + \cdots + \lambda_0)\iota_{2r+1} \otimes \sigma \right)$$

where $\pi : spin^c(2r+1) \to SO(2r+1) \times SO(2)$ *is the double covering,* $\iota_{2r+1} : SO(2r+1) \to U(2r+1)$ *is the inclusion representation and* $\sigma : SO(2) \approx U(1)$ *is the representation identifying* $SO(2)$ *with* $U(1)$.

PROOF: The representation $\Delta_{2r+1}^c \otimes \Delta_{2r+1}^c$ takes $((a,b))$ and $((-a,b))$ to the same image, and so is the pullback under π^* of a certain representation of $SO(2r+1) \times SO(2)$. But, considered as a representation of spin$(2r+1) \times$ spin(2) it is just

$$(\Delta_{2r+1} \otimes \Delta_2^+) \otimes (\Delta_{2r+1} \otimes \Delta_2^+) = (\Delta_{2r+1} \otimes \Delta_{2r+1}) \otimes (\Delta_2^+)^2 .$$

The proposition now follows from (#), together with $(\Delta_2^+)^2 = \rho_2$, and the injectivity of the representation ring of $SO(r)$ into that of spin(r) under the double cover.

PROPOSITION (2.2). *Under the second exterior power operation* λ_2 *one has*

$$\lambda_2(\Delta_{2r+1}^c) = \pi^* \left(\sum \lambda_i(\iota_{2r+1}) \otimes \sigma \right), \ i \equiv r+2 \text{ or } r+3 (\mathrm{mod}\, 4) \text{ and } 1 \le i \le r-1 .$$

PROOF: For the same reason as in Proposition (2.1), $\lambda_2(\Delta_{2r+1}^c)$ is a pullback representation under π^*. The proof is now completed by noting that $\lambda_2(\Delta_{2r+1}^c)$ sends $((a,b))$ into

$$(\text{the second exterior matrix of } \Delta_{2r+1}(a)) \cdot (\Delta_2^+(b))^2 ,$$

and by appealing to (#).

Remark. There are, of course, similar results for $(\Delta^c_{2r})^+$ but we shall not need them in this note.

§3. Proof of Theorems 1, 2 and 3.

We recall that $\widetilde{KO}(RP^n)$ is a cyclic group of order $2^{\phi(n)}$ with $\phi(n)$ as in Theorem 2, and $\tilde{K}(RP^n)$ is cyclic of order $2^{[n/2]}$. The first cyclic group is generated by $x = \xi - \epsilon$ and the second by $y = \xi \otimes \mathbf{C} - \epsilon_{\mathbf{C}}$. As a stable class, y is represented by the complex line bundle $H = \xi \otimes \mathbf{C}$. As is well-known, $x^2 = -2x$ and $y^2 = -2y$.

To prove Theorem 1, suppose that over RP^n

$$4k\xi_n = (4k-6)\epsilon \oplus \eta$$

where η is a 6-dimensional bundle. Since $\omega_1(\eta) = 0 = \omega_2(\eta)$, η has a spin(6) structure, so that there is a 4- dimensional complex vector bundle $\Delta_6^+(\eta)$ associated with the spinor representation $\Delta_6^+ : \text{spin}(6) \to U(4)$. As an element in $K(RP^n)$, we have $\Delta_6^+(\eta) = my + 4$ for some m. By (#), $\lambda_2(\Delta_6^+(\eta)) = \lambda_1(\eta \otimes \mathbf{C}) = \eta \otimes \mathbf{C} = 4ky + 6$. Since

$$\begin{aligned}\lambda_2(my+4) &= \lambda_2(my) + \lambda_1(my)\lambda_1(4) + \lambda_2(4)\\ &= -m^2y + 4my + 6\,,\end{aligned}$$

we deduce

$$-m(m-4) \equiv 4k \qquad (\text{mod}\, 2^{[n/2]})\,.$$

In particular, m must be even.

We now invoke the consideration of K theory Pontryagin classes, by a procedure similar to Atiyah's in [3]. The equation $\Delta_6^+(\eta) = my + 4$ means that the complex bundle mH has <u>complex</u> geometric dimension ≤ 4, and standard arguments in [3] lead to the conclusion that $(1+yt)^m$ is a polynomial in t of degree ≤ 4 having coefficients in $K(RP^n)$. In particular,

$$y^5\binom{m}{5} = 16\binom{m}{5}y = \frac{2}{15}m(m-1)(m-2)(m-3)(m-4)y = 0\,,$$

so we have another congruence

$$2m(m-2)(m-4) \equiv 0 \qquad (\mathrm{mod}\, 2^{[n/2]}) .$$

Combining with the previous congruence, one gets

$$2(m-2)(4k) \equiv 0 \qquad (\mathrm{mod}\, 2^{[n/2]}) .$$

Let $2^{\nu_2(h)}$ denote the highest power of 2 dividing h. If k is even, then we conclude from the first congruence that m must be divisible by 4, so that $\nu_2(m-2) = 1$. Hence the last congruence simplifies to

$$16k \equiv 0 \qquad (\mathrm{mod}\, 2^{[n/2]})$$

completing (i). If k is odd, then $\nu_2(4k) = 2$ and we must have

$$\nu_2(m-2) \geq [n/2] - 3 .$$

Therefore,

$$4k - 4 \equiv -(m^2 - 4m + 4) \equiv -(m-2)^2 \equiv 0 \qquad (\mathrm{mod}\, 2^{[n/2]})$$

because $\nu_2(m-2)^2 = 2\nu_2(m-2) \geq 2[n/2] - 6 \geq [\frac{n}{2}]$ since $n \geq 12$, and part (ii) is now proved.

We now move on to prove Theorem 2. Suppose that, over RP^n, there is a 7-dimensional vector bundle η satisfying

$$4k\xi_n = (4k-7)\epsilon \oplus \eta .$$

If, for any vector bundle ζ, we write a formal power series

$$\lambda_t(\zeta) = 1 + \lambda_1(\zeta)t + \lambda_2(\zeta)t^2 + \cdots ,$$

then $\lambda_t(\zeta \oplus \zeta') = \lambda_t(\zeta)\lambda_t(\zeta')$. In particular,

$$\begin{aligned} \lambda_t(\eta) &= \lambda_t(4k\xi_n)\lambda_t((4k-7)\epsilon)^{-1} \\ &= (1 + (x+1)t)^{4k}(1+t)^{7-4k} , \end{aligned}$$

and the exterior powers $\lambda_1(\eta), \lambda_2(\eta), \ldots$, as elements in $KO(RP^n)$, can be read out from the various coefficients in this formal power series of t. Specifically, one has

$$\begin{aligned}\lambda_1(\eta) &= 4kx + 7\\ \lambda_2(\eta) &= -4k(4k-7)x + 21\\ \lambda_3(\eta) &= \left[\binom{4k}{3} + 4k(4k-7)(2k-3)\right] x + 35\,.\end{aligned}$$

We now observe that the spinor representation $\Delta_7 : \mathrm{spin}(7) \to U(8)$ is <u>real</u> in the sense that it factorizes as

$$\mathrm{spin}(7) \to SO(8) \subset U(8)\,.$$

Hence we can consider the associated <u>real</u> vector bundle $\Delta_7(\eta)$ and write it as an element $mx + 8$ in $KO(RP^n)$ for some integer m. Similar to the arguments in the proof of Theorem 1, the relation $\lambda_2\Delta_7 = \lambda_2 + \lambda_1$ over R leads to the congruence

$$\begin{aligned}-m^2 + 8m &\equiv -4k(4k-7) + 4k \qquad (\mathrm{mod}\, 2^{\phi(n)})\\ &\equiv -16k(k-2) \quad (\mathrm{mod}\, 2^{\phi(n)})\,.\end{aligned}$$

Also, $\Delta_7(\eta) \otimes_R \Delta_7(\eta) = (mx+8)^2 = -2m^2x + 16mx + 64$. Now the relation $\Delta_7 \otimes_R \Delta_7 = \lambda_3 + \lambda_2 + \lambda_1 + \lambda_0$ over R can be applied and, after some calculation, leads to the congruence

$$-2m^2 + 16m \equiv -16k(k-2) + \binom{4k}{3} + 4k(4k-7)(2k-3) \quad (\mathrm{mod}\, 2^{\phi(n)})\,.$$

Multiplying the first congruence by -2 and adding the result to the second congruence, we eliminate m to obtain

$$16k(k-2) + \binom{4k}{3} + 4k(4k-7)(2k-3) \equiv 0 \quad (\mathrm{mod}\, 2^{\phi(n)})\,,$$

or, after simplication

$$\frac{1}{3}\cdot 4k \cdot 8(4k-5)(k-1) \equiv 0 \quad (\mathrm{mod}\, 2^{\phi(n)})\,.$$

Theorem 2 is now proved by dropping the odd factors $\frac{1}{3}$ and $(4k-5)$ from this congruence.

To prove Theorem 3, suppose that over RP^n, there is a 7-dimensional vector bundle η such that

$$(4k+2)\xi_n = (4k-5)\epsilon \oplus \eta \, .$$

Note that $\omega_2(\eta) \neq 0$ and there is no spin(7) structure for η. Nevertheless, $\eta \oplus 2\xi_n$ has a spin(9) structure, and the pull-back diagram of Lie group homomorphisms

$$\begin{array}{ccc} \mathrm{spin}(7) \times_{\pi} \mathrm{spin}(2) & \longrightarrow & \mathrm{spin}(9) \\ \downarrow \pi & & \downarrow \\ SO(7) \times SO(2) & \longrightarrow & SO(9) \end{array}$$

shows that $\eta \oplus 2\xi_n$ has a spin$^c(7)$ structure. Therefore, there is a $U(8)$ bundle $\Delta_7^c(\eta \oplus 2\xi_n)$ associated to the representation

$$\Delta_7^c : \mathrm{spin}^c(7) \to U(8)$$

and we can write it as $my + 8$ as an element in $K(RP^n)$.

We now exploit the "twisted" relations over $\mathbf{C}$

$$\lambda_2(\Delta_7^c(\eta \oplus 2\xi_n)) = (\lambda_2(\eta) + \lambda_1(\eta)) \otimes H \, ,$$
$$\Delta_7^c(\eta \oplus 2\xi_n) \otimes \Delta_7^c(\eta \oplus 2\xi_n) = (\lambda_3(\eta) + \lambda_2(\eta) + \lambda_1(\eta) + 1) \otimes H \, ,$$

as established in §2. After direct calculations as in the proof of Theorem 2, they lead to the following congruences:

$$-m^2 + 8m \equiv 16k^2 - 16k + 16 \pmod{2^{[n/2]}} \, ,$$
$$-2m^2 + 16m \equiv -\frac{16}{3}(4k+2)(2k^2-4k+3) + 64 \pmod{2^{[n/2]}} \, .$$

Multiplying the first congruence by -2 and adding the result to the second congruence, we eliminate m to obtain

$$-\frac{16}{3} \cdot 2k(4k+1)(k-1) \equiv 0 \qquad \pmod{2^{[n/2]}} \, .$$

The conclusion of Theorem 3 now follows by dropping the odd factors $-\frac{1}{3}$ and $(4k+1)$ from this congruence.

§4. An illustrative example. The strength of the theorems in this note can be seen through the example of $32\xi_{16}$ over RP^{16}. By Theorem 1, it cannot have geometric dimension ≤ 6. On the other hand, one has

Proposition. *$32\xi_{16}$ has 25 independent sections, so that its geometric dimension is precisely equal to 7.*

Proof: We start with the well-known fact that over RP^{15}, $16\xi_{15}$ has 9 independent sections. Thus there is a 7-dimensional vector bundle η over RP^{15} such that

$$16\xi_{15} = 9\epsilon \oplus \eta\,.$$

This η, being stably trivial over RP^8, is actually trivial over RP^8 by the results of James and Thomas in [9, Cor.1.12]. This means that there is a 7-dimensional vector bundle $\tilde{\eta}$ over the truncated projective space $RP_9^{15} = RP^{15}/RP^8$ such that $\eta = c^*(\tilde{\eta})$, where $c : RP^{15} \to RP_9^{15}$ is the collapsing map. Consider now the map $f : RP_9^{16} \to RP_9^{16}$ which is twice the identity map of the (desuspendable) space RP_9^{16}. It is easy to see that homotopically f compresses into the 15 skeleton, namely that f factors up to homotopy into

$$RP_9^{16} \xrightarrow{f_0} RP_9^{15} \xrightarrow{\text{incl.}} RP_9^{16}\,.$$

Denoting the collapsing map $RP^{16} \to RP_9^{16}$ again by c, we see clearly that $c^* f_0^*(\tilde{\eta})$ is a 7-dimensional vector bundle in the stable class of $32\xi_{16}$. Thus $32\xi_{16} = 25\epsilon \oplus c^* f_0^*(\tilde{\eta})$ and the proposition follows.

While this proposition is only an isolated result, it does help to produce an infinite sequence of results on geometric dimension by the use of "KO- periodic maps" (cf. §5)

$$\cdots \to RP_{8\ell+9}^{8\ell+16} \to \cdots \to RP_{17}^{24} \to RP_9^{16}\,,$$

namely, we have

Corollary. *The geometric dimension of $2^{4\ell+5}\xi_{8\ell+16}$ is precisely equal to 7 for all $\ell \geq 0$.*

Proof: By Theorem 1, the geometric dimension in question can't be 6 or less. For the manufacturing of a 7-dimensional vector bundle over $RP^{8\ell+16}$ in the stable class of $2^{4\ell+5}\xi_{8\ell+16}$, we use the above sequence of periodic maps to pull-back $\tilde{\eta}$. Details are the same as in [10].

If the result of Davis [5, Thm. 1.1] is applied to the present case, the conclusion would only be that the geometric dimension in question is ≥ 6. Thus the use of spinor representations in this case provides an improvement by one unit, which is the crucial unit required to settle the geometric dimension, and yet avoids the heavy machinery used in [6]. The result of Theorem 3, on the other hand, is about as strong as Theorem 1, and has little overlap with [5] or [6].

§5. Constructing the KO-periodic map $RP_{17}^{24} \to RP_9^{16}$.

Recall that a KO-periodic map (or KO-equivalence [10]) is one inducing an isomorphism in KO cohomology theory. The KO-periodic maps

$$\cdots \to RP_{8\ell+9}^{8\ell+16} \to \cdots \to RP_{25}^{32} \to RP_{17}^{24}$$

used in §4 can be constructed by using Theorem 3.3 of [10]. However, the map $f : RP_{17}^{24} \to RP_9^{16}$ cannot be so constructed because the connectivity of RP_9^{24} is not high enough to meet the hypothesis of that theorem. Fortunately there is an independent approach due to J.F. Adams which in the case of our particular f would run as follows.

Step 1. Let $X = S^8 \bigcup_{16} e^9$ be the Moore space obtained from the 8-dimensional sphere by attaching a 9-cell using a self-map of degree sixteen. Then there is an "Adams map"

$$A : \Sigma^8 X \to \Sigma^8 X$$

inducing isomorphism in complex K^* cohomology theory. Details for constructing this A will be given in Proposition (5.1) below.

Step 2. Recall that the truncated projective space RP_9^{16} is the Thom complex of $9\xi_7$. But, over RP^7

$$9\xi_7 = 8\xi_7 \oplus \xi_7 = 8\epsilon \oplus \xi_7 ,$$

so the Thom complex in question is homeomorphic to $\Sigma^8 RP^8$, with RP^8 itself being the Thom complex of ξ_7. Similarly, $RP_{17}^{24} \approx \Sigma^{16} RP^8$. We now "smash A with the identity map of RP^8" to obtain a map

$$h = A \wedge id : \left(S^{16} \bigcup_{16} e^{17}\right) \wedge RP^8 \to \left(S^8 \bigcup_{16} e^9\right) \wedge RP^8$$

which on account of the Künneth formula also induces isomorphism in K^* theory.

Step 3. While the suspension order of RP^8 has not been precisely determined, one does know from Mukai [11,p.62] that the suspension order of $\Sigma^5 RP^8$ is equal to 16. Because of this,

$$\left(S^8 \bigcup_{16} e^9\right) \wedge RP^8 = \left(S^2 \bigcup_{16} e^3\right) \wedge \Sigma^6 RP^8$$

splits into a wedge $\Sigma^8 RP^8 \vee \Sigma^9 RP^8$, and there is a similar splitting of $(S^{16} \bigcup_{16} e^{17}) \wedge RP^8$ into the wedge $\Sigma^{16} RP^8 \vee \Sigma^{17} RP^8$. Since $\tilde{K}(\Sigma^9 RP^8) = 0 = \tilde{K}(\Sigma^{17} RP^8)$, the component $f : \Sigma^{16} RP^8 \to \Sigma^8 RP^8$ of h induces an isomorphism in K-cohomology groups. This f must then induce isomorphism in KO cohomology as well, yielding the map $RP^{24}_{17} \to RP^{16}_9$ as desired in §4.

All that remains now is to prove

PROPOSITION (5.1). *Let $X = S^8 \bigcup_{16} e^9$. There is a map*

$$A : \Sigma^8 X \to X$$

inducing an isomorphism in K^* cohomology theory.

PROOF: This is basically Adams' theorem [2, Lemma 12.5], the difficulty being that one is outside the stable range, i.e., $\dim \Sigma^8 X$ exceeds twice the connectivity of X by 3. For large N one can certainly use Adams' theorem to get a map $A' : \Sigma^{8N+8} X \to \Sigma^{8N} X$ inducing isomorphism in K^* cohomology. The composite

$$S^{8N+16} \xrightarrow{i} \Sigma^{8N+8} X \xrightarrow{A'} \Sigma^{8N} X ,$$

where i denotes the inclusion of the bottom cell, can be desuspended by Freudenthal's theorem into a map $g_1 : S^{17} \to \Sigma X$. Consider the homotopy exact sequence of the pair $(\Omega\Sigma X, X)$:

$$\to \pi_{16}(X) \xrightarrow{\Sigma} \pi_{16}(\Omega\Sigma X) \to \pi_{16}(\Omega\Sigma X, X) \xrightarrow{\Delta} \pi_{15}(X) \xrightarrow{\Sigma} \pi_{15}(\Omega\Sigma X) \cdots$$

A standard calculation with the Hurewicz isomorphism theorem shows that $\pi_{16}(\Omega\Sigma X, X) \approx H_{16}(\Omega\Sigma X, X; Z) = Z_{16}$. On the other hand, a routine use of the Pontryagin square operation as given in Thomas [12] shows that the

Whitehead product element $[\iota_8, \iota_8]$, considered as an element in $\pi_{15}(X)$, has order 16. Since $\Sigma[\iota_8, \iota_8] = 0$, we conclude by exactness that Δ is one to one and $\Sigma : \pi_{16}(X) \to \pi_{16}(\Omega\Sigma X)$ is onto, so that g_1 can be further desuspended into a map $g : S^{16} \to X$.

If $c : X \to S^9$ denotes the map collapsing the bottom 8-sphere of X to a point, then it follows from Adams' work quoted above that the composite $cg : S^{16} \to S^9$ represents the element $\pm 15\sigma$ in the group $\pi_{16}(S^9) = Z_{240}$, with σ denoting the generator of this group (i.e., represented by the suspension of the Hopf invariant one map $S^{15} \to S^8$). In particular, the element $[g] \in \pi_{16}(X)$ has order divisible by 16. From Lemma (5.2) below we shall see that it has order precisely 16, so that g extends to a map $A : S^{16} \bigcup_{16} e^{17} \to X$, or $\Sigma^8 X \to X$. That this map induces isomorphism in K^* theory is checked in the same way as in [2, Lemma 12.5].

To bring this section to a close, we present

LEMMA (5.2). *For <u>any</u> map $g : S^{16} \to X$ the composite*

$$g \circ 16 : S^{16} \xrightarrow{16} S^{16} \xrightarrow{g} X$$

is homotopically trivial.

PROOF: First, observe that if an integer m is divisible by 16, then the composite

$$S^{16} \xrightarrow{g} X \xrightarrow{m.id_X} X$$

of g with m times the identity map of X, abbreviated as $m \circ g$, is homotopically trivial by Toda [14, Thm. 4.4]. However, since $g \circ m \neq m \circ g$ in general, this is not yet a proof for the lemma. We must start by analyzing the long homotopy exact sequence of the pair (X, S^8):

$$\cdots \longrightarrow \pi_{16}(S^8) \xrightarrow{i_*} \pi_{16}(X) \xrightarrow{j_*} \pi_{16}(X, S^8) \xrightarrow{\partial} \pi_{15}(S^8) \longrightarrow \cdots$$

According to Sasao [13], there is a split exact sequence

$$0 \longrightarrow Z \longrightarrow \pi_{16}(X, S^8) \xrightarrow{c'_*} \pi_{16}(S^9) \longrightarrow 0$$

where $c' : (X, S^8) \to (S^9, *)$ is collapsing, and the group Z inside $\pi_{16}(X, S^8)$ is generated by the relative Whitehead product $[\chi, \iota_8]_r$, with $\chi : (I^9, \partial I^9) \to (X, S^8)$ being the characteristic map for the 9-cell of X. Clearly, $\partial[\chi, \iota_8]_r = 16[\iota_8, \iota_8] \in \pi_{15}(S^8) = Z \oplus Z_{120}$. Because $\pi_{16}(S^8) = Z_2 \oplus Z_2 \oplus Z_2 \oplus Z_2$ (see [15]), we conclude that $\pi_{16}(X)$ is a torsion group mapped by j_* into the torsion subgroup of $\pi_{16}(X, S^8)$, which in turn is isomorphic to $\pi_{16}(S^9)$ under c'_*. The net outcome of all these is that Kernel c_* = Image i_* in the following commutative diagram

$$\begin{array}{ccccccc} \longrightarrow & \pi_{16}(S^8) & \xrightarrow{i_*} & \pi_{16}(X) & \xrightarrow{j_*} & \pi_{16}(X, S^8) & \xrightarrow{\Delta} \\ & & & & \searrow^{c_*} & \downarrow c'_* & \\ & & & & & \pi_{16}(S^9) & \end{array}$$

For any map $g : S^{16} \to X$, the element represented by $2 \circ g - g \circ 2$ is in Kernel c_*, and is thus of the form $i_*(u)$ for some $u \in \pi_{16}(S^8)$.

It follows that, by slight abuse of notation,

$$(2 \circ g - g \circ 2) \circ 2 = i_*(u) \circ 2 = i_*(u) + i_*(u) = 0 ,$$

or $2 \circ g \circ 2 = g \circ 4$. Using this fact repeatedly, we obtain

$$\begin{aligned} g \circ 16 &= (g \circ 4) \circ 4 = (2 \circ g \circ 2) \circ 4 = 2 \circ (g \circ 4) \circ 2 \\ &= 2 \circ (2 \circ g \circ 2) \circ 2 = 4 \circ (g \circ 4) = 4 \circ (2 \circ g \circ 2) \\ &= 8 \circ g \circ 2 = 8 \circ (2 \circ g - i_*(u)) = 16 \circ g - i_*(8 \circ u) \\ &= -i_*(8 \circ u) . \end{aligned}$$

Invoking Toda's description [15, p.61] of the elements of $\pi_{16}(S^8)$, one can now do a case by case check to confirm that $8 \circ u = 0$ for all $u \epsilon \pi_{16}(S^8)$. We leave out the details, but point out cheerfully that this completes the proof of Lemma (5.2).

REFERENCES

[1] J. F. Adams, Geometric dimension of vector bundles over RP^n, *Proc. Int. Conf. on Prospects of Math.*, Kyoto 1973, p.1–17.

[2] J. F. Adams, On the groups $J(X) - IV$, *Topology* **5** (1966), p.21–72.

[3] M. F. Atiyah, Immersions and embeddings of manifolds, *Topology* **1** (1962), p.291–310.

[4] R. Bott, *Lectures on $K(X)$*, Benjamin, New York, 1969.

[5] D. M. Davis, Generalized homology and the generalized vector field problem, *Quart. J. Math. Oxford* **23** (1974), p.169–193.

[6] D. M. Davis, S. Gitler and M. Mahowald, The stable geometric dimension of vector bundles over real projective spaces, *Tran. Amer. Math. Soc.*, **268** (1981), p. 39–61; Corrections 280 (1983), p.841–843.

[7] S. Gitler and K. Y. Lam, The generalized vector field problem and bilinear maps, *Bol. Soc. Matem. Mexicana* **14** (1969), p.65–69.

[8] D. Hüsemoller, *Fiber Bundles*, second edition, Springer- Verlag, New York, 1975.

[9] I. James and E. Thomas, An approach to the enumeration problem for nonstable vector bundles, *J. of Math. and Mechanics* **14** (1965), p.485–506.

[10] K. Y. Lam, KO-equivalence and existence of nonsingular bilinear maps, *Pacif J. Math.* **82** (1979), p.145–153.

[11] J. Mukai, A note on the Kahn-Priddy map, *J. Math. Soc. Japan* **40** (1988), p.53–63.

[12] E. Thomas, A generalization of the Pontryagin square cohomology operation, *Proc. Nat. Acad. Sci. U.S.A.* **42** (1956), p.266-269.

[13] S. Sasao, On homotopy groups $\pi_{2n}(K^n_m S^n)$, *Proc. Japan Acad.* **39** (1963), p.557–558.

[14] H. Toda, Order of the identity class of a suspension space, *Ann. of Math.* **78** (1963), p.300–325.

[15] H. Toda, Composition methods in homotopy groups of spheres, Princeton 1962.

K. Y. Lam
Department of Mathematics
University of British Columbia
Vancouver, B.C. V6T 1Y4
Canada

D. Randall
Department of Mathematics
Loyola University
New Orleans, Lousiana, 70018
U.S.A.

The characteristic classes for the exceptional Lie groups

Mamoru MIMURA

§0. Introduction.

The compact, simply connected, simple Lie groups are classified as follows:

$$A_n \text{ for } n \geq 1, \quad B_n \text{ for } n \geq 2, \quad C_n \text{ for } n \geq 3, \quad D_n \text{ for } n \geq 4,$$
$$E_j \text{ for } j = 6, 7, 8, \quad F_4, \quad G_2.$$

The first four are called classical and the last five are called exceptional.

For each such a group G there exists the universal G-bundle with the base space BG and the total space acyclic. The base space BG is called the classifying space. In fact, BG classifies fibre bundles in the following sense.

As is well known, the equivalence classes of fibre bundles with fibre F on which G acts effectively are in a one to one correspondence with the equivalence classes of associated principal G-bundles. Here we have the well known theorem due to Steenrod.

Theorem. *For a (para)compact CW-complex X, there is a one to one correspondence between the set of homotopy classes $[X, BG]$ and $\{$the equivalence classes of principal G-bundles over $X\}$.*

The correpondence is given by assigning $[f]$ the bundle induced by f from the universal bundle. So, for any G-bundle over X, we have a homomorphism $f^* : H^*(BG; R) \to H^*(X; R)$ called the characteristic map of the bundle, where R is a coefficient field. The $\mathrm{Im}\, f^*$ is the characteristic ring of the bundle. For given two maps f and g, if $\mathrm{Im}\, f^*$ and $\mathrm{Im}\, g^*$ are different, then f is not homotopic to g and hence the fibre bundles corresponding to f and g are not equivalent. Thus, to classify fibre bundles, it is significant to determine $H^*(BG; R)$.

The purpose of this paper is to describe methods how to compute $H^*(BG; Z/p)$, the characteristic classes for exceptional Lie groups G, mainly by making use of the Eilenberg-Moore spectral sequence. The paper is organized as follows. In §1 we recollect the Hopf algebra structure of $H^*(G;Z/p)$ for later use. In §2 we prove the Borel theorem using the Eilenberg-Moore spectral sequence. Then in §3 we observe the relation between the invariant subalgebra of the Weyl group and the characteristic classes. In §4 we recollect the results on $\mathrm{Cotor}_A(Z/p, Z/p)$ and observe that the Eilenberg-Moore spectral sequence collapses for all the cases except for $(G,p) = (E_8, 2)$ and $(AdE_6, 3)$. In §5 we construct a small $Z/2$-resolution over $H^*(E_8;Z/2)$. In the last section, §6, we construct a small $Z/3$-resolution over $H^*(AdE_6;Z/3)$.

The details of the computations of $\mathrm{Cotor}_A(Z/2, Z/2)$ with $A = H^*(E_8; Z/2)$ and $H^*(AdE_6;Z/3)$ and those of $H^*(BE_8;Z/2)$ and $H^*(BAdE_6;Z/3)$ will appear elsewhere.

We would like to thank our joint authors Akira Kono, Nobuo Shimada and Yuriko Sambe.

§1. The Hopf algebra structure.

For later use let us recall ([A1,2,3], [AS], [Bo2,3,4], [IKT], [K1], [K2], [KM2], [Th]) the following results on the Hopf algebra structure over the Steenrod algebra $\mathcal{A}_p$ of the cohomology mod p of the exceptional Lie groups. In the below $\overline{\phi}$ is the reduced diagonal map induced from the multiplication on the group.

$(G_2.2)$ $$H^*(G_2;Z/2) = Z/2[x_3]/(x_3^4) \otimes \Lambda(x_5)$$

with all x_i primitive, where $Sq^2 x_3 = x_5$.

$(G_2.p)$ $$H^*(G_2;Z/p) = \Lambda(x_3, x_{11}) \quad \text{for} \quad p > 2$$

with all x_i primitive, where $\wp^1 x_3 = x_{11}$ for $p = 5$.

$(F_4.2)$ $$H^*(F_4;Z/2) = Z/2[x_3]/(x_3^4) \otimes \Lambda(x_5, x_{15}, x_{23})$$

with all x_i primitive, where $Sq^2 x_3 = x_5$ and $Sq^8 x_{15} = x_{23}$.

$(F_4.3)$ $$H^*(F_4;Z/3) = Z/3[x_8]/(x_8^3) \otimes \Lambda(x_3, x_7, x_{11}, x_{15})$$

with x_i primitive for $i = 3, 7, 8$, and

$$\overline{\phi}(x_j) = x_8 \otimes x_{j-8} \quad \text{for} \quad j = 11, 15,$$

where $\wp^1 x_3 = x_7$, $\beta x_7 = x_8$, $\wp^1 x_{11} = x_{15}$.

$(F_4.p)$ $$H^*(F_4;Z/p) = \Lambda(x_3, x_{11}, x_{15}, x_{23}) \quad \text{for} \quad p > 3$$

with all x_i primitive, where $\wp^1 x_3 = x_{11}$, $\wp^1 x_{15} = x_{23}$ for $p = 5$; $\wp^1 x_3 = x_{15}$, $\wp^1 x_{11} = x_{23}$ for $p = 7$; $\wp^1 x_3 = x_{23}$ for $p = 11$.

$(E_6.2)$ $$H^*(E_6; Z/2) = Z/2[x_3]/(x_3^4) \otimes \Lambda(x_5, x_9, x_{15}, x_{17}, x_{23})$$

with x_i primitive for $i = 3, 5, 9, 17$, and

$$\overline{\phi}(x_j) = x_3^2 \otimes x_{j-6} \quad \text{for} \quad j = 15, 23,$$

where $Sq^2 x_3 = x_5$, $Sq^4 x_5 = x_9$, $Sq^8 x_9 = x_{17}$, $Sq^8 x_{15} = x_{23}$.

$(E_6.3)$ $$H^*(E_6; Z/3) = Z/3[x_8]/(x_8^3) \otimes \Lambda(x_3, x_7, x_9, x_{11}, x_{15}, x_{17})$$

with x_i primitive for $i = 3, 7, 8, 9$, and

$$\overline{\phi}(x_j) = x_8 \otimes x_{j-8} \quad \text{for} \quad j = 11, 15, 17,$$

where $\wp^1 x_3 = x_7$, $\beta x_7 = x_8$, $\wp^1 x_{11} = x_{15}$.

$(E_6.p)$ $$H^*(E_6; Z/p) = \Lambda(x_3, x_9, x_{11}, x_{15}, x_{17}, x_{23}) \quad \text{for} \quad p > 3$$

with all x_i primitive, where $\wp^1 x_3 = x_{11}$, $\wp^1 x_9 = x_{17}$, $\wp^1 x_{15} = x_{23}$ for $p = 5$; $\wp^1 x_3 = x_{15}$, $\wp^1 x_{11} = x_{23}$ for $p = 7$; $\wp^1 x_3 = x_{23}$ for $p = 11$.

$(E_7.2)$ $$H^*(E_7; Z/2) = Z/2[x_3, x_5, x_9]/(x_3^4, x_5^4, x_9^4) \otimes \Lambda(x_{15}, x_{17}, x_{23}, x_{27})$$

with x_i primitive for $i = 3, 5, 9, 17$, and

$$\begin{aligned} \overline{\phi}(x_{15}) &= x_5 \otimes x_5^2 + x_9 \otimes x_3^2, \\ \overline{\phi}(x_{23}) &= x_5 \otimes x_9^2 + x_{17} \otimes x_3^2, \\ \overline{\phi}(x_{27}) &= x_9 \otimes x_9^2 + x_{17} \otimes x_5^2, \end{aligned}$$

where $Sq^2 x_3 = x_5$, $Sq^4 x_5 = x_9$, $Sq^8 x_9 = x_{17}$, $Sq^8 x_{15} = x_{23}$, $Sq^4 x_{23} = x_{27}$.

$(E_7.3)$ $$H^*(E_7; Z/3) = Z/3[x_8]/(x_8^3) \otimes \Lambda(x_3, x_7, x_{11}, x_{15}, x_{19}, x_{27}, x_{35})$$

with x_i primitive for $i = 3, 7, 8, 19$, and

$$\begin{aligned} \overline{\phi}(x_j) &= x_8 \otimes x_{j-8} \qquad \text{for} \quad j = 11, 15, 27, \\ \overline{\phi}(x_{35}) &= x_8 \otimes x_{27} - x_8^2 \otimes x_{19}, \end{aligned}$$

where $\wp^1 x_3 = x_7$, $\beta x_7 = x_8$, $\wp^1 x_{11} = x_{15}$, $\wp^3 x_7 = x_{19}$, $\wp^3 x_{15} = x_{27}$, $\wp^1 x_{15} = \varepsilon x_{19}$ $(\varepsilon = \pm 1)$.

$(E_7.p)$ $$H^*(E_7; Z/p) = \Lambda(x_3, x_{11}, x_{15}, x_{19}, x_{23}, x_{27}, x_{35}) \quad \text{for} \quad p > 3$$

with all x_i primitive, where $\wp^1 x_3 = x_{11}$, $\wp^1 x_{15} = x_{23}$, $\wp^1 x_{19} = x_{27}$, $\wp^1 x_{27} = x_{35}$ for $p = 5$; $\wp^1 x_3 = x_{15}$, $\wp^1 x_{11} = x_{23}$, $\wp^1 x_{23} = x_{35}$ for $p = 7$; $\wp^1 x_3 = x_{23}$, $\wp^1 x_{15} = x_{35}$ for $p = 11$; $\wp^1 x_3 = x_{27}$, $\wp^1 x_{11} = x_{35}$ for $p = 13$; $\wp^1 x_3 = x_{35}$ for $p = 17$.

$(E_8.2)$ $$H^*(E_8; Z/2) = Z/2[x_3, x_5, x_9, x_{15}]/(x_3^{16}, x_5^8, x_9^4, x_{15}^4) \otimes \Lambda(x_{17}, x_{23}, x_{27}, x_{29})$$

with x_i primitive for $i = 3, 5, 9, 17$, and

$$\begin{aligned}
\overline{\phi}(x_{15}) &= x_3 \otimes x_3^4 + x_5 \otimes x_5^2 + x_3^2 \otimes x_9, \\
\overline{\phi}(x_{23}) &= x_3 \otimes x_5^4 + x_5 \otimes x_9^2 + x_3^2 \otimes x_{17}; \\
\overline{\phi}(x_{27}) &= x_3 \otimes x_3^8 + x_9 \otimes x_9^2 + x_5^2 \otimes x_{17}, \\
\overline{\phi}(x_{29}) &= x_5 \otimes x_3^8 + x_9 \otimes x_5^4 + x_3^4 \otimes x_{17}, \\
\overline{\phi}(x_{30}) &= x_3^2 \otimes x_3^8 + x_5^2 \otimes x_5^4 + x_3^4 \otimes x_9^2,
\end{aligned}$$

where $Sq^2 x_3 = x_5$, $Sq^4 x_5 = x_9$, $Sq^8 x_9 = x_{17}$, $Sq^8 x_{15} = x_{23}$, $Sq^4 x_{23} = x_{27}$, $Sq^2 x_{27} = x_{29}$.

$(E_8.3)$ $$H^*(E_8; Z/3) = Z/3[x_8, x_{20}]/(x_8^3, x_{20}^3) \otimes \Lambda(x_3, x_7, x_{15}, x_{19}, x_{27}, x_{35}, x_{39}, x_{47})$$

with x_i primitive for $i = 3, 7, 8, 19, 20$, and

$$\begin{aligned}
\overline{\phi}(x_{15}) &= x_8 \otimes x_7, \\
\overline{\phi}(x_{27}) &= x_8 \otimes x_{19} + x_{20} \otimes x_7, \\
\overline{\phi}(x_{35}) &= x_8 \otimes x_{27} - x_8^2 \otimes x_{19} + x_{20} \otimes x_{15} + x_{20} x_8 \otimes x_7, \\
\overline{\phi}(x_{39}) &= x_{20} \otimes x_{19}, \\
\overline{\phi}(x_{47}) &= x_8 \otimes x_{39} + x_{20} \otimes x_{27} + x_{20} x_8 \otimes x_{19} - x_{20}^2 \otimes x_7,
\end{aligned}$$

where $\wp^1 x_3 = x_7$, $\beta x_7 = x_8$, $\wp^1 x_{15} = \varepsilon x_{19}$, $\beta x_{19} = x_{20}$, $\wp^1 x_{35} = \varepsilon x_{39}$, $\wp^3 x_7 = x_{19}$, $\wp^3 x_8 = x_{20}$, $\wp^3 x_{15} = x_{27}$, $\wp^3 x_{27} = -x_{39}$, $\wp^3 x_{35} = x_{47}$ with $\varepsilon = \pm 1$ simultaneously.

$(E_8.5)$ $$H^*(E_8; Z/5) = Z/5[x_{12}]/(x_{12}^5) \otimes \Lambda(x_3, x_{11}, x_{15}, x_{23}, x_{27}, x_{35}, x_{39}, x_{47})$$

with x_i primitive for $i = 3, 11, 12$, and

$$\begin{aligned}
\overline{\phi}(x_j) &= x_{12} \otimes x_{j-12} && \text{for } j = 15, 23, \\
\overline{\phi}(x_k) &= 2x_{12} \otimes x_{k-12} + x_{12}^2 \otimes x_{k-24} && \text{for } k = 27, 35, \\
\overline{\phi}(x_l) &= 3x_{12} \otimes x_{l-12} + 3x_{12}^2 \otimes x_{l-24} + x_{12}^3 \otimes x_{l-36} && \text{for } l = 39, 47,
\end{aligned}$$

where $\wp^1 x_3 = x_{11}$, $\beta x_{11} = x_{12}$, $\wp^1 x_{15} = x_{23}$, $\wp^1 x_{27} = x_{35}$, $\wp^1 x_{39} = x_{47}$.

$(E_8.p)$ $\quad H^*(E_8;Z/p) = \Lambda(x_3, x_{15}, x_{23}, x_{27}, x_{35}, x_{39}, x_{47}, x_{59})$ for $p > 5$

with all x_i primitive, where $\wp^1 x_3 = x_{15}$, $\wp^1 x_{23} = x_{35}$, $\wp^1 x_{27} = x_{39}$, $\wp^1 x_{35} = x_{47}$, $\wp^1 x_{47} = x_{59}$ for $p = 7$; $\wp^1 x_3 = x_{23}$, $\wp^1 x_{15} = x_{35}$, $\wp^1 x_{27} = x_{47}$, $\wp^1 x_{39} = x_{59}$ for $p = 11$; $\wp^1 x_3 = x_{27}$, $\wp^1 x_{15} = x_{39}$, $\wp^1 x_{23} = x_{47}$, $\wp^1 x_{35} = x_{59}$ for $p = 13$; $\wp^1 x_3 = x_{35}$, $\wp^1 x_{15} = x_{47}$, $\wp^1 x_{27} = x_{59}$ for $p = 17$; $\wp^1 x_3 = x_{39}$, $\wp^1 x_{23} = x_{59}$ for $p = 19$; $\wp^1 x_3 = x_{47}$, $\wp^1 x_{15} = x_{59}$ for $p = 23$; $\wp^1 x_3 = x_{59}$ for $p = 29$.

Remark. *For each pair of the inclusions*

$$G_2 \subset F_4 \subset E_6 \subset E_7 \subset E_8$$

the smaller group is mod 2 totally non-homologous to zero in the larger group.

As is well known, among the exceptional groups, only E_6 and E_7 have the center $Z/3$ and $Z/2$ respectively; the quotient groups are denoted by AdE_6 and AdE_7.

$(AdE_6.3)$ $\quad H^*(AdE_6;Z/3) = Z/3[x_2, x_8]/(x_2^9, x_8^3) \otimes \Lambda(x_1, x_3, x_7, x_9, x_{11}, x_{15})$

with x_i primitive for $i = 1, 2$, and

$$\begin{aligned}
\overline{\phi} x_3 &= x_2 \otimes x_1, \\
\overline{\phi} x_7 &= x_2^3 \otimes x_1, \\
\overline{\phi} x_8 &= x_2^3 \otimes x_2, \\
\overline{\phi} x_9 &= x_2 \otimes x_7 - x_2^3 \otimes x_3 + x_8 \otimes x_1 + x_2^4 \otimes x_1, \\
\overline{\phi} x_{11} &= x_2 \otimes x_9 - x_2^2 \otimes x_7 + x_8 \otimes x_3 - x_2^4 \otimes x_3 + x_8 x_2 \otimes x_1 - x_2^5 \otimes x_1, \\
\overline{\phi} x_{15} &= x_2^3 \otimes x_9 + x_8 \otimes x_7 + x_2^6 \otimes x_3 + x_8 x_2^3 \otimes x_1,
\end{aligned}$$

where $x_2 = \beta x_1$, $x_7 = \wp^1 x_3$, $x_8 = \beta x_7$, $x_{15} = \wp^1 x_{11}$, $\wp^1 x_8 = -x_2^6$.

$(AdE_6.p)$ $\quad H^*(AdE_6;Z/p) = H^*(E_6;Z/p)$
as Hopf algebra over $\mathcal{A}_p$ for $p \neq 3$.

$(AdE_7.2)$ $\quad H^*(AdE_7;Z/2) = Z/2[x_1, x_5, x_9]/(x_1^4, x_5^4, x_9^4)$
$\otimes \Lambda(x_6, x_{15}, x_{17}, x_{23}, x_{27})$

with x_i primitive for $i = 1, 3, 5, 6, 9, 17$, and

$$\begin{aligned}
\overline{\phi} x_{15} &= x_6 \otimes x_9 + x_5^2 \otimes x_5, \\
\overline{\phi} x_{23} &= x_9^2 \otimes x_5 + x_6 \otimes x_{17}, \\
\overline{\phi} x_{27} &= x_9^2 \otimes x_9 + x_5^2 \otimes x_{17}.
\end{aligned}$$

$(AdE_7.p)$ $\quad H^*(AdE_7; Z/p) = H^*(E_7; Z/p)$
as Hopf algebra over $\mathcal{A}_p$ for $p \neq 2$.

§2. The Borel theorem and the Eilenberg-Moore spectral sequence.

Let G be a compact, connected Lie group. Consider the following three statements:

(a) $H^*(G; Z)$ is p-torsion free.
(b) $H^*(G; Z/p)$ is generated by (universally) transgressive elements.
(c) $H^*(G; Z/p)$ is generated by primitive elements.

As is well known, (a) is equivalent to

(a)' $H^*(G; Z/p) = \Lambda(x_1, \ldots, x_l)$ with $|x_i| =$ odd.

Borel [Bo2] showed that (a)' implies (b), which implies (c). On the other hand, Browder [Bd] showed that (c) implies (a), if p is odd.

More precisely, Borel showed

Theorem 2.1. *If $H^*(G; Z/p)$ is an exterior algebra generated by the elements of odd degrees $r_1, \ldots, r_l$, then*

(a) $H^*(G; Z/p) \cong \Lambda(x_1, \ldots, x_l)$ *with* $|x_i| = r_i$,
where x_i is (universally) transgressive;
(b) $H^*(BG; Z/p) \cong Z/p[y_1, \ldots, y_l]$ *with* $|y_i| = r_i + 1$,
where y_i is the transgression image of x_i.

Theorem 2.2. *If $H^*(G; Z/2)$ has a simple system of (universally) transgressive elements $x_1, \ldots, x_l$, then*

$$H^*(BG; Z/2) \cong Z/2[y_1, \ldots, y_l] \quad \textit{with} \quad |y_i| = |x_i| + 1,$$

where y_i is the transgression image of x_i.

Now we recall the Eilenberg-Moore spectral sequence (EMSS) $\{E_r(G), d_r\}$ for (G, p) such that

$$E_2(G) = \mathrm{Cotor}_A(Z/p, Z/p) \quad \text{with} \quad A = H^*(G; Z/p),$$
$$E_\infty(G) = Gr H^*(BG; Z/p).$$

For details see [RS1,2].

In [Mo] Moore pointed out that EMSS gives an easy proof of Theorem 2.1. In fact, by a Koszul resolution one can show that $\mathrm{Cotor}_A(Z/p, Z/p)$ is a polynomial algebra on corresponding generators of one higher degrees if A is an exterior algebra with generators of odd degrees. Then all terms of E_2 of odd total degree are trivial and hence all $d_r = 0$ which implies $E_2 = E_\infty$. Since E_∞ is a polynomial algebra, it is algebraically free and so $E_\infty = H^*(BG; Z/p)$ as an algebra.

The proof of Theorem 2.2 is quite similar. In fact, $\mathrm{Cotor}_A(Z/2, Z/2)$ with $A = H^*(G; Z/2)$ is seen to be isomorphic to $Z/2[y_1, \ldots, y_l]$, since A is primitively generated.

Immediate consequences of $(G_2 . p)$, $(F_4 . p)$, $(E_6 . p)$, $(E_7 . p)$ and $(E_8 . p)$ are the following

Theorem 2.3. *The Eilenberg-Moore spectral sequence of G with Z/p -coefficient collapses for (G_2, p) for $p > 2$, (F_4, p), (E_6, p), (E_7, p) for $p > 3$ and (E_8, p) for $p > 5$; and*

$$\begin{array}{lll} H^*(BG_2; Z/p) \cong Z/p[y_4, y_{12}] & \text{for} & p > 2, \\ H^*(BF_4; Z/p) \cong Z/p[y_4, y_{12}, y_{16}, y_{24}] & \text{for} & p > 3, \\ H^*(BE_6; Z/p) \cong Z/p[y_4, y_{10}, y_{12}, y_{16}, y_{18}, y_{24}] & \text{for} & p > 3, \\ H^*(BE_7; Z/p) \cong Z/p[y_4, y_{12}, y_{16}, y_{20}, y_{24}, y_{28}, y_{36}] & \text{for} & p > 3, \\ H^*(BE_8; Z/p) \cong Z/p[y_4, y_{16}, y_{24}, y_{28}, y_{36}, y_{40}, y_{48}, y_{60}] & \text{for} & p > 5. \end{array}$$

By Théorèmes 17.3 and 19.2 of [Bo2] the generators of $H^*(G; Z/2)$ for $G = G_2$ and F_4 can be taken to be universally transgressive and so the following follows from Theorem 2.2.

Theorem 2.4. *The Eilenberg-Moore spectral sequences of G_2 and F_4 with $Z/2$-coefficient collapse; and*

$$H^*(BG_2; Z/2) \cong Z/2[y_4, y_6, y_7],$$
$$H^*(BF_4; Z/2) \cong Z/2[y_4, y_6, y_7, y_{12}, y_{16}].$$

Thus the cases not covered by the Borel theorem are the following

$$\begin{array}{llll} (G,p) = (F_4, 3), & & & \\ & (E_6, 2), & (E_6, 3), & (AdE_6, 3), \\ & (E_7, 2), & (AdE_7, 2), & (E_7, 3), \\ & (E_8, 2), & (E_8, 3), & (E_8, 5). \end{array}$$

§3. The invariant subalgebra of the Weyl group.

Let T be a maximal torus of G and $N(T)$ the normalizer of T in G. Then the quotient $W(G) = N(T)/T$ is a finite group, called the Weyl group of G. $W(G)$ acts on BT and hence on $H^*(BT;R)$, where R is a coefficient field. The invariant subalgebra $H^*(BT;R)^{W(G)}$ under the action of $W(G)$ contains the image of

$$i^* : H^*(BG;R) \to H^*(BT;R)$$

where $i : T \to G$ is the inclusion. Borel showed in [Bo2] that

$$i^* : H^*(BG;R) \cong H^*(BT;R)^{W(G)}$$

for $G = U(n)$, $SU(n)$, $Sp(n)$ for any R and for $G = SO(n)$ if char $R \neq 2$.

Along this line Toda [T1] calculated $H^*(BT;Z/3)^{W(F_4)}$ as follows, where T is a maximal torus of $Spin(9)$ which is a maximal subgroup of F_4.

As is well known, the natural map $\rho : BT \to BG$ induces a homomorphism

$$\rho^* : H^*(BG;Z/3) \to H^*(BT;Z/3)$$

such that the image of ρ^* is contained in the subalgebra $H^*(BT;Z/3)^{W(G)}$ which consists of the elements invariant under the action of $W(G)$, where $G = Spin(9)$ or F_4.

For $G = Spin(9)$, ρ^* is injective and the image coincides with the invariant subalgebra which is a polynomial algebra on the Pontrjagin classes $p_i \in H^{4i}$. Thus one may identify as follows:

$$H^*(BSpin(9);Z/3) = H^*(BT;Z/3)^{W(Spin(9))} = Z/3[p_1, p_2, p_3, p_4].$$

Defining the elements

$$\begin{aligned}
\overline{p}_2 &= p_2 - p_1^2, \\
\overline{p}_5 &= p_4 p_1 + p_3 \overline{p}_2, \\
\overline{p}_9 &= p_3^3 - p_4 p_3 p_1^2 + p_3^2 \overline{p}_2 p_1 - p_4 \overline{p}_2 p_1^3, \\
\overline{p}_{12} &= p_4^3 + p_4^2 \overline{p}_2^2 + p_4 \overline{p}_2^4,
\end{aligned}$$

Toda obtained

$$H^*(BT;Z/3)^{W(F_4)} = Z/3[p_1, \overline{p}_2, \overline{p}_5, \overline{p}_9, \overline{p}_{12}]/(r_{15})$$

where $r_{15} = \overline{p}_5^3 + \overline{p}_5^2 \overline{p}_2^2 p_1 - \overline{p}_{12} p_1^3 - \overline{p}_9 \overline{p}_2^3$.

Then by making use of the Serre spectral sequence associated with the bundle

$$F_4/Spin(9) \to BSpin(9) \to BF_4,$$

he obtained

Theorem 3.1. *As an algebra*

$$H^*(BF_4;Z/3) \cong (Z/3[t_4,t_8,t_{20},t_{26},t_{36},t_{48}]\otimes\Lambda(t_9,t_{21},t_{25}))/R,$$

where R is the ideal generated by the following

$$t_9t_4,\quad t_9t_8,\quad t_{21}t_4,\quad t_{21}t_8+t_{20}t_9,\quad t_{25}t_4+t_{20}t_9,\quad t_{26}t_4+t_{21}t_9,\quad t_{25}t_8,$$
$$t_{26}t_8-t_{25}t_9,\quad t_{21}t_{20},\quad t_{25}t_{20},\quad t_{26}t_{20}-t_{25}t_{21},\quad t_{20}^3-t_{48}t_4^3-t_{36}t_8^3+t_{20}^2t_8^2t_4.$$

Further Toda proceeded to the case $(E_6,3)$ as follows. Let $U=Spin(10)\cdot S^1$ be the isotropy subgroup of the symmetric space $EIII=E_6/U$ and let T be a maximal torus of U. For a suitable choice of a basis $\{t, t_1,\ldots,t_5\}$ of $H^2(BT;Z/3)$, $W(U)$ operates trivially on t and similarly on t_i to $W(Spin(10))$. Then one has

$$H^*(BT;Z/3)^{W(E_6)}\subset H^*(BT;Z/3)^{W(U)}=Z/3[t,p_1,p_2,c_5,p_3,p_4],$$

where $c_i\in H^{2i}$ is the i-th Chern class and $p_i\in H^{4i}$ is the i-th Pontrjagin class. Put

$$x_4=p_1,\qquad x_8=p_2-p_1^2,\qquad y_{10}=c_5-(p_2-p_1^2)t-p_1t^3,$$
$$x_{12}=p_3+c_5t+x_8t^2+p_1t^4,\qquad x_{16}=p_4+c_5t^3+x_8t^4+p_1t^6.$$

Then obviously

$$H^*(BT;Z/3)^{W(U)}=Z/3[t,x_4,x_8,y_{10},x_{12},x_{16}].$$

One sees that x_4, x_8 and y_{10} are invariant under $W(E_6)$ and that so are the following two elements:

$$x_{20}=x_{16}x_4+x_{12}x_8=y_{10}^2+\wp^3x_8+x_8^2x_4,$$
$$y_{22}=\wp^3y_{10}=x_{12}y_{10}-x_4f$$

where $f=(x_{16}-x_8^2)t-(x_{12}-x_8x_4)t^3+x_8t^5-x_4t^7+t^9$.

Let $\lambda: BE_6\to BU(27)$ be a map induced from the representation of E_6 corresponding to a fundamental weight. By inspecting $\lambda^*(c_{18})$ mod 3 one can find an invariant

$$x_{36}=x_{12}^3-x_{16}x_{12}x_4^2+x_{12}^2x_8x_4-x_{16}x_8x_4^3-x_4^3g,$$

where $g=-(x_{12}+x_8x_4)y_{10}t+\cdots$.

Then the following lemma is deduced from the fact that the field $Z/3(t,x_4,x_8,y_{10},x_{12},x_{16})$ is an extension of degree 27 of the subfield $Z/3(x_4,x_8,y_{10},x_{20},y_{22},x_{36})$.

Lemma 3.2. (1) $Z/3(t, t_1, \ldots, t_5)^{W(E_6)} = Z/3(x_4, x_8, y_{10}, x_{20}, y_{22}, x_{36})$.

(2) $H^*(BT; Z/3)^{W(E_6)} = Z/3[t, x_4, x_8, y_{10}, x_{12}, x_{16}]$
$\cap Z/3(x_4)[x_8, y_{10}, x_{20}, y_{22}, x_{36}]$.

More concretely, the subalgebra $H^*(BT; Z/3)^{W(E_6)}$ is generated by x_{4i} $(i = 1, 2, 5, 9, 12)$ and y_{2j} $(j = 5, 11, 13, 27, 29, 30, 32, 38)$, where these new generators are given by fractions over x_4:

$$
\begin{aligned}
y_{26} &= (x_{20}y_{10} - y_{22}x_8)/x_4,\\
x_{48} &= (-x_{36}x_8^3 + x_{20}^3 + x_{20}^2x_4x_8^2 + x_{20}x_4^2x_8^4)/x_4^3,\\
x_{54} &= (x_{36}y_{10}^3 - y_{22}^3 + x_{20}y_{22}x_4y_{10}^2 + y_{22}^2x_4x_8y_{10} - y_{22}x_4^2x_8^2y_{10}^2)/x_4^3,\\
y_{58} &= (x_{36}x_8^2y_{10} - x_{20}^2y_{22})/x_4,\\
y_{60} &= (x_{36}x_8y_{10}^2 - x_{20}y_{22}^2)/x_4,\\
y_{64} &= (y_{58}y_{10} - x_{20}y_{22}y_{26} - y_{22}^2x_8^3)/x_4,\\
y_{76} &= (y_{58}y_{22} + y_{60}x_{20} + x_{20}y_{22}^2x_8^2)/x_4.
\end{aligned}
$$

Based on the result of Mimura [MS1] on EMSS for $(E_6, 3)$ (see Theorem 4.12), Toda announced the following (cf. Theorem 4.9) in [T2]:

Theorem 3.3. *The Eilenberg-Moore spectral sequence with $Z/3$-coefficient for E_6 collapses. The natural homomorphism ρ^* maps $H^*(BE_6; Z/3)$ onto the invariant subalgebra $H^*(BT; Z/3)^{W(E_6)}$ and its kernel is the ideal generated by x_9, x_{21}, x_{25}, x_{26} and y_{27}.*

The details of computation has not yet appeared.

§4. The twisted tensor product and the E_2-term $\mathrm{Cotor}_A(Z/p, Z/p)$ with $A = H^*(G; Z/p)$.

Immediately after Toda had determined $H^*(BF_4; Z/3)$, Shimada [KMS1] pointed out that one can calculate at least the module structure of $H^*(BF_4; Z/3)$ quite easily by making use of EMSS.

In general there are two cruxes in EMSS, namely

(1) computing $E_2(G) = \mathrm{Cotor}_A(Z/p, Z/p)$ with $A = H^*(G; Z/p)$,

(2) computing differentials d_r.

In order to calculate $\mathrm{Cotor}_A(Z/p, Z/p)$, one needs to construct a Z/p-resolution over A. Theoretically we have a cobar construction, which is, from the practical view point, too large to calculate. As for $A = H^*(F_4; Z/3)$ Shimada constructed a fairly small $Z/3$-resolution over A by making use of the twisted tensor product, a notion due to Brown [Br]. He calculated $E_2(F_4) = \mathrm{Cotor}_A(Z/3, Z/3)$ and showed that EMSS collapses for dimensional reasons.

His idea, namely constructing a small Z/p-resolution over $A = H^*(G; Z/p)$ by making use of the twisted tensor product, should work for the other cases.

In this section we study $H^*(BG; Z/p)$ for the following eight cases:

$$\begin{aligned}(G,p) = (E_6, 2),\quad &(E_7, 2),\quad (AdE_7, 2),\\ (F_4, 3),\quad &(E_6, 3),\quad (E_7, 3),\quad (E_8, 3),\\ (E_8, 5).\end{aligned}$$

Computation of the E_2-term $\mathrm{Cotor}_A(Z/p, Z/p)$ of EMSS turns out to be easy for the first five cases but not so straightforward for the last three cases. We recollect the results on $\mathrm{Cotor}_A(Z/p, Z/p)$ and explain how to show the collapsing of EMSS.

To begin with let us recall a notion of the twisted tensor product due to Brown [Br]. Let A be a (graded) differential augmented coalgebra over Z/p with differential d_A and coproduct ϕ_A and C a (graded) differential augmented algebra over Z/p with differential d_C and product ψ_C and augmentation ε. Let $\theta : A \to C$ be a Z/p-linear map of degree 1 satisfying the conditions

$$\varepsilon\circ\theta = 0 \qquad \text{and} \qquad d_C\circ\theta + \theta\circ d_A + \psi_C\circ(\theta\otimes\theta)\circ\phi_A = 0.$$

Then the twisted tensor product $A\otimes C$ with respect to θ is a differential Z/p -module with the differential operator

$$d = d_A\otimes 1 + 1\otimes d_C + (1\otimes\psi_C)(1\otimes\theta\otimes 1)(\phi_A\otimes 1).$$

Now we apply this notion as follows for $A = H^*(G; Z/p)$.

Take $A = H^*(G; Z/p)$ with coproduct ϕ induced from the obvious multiplication of G (and $d_A = 0$). Let L be a Z/p-subspace of $\tilde{H}^*(G; Z/p)$, $\iota : L \to A$ the inclusion and $\theta : A \to L$ a map such that $\theta\circ\iota = 1_L$. Let $s : L \to sL$ be a suspension. Define $\overline{\theta} : A \to sL$ by $\overline{\theta} = s\circ\theta$ and $\overline{\iota} : sL \to A$ by $\overline{\iota} = \iota\circ s^{-1}$. Construct the tensor algebra $T(sL)$ and denote by ψ the product in $T(sL)$. Let I be the ideal of $T(sL)$ generated by $\mathrm{Im}(\psi\circ(\overline{\theta}\otimes\overline{\theta})\circ\phi)\circ(\ker\overline{\theta})$. Put $C = T(sL)/I$. Then the map $\overline{\theta} : A \to sL$ induces a map $A \to T(sL)$ which is again denoted by $\overline{\theta}$. We define a map $d_C = -\psi\circ(\overline{\theta}\otimes\overline{\theta})\circ\phi\circ\overline{\iota} : sL \to T(sL)$ and extend it naturally over $T(sL)$ (as derivation). Since $d_C\circ\phi = \psi\circ(d_C\otimes 1 + 1\otimes d_C)$ holds, we deduce $d_C(I) \subset I$. So d_C induces a map $C \to C$, which is again denoted by $d_C : C \to C$. Then it is easy to see $d_C\circ d_C = 0$. This shows that C is a differential algebra over Z/p. Since the relation

$$d_C\circ\overline{\theta} + \psi\circ(\overline{\theta}\otimes\overline{\theta})\circ\phi = 0$$

holds, we can now construct the twisted tensor product $W = A \otimes C$ with respect to $\overline{\theta}$. That is, $W = A \otimes C$ is an A-comodule with a differential operator

$$d_W = 1 \otimes d_C + (1 \otimes \psi) \circ (1 \otimes \overline{\theta} \otimes 1) \circ (\phi \otimes 1).$$

In the below we denote d_W and d_C simply by d by abuse of notation. We will choose L suitably so that $A \otimes C$ is acyclic and hence it is a Z/p-resolution over A. Consequently we have

$$H(C : d) = \mathrm{Cotor}_A(Z/p, Z/p).$$

(I) The case $(E_6, 2)$. [KM1]

We choose $L = \{x_3, x_3^2, x_5, x_9, x_{17}, x_{15}, x_{23}\}$. Then $A \otimes C$ becomes acyclic so that

$$H(C : d) = \mathrm{Cotor}_A(Z/2, Z/2).$$

Theorem 4.1. *As an algebra $\mathrm{Cotor}_A(Z/2, Z/2)$ with $A = H^*(E_6; Z/2)$ is isomorphic to M'/R' where*

$$M' = Z/2[m_i \mid i = 4, 6, 7, 10, 18, 32, 34, 48]$$

and R' is the ideal generated by

$$m_7 m_{10}, \quad m_7 m_{18}, \quad m_7 m_{34}, \quad m_{34}^2 + m_{10}^2 m_{48} + m_{18}^2 m_{32}.$$

The collapsing of EMSS for E_6 with $Z/2$-coefficient can be shown by using the naturality with respect to the obvious inclusion $f : F_4 \to E_6$ and by the fact that $E_2(F_4) \cong E_\infty(F_4) \cong Z/2[y_i \mid i = 4, 6, 7, 12, 16]$ by Theorem 2.4. Thus we have

Theorem 4.2. *As a module*

$$H^*(BE_6; Z/2) \cong \mathrm{Cotor}_A(Z/2, Z/2) \cong M'/R'.$$

Now observe that E_6 contains a subgroup $Spin(10)$. To determine the algebra structure of $H^*(BE_6; Z/2)$ we consider the induced map $j : BSpin(10) \to BE_6$ for which we have

$$j^* m_4 = w_4, \quad j^* m_6 = Sq^2 w_4 = w_6, \quad j^* m_7 = Sq^3 w_4 = w_7,$$

where $H^*(BSpin(10); Z/2) \cong Z/2[w_4, w_6, w_7, w_8, w_{10}]/(w_7 w_{10}) \otimes Z/2[e_{32}]$.

We define new elements

$$\overline{m}_{10} = Sq^4 m_6 + m_6 m_4, \quad \overline{m}_{18} = Sq^8 \overline{m}_{10}, \quad \overline{m}_{34} = Sq^{16} \overline{m}_{18}$$

so that we have

$$\overline{m}_{10} m_7 = \overline{m}_{18} m_7 = \overline{m}_{34} m_7 = 0.$$

Now studying the homomorphisms $f^* : H^*(BE_6; Z/2) \to H^*(BF_4; Z/2)$ and $i^* : H^*(BE_6; Z/2) \to H^*(BT^6; Z/2)$, where $i : T^6 \to E_6$ is the inclusion of a maximal torus, we can show

Theorem 4.3. *As an algebra*

$$H^*(BE_6; Z/2) \cong Z/2[m_4, m_6, m_7, \overline{m}_{10}, \overline{m}_{18}, m_{32}, \overline{m}_{34}, m_{48}]/R$$

where R is the ideal generated by

$$m_7\overline{m}_{10},\ m_7\overline{m}_{18},\ m_7\overline{m}_{34} \text{ and } r_{68} = \overline{m}_{34}^2 + \overline{m}_{10}^2 m_{48} + \overline{m}_{18}^2 m_{32} + (\textit{higher term}).$$

(II) The cases $(E_7, 2)$ and $(AdE_7, 2)$. [KMS2], [KM3]

We choose $L = \{x_3, x_3^2, x_5, x_5^2, x_9, x_9^2, x_{17}, x_{15}, x_{23}, x_{27}\}$. Then $A \otimes C$ becomes acyclic so that

$$H(C : d) = \mathrm{Cotor}_A(Z/2, Z/2).$$

Theorem 4.4. *As an algebra $Cotor_A(Z/2, Z/2)$ with $A = H^*(E_7; Z/2)$ is isomorphic to M/R, where*

$$M = Z/2[m_i \mid i = 4, 6, 7, 10, 11, 18, 19, 34, 35, 66, 67, 64, 96, 112]$$

and R is the ideal generated by

$$m_6 m_{11} + m_{10} m_7, \quad m_6 m_{19} + m_7 m_{18}, \quad m_{10} m_{19} + m_{11} m_{18}, \quad m_{11}^3 + m_7^2 m_{19},$$
$$m_6 m_{35} + m_7 m_{34}, \quad m_{10} m_{35} + m_{11} m_{34}, \quad m_{11} m_{19}^2, \quad m_{18} m_{35} + m_{19} m_{34},$$
$$m_{19}^3, \quad m_6 m_{67} + m_7 m_{66}, \quad m_{10} m_{67} + m_{11} m_{66}, \quad m_{18} m_{67} + m_{19} m_{66},$$
$$m_7^2 m_{67} + m_{11} m_{35}^2, \quad m_{11}^2 m_{67} + m_{19} m_{35}^2, \quad m_{34} m_{67} + m_{35} m_{66}, \quad m_{19}^2 m_{67},$$
$$m_{66}^2 + m_{10}^2 m_{112} + m_{18}^2 m_{96}, \quad m_{66} m_{67} + m_{10} m_{11} m_{112} + m_{18} m_{19} m_{96},$$
$$m_{67}^2 + m_{11}^2 m_{112} + m_{19}^2 m_{96}, \quad m_{35}^2 m_{67} + m_7^2 m_{11} m_{112} + m_{11}^2 m_{19} m_{96},$$
$$m_{34}^2 + m_6^4 m_{112} + m_{10}^4 m_{96} + m_{18}^4 m_{64},$$
$$m_{34}^3 m_{35} + m_6^3 m_7 m_{112} + m_{10}^3 m_{11} m_{96} + m_{18}^3 m_{19} m_{64},$$
$$m_{34}^2 m_{35}^2 + m_6^2 m_7^2 m_{112} + m_{10}^2 m_{11}^2 m_{96} + m_{18}^2 m_{19}^2 m_{64},$$
$$m_{34} m_{35}^3 + m_6 m_7^3 m_{112} + m_7^2 m_{11} m_{18} m_{96}, \quad m_{35}^4 + m_7^4 m_{112} + m_{11}^4 m_{96}.$$

We thank M. Nakamura for pointing out some missing terms in R in [KMS2], [MM].

Now observe that E_7 contains the subgroup $Spin(12)$. With the aid of the algebra structure of $H^*(BSpin(12); Z/2)$ we define new elements of $H^*(BE_7; Z/2)$ as follows:

$$\overline{m}_6 = Sq^2\overline{m}_4, \quad \overline{m}_7 = Sq^1\overline{m}_6, \quad \overline{m}_{10} = Sq^4\overline{m}_6 + \overline{m}_6\overline{m}_4, \quad \overline{m}_{11} = Sq^1\overline{m}_{10},$$
$$\overline{m}_{18} = Sq^8\overline{m}_{10}, \quad \overline{m}_{19} = Sq^1\overline{m}_{18}, \quad \overline{m}_{34} = Sq^{16}\overline{m}_{18}, \quad \overline{m}_{35} = Sq^1\overline{m}_{34}.$$

Then we see that $\overline{m}_i$ represents m_i in $E_\infty(E_7)$ for $i = 4, 6, 7, 10, 11, 18, 19, 34, 35$, and hence EMSS collapses for degrees < 64. Further we define

$$\overline{m}_{66} = Sq^{32}(\overline{m}_{34} + \overline{m}_{19}\overline{m}_{11}\overline{m}_4 + \overline{m}_{10}^3\overline{m}_4 + \overline{m}_{18}\overline{m}_{10}\overline{m}_6),$$
$$\overline{m}_{67} = Sq^1\overline{m}_{66}$$

so that the elements $\overline{m}_{66}$ and $\overline{m}_{67}$ cannot be a sum of monomials of $\overline{m}_i$ for $i \leq 35$.

Consider the complex representation $\lambda(56) : E_7 \to U(56)$ corresponding to the dominant root (weight) of E_7 and consider the induced homomorphism $\lambda(56)^* : H^*(BU(56); Z/2) \to H^*(BE_7; Z/2)$. It is easy to see that m_{64} in $E_2(E_7)$ is a permanent cycle and represented by $\lambda(56)^*c_{32}$ and that the elements m_{66} and m_{67} are also permanent cycles and represented by $\overline{m}_{66}$ and $\overline{m}_{67}$ respectively. Further one can see that the elements m_{96} and m_{112} are permanent cycles and represented by $\lambda(56)^*c_i$ for $i = 48$ and 56, respectively.

Thus we obtain

Theorem 4.5. *As a module*

$$H^*(BE_7; Z/2) \cong \mathrm{Cotor}_A(Z/2, Z/2) \cong M/R.$$

Now we put $B = H^*(E_7; Z/2)//(x_3)$. Since x_3 is primitive, B is also a Hopf algebra. Then it follows from $(E_7.2)$ and $(AdE_7.2)$ that

$$H^*(E_7; Z/2) \cong \Lambda(s_3) \otimes B^*,$$
$$H^*(AdE_7; Z/2) \cong \Lambda(s_1, s_2) \otimes B^*,$$

where $\deg s_i = i$ and B^* is the dual of B. Then as an algebra

$$\mathrm{Ext}_{A^*}(Z/2, Z/2) \cong M/R$$

where $A^* = H_*(E_7; Z/2)$. Thus we have

Theorem 4.6. *As an algebra $\mathrm{Cotor}_A(Z/2, Z/2)$ with $A = H^*(AdE_7; Z/2)$ is isomorphic to M'/R, where*

$$M' = Z/2[m_i \mid i = 2, 3, 6, 7, 10, 11, 18, 19, 34, 35, 66, 67, 64, 96, 112]$$

and R is the same as in Theorem 4.1.

Using a representation $\overline{\mu} : AdE_7 \to O(133)$ induced from the adjoint representation $\mu : E_7 \to O(133)$, where $\ker \mu$ is the center of E_7, one can see that there exist indecomposable elements of $H^*(BAdE_7; Z/2)$ in the corresponding degrees of $\mathrm{Cotor}_A(Z/2, Z/2)$ in Theorem 4.6, and hence one obtains

Theorem 4.7. *As a module*

$$H^*(BAdE_7; Z/2) \cong Cotor_A(Z/2, Z/2) \cong M'/R,$$

where R is the same as in Theorem 4.4 in the above.

In general ([S], [M]) one can introduce cohomology operations in EMSS which commute with differentials of the spectral sequence. Using this one can also show [MM] that EMSS collapses for the cases $(E_6, 2)$ and $(E_7, 2)$. Thus we have

Corollary 4.8. (1) *As a module*

$$H^*(BX_6; Z/2) \cong Cotor_A(Z/2, Z/2) \quad \text{with} \quad A = H^*(X_6; Z/2)$$

for any associative H-space X_6 such that $H^*(X_6; Z/2) \cong H^*(E_6; Z/2)$ as a Hopf algebra over $\mathcal{A}_2$.

(2) *As a module*

$$H^*(BX_7; Z/2) \cong Cotor_A(Z/2, Z/2) \quad \text{with} \quad A = H^*(X_7; Z/2)$$

for any associative H-space X_7 such that $H^*(X_7; Z/2) \cong H^*(E_7; Z/2)$ as a Hopf algebra over $\mathcal{A}_2$.

(III) The cases $(F_4, 3)$ and $(E_6, 3)$. [KMS1], [MS1], [KM4]

For the case $(F_4, 3)$ we choose $L = \{x_3, x_7, x_8, x_8^2, x_{11}, x_{15}\}$. Then $A \otimes C$ becomes acyclic so that $H(C : d) = \mathrm{Cotor}_A(Z/3, Z/3)$.

Theorem 4.9. *As an algebra, $Cotor_A(Z/3, Z/3)$ with $A = H^*(F_4; Z/3)$ is isomorphic to M/R, where*

$$M = Z/3[m_i \mid i = 4, 8, 9, 20, 21, 25, 26, 36, 48]$$

and R is the ideal generated by

$$m_4m_9, \quad m_8m_9, \quad m_9^2, \quad m_4m_{21}, \quad m_8m_{25}, \quad m_4m_{25} + m_8m_{21}, \quad m_{20}m_{21},$$
$$m_{20}m_{25}, \quad m_{21}^2, \quad m_{25}^2, \quad m_9m_{20} - m_4m_{25} + m_8m_{21}, \quad m_{20}^3 - m_4^3m_{48} + m_8^3m_{36},$$
$$m_{26}m_4 + m_{21}m_9, \quad m_{26}m_8 + m_{25}m_9, \quad m_{26}m_{20} - m_{21}m_{25}.$$

Then for dimensional reasons, one can show the collapsing of EMSS, and hence

Theorem 4.10. *As a module*

$$H^*(BF_4; Z/3) \cong \mathrm{Cotor}_A(Z/3, Z/3) \cong M/R.$$

Corollary 4.11. *As a module*

$$H^*(BX_4; Z/3) \cong \mathrm{Cotor}_A(Z/3, Z/3) \quad \text{with} \quad A = H^*(X_4; Z/3)$$

for any associative H-space X_4 such that $H^*(X_4; Z/3) \cong H^*(F_4; Z/3)$ as an algebra over $\mathcal{A}_3$.

For the case $(E_6, 3)$ we choose $L = \{x_3, x_7, x_8, x_9, x_{11}, x_{15}, x_{17}, x_8^2\}$. Then $A \otimes C$ becomes acyclic so that $H(C : d) = \mathrm{Cotor}_A(Z/3, Z/3)$.

Theorem 4.12. *As an algebra $\mathrm{Cotor}_A(Z/3, Z/3)$ with $A = H^*(E_6; Z/3)$ is isomorphic to M/R, where*

$$M = Z/3[m_i \mid i = 4, 8, 9, 10, 20, 21, 22, 25, 26, 26', 27, 36, 48, 54, 58, 60, 64, 76]$$

and R is the ideal generated by some elements of M.

(For details see Theorem 5.20 of [MS1].)

Note that m'_{26} is an element of degree 26, different from m_{26}.

By the naturality of EMSS with respect to the obvious inclusion $f : F_4 \to E_6$ and by the triviality of EMSS for F_4 by Theorem 4.10 one can show that EMSS for E_6 collapses for degrees < 36. Then, for dimensional reasons, one sees that the elements m_i for 36, 48, 54, 58, 60, 64, 76 are permanent cycles. Thus one has

Theorem 4.13. *As a module*

$$H^*(BE_6; Z/3) \cong \mathrm{Cotor}_A(Z/3, Z/3) \cong M/R.$$

As in the cases $(E_6, 2)$, $(E_7, 2)$ one can show ([MS4]) the collapsing of EMSS for $(E_6, 3)$ by introducing the cohomology operations in EMSS. Thus we have

Corollary 4.14. *As a module*

$$H^*(BX_6; Z/3) \cong \mathrm{Cotor}_A(Z/3, Z/3) \quad \text{with} \quad A = H^*(X_6; Z/3)$$

for any associative H-space X_6 *such that* $H^*(X_6; Z/3) \cong H^*(E_6; Z/3)$ *as a Hopf algebra over* $\mathcal{A}_3$.

(IV) The cases $(E_7, 3)$ and $(E_8, 3)$. [MS 1,2,3]

As for the cases $(E_7, 3)$ and $(E_8, 3)$ the following are the only results appeared so far concerning EMSS.

For the case $(E_7, 3)$ we choose $L = \{x_3, x_7, x_8, x_{19}, x_{11}, x_{15}, x_{27}, x_8^2, x_{35}\}$. Then $A \otimes C$ becomes acyclic so that $H(C : d) = \mathrm{Cotor}_A(Z/3, Z/3)$.

Theorem 4.15. *As an algebra* $\mathrm{Cotor}_A(Z/3, Z/3)$ *with* $A = H^*(E_7; Z/3)$ *is isomorphic to* M/R, *where* $M = Z/3[m_i \mid i =$ 4, 8, 9, 20, 20′, 21, 25, 26, 32, 36, 36′, 44, 48, 48′, 52, 56, 60, 64, 68, 68′, 72, 76, 80, 80’, 84, 84′, 84″, 88, 96, 96′, 96″, 100, 100′, 104, 108, 108′, 112, 112′, 116, 116′, 120, 120′, 124, 128, 132, 140, 152, 156, 168] *and* R *is the ideal generated by some elements of* M.

(For details see Theorem 4.10 of [MS1].)

For the case $(E_8, 3)$ we choose $L = \{x_3, x_7, x_8, x_{19}, x_{20}, x_8^2, x_{20}^2, x_{15}, x_{39}, x_{27}, x_{35}, x_{47}\}$. Then $A \otimes C$ becomes acyclic so that $H(C : d) = \mathrm{Cotor}_A(Z/3, Z/3)$.

Theorem 4.16. *As an algebra* $\mathrm{Cotor}_A(Z/3, Z/3)$ *with* $A = H^*(E_8; Z/3)$ *is isomorphic to* M/R, *where* $M = Z/3[m_i \mid i = 4, 8, 9, 20, 21, 25, 26, 48, 52, 56, 56', 57, 61, 62, 84, 88, 88', 92, 100, 104, 108, 120, 124, 128, 136, 140, 144, 152, 168]$ *and* R *is the ideal generated by some elements of* M.

(For details see Propositions 12.1, 12.2 and 12.3 of [MS3].)

In the above theorems m_i' and m_i'' mean elements of degree i, each different from m_i.

Then one can prove the collapsing of EMSS for these two cases by introducing cohomology operations in EMSS ([S], [M]) and also by using the results on $H^*(B\tilde{E}_j; Z/3)$, $j = 7, 8$ ([HK]), where $\tilde{E}_j$ is the 3-connective cover of E_j. In particular, for the case $(E_7, 3)$, one can show ([MS4]) that all elements but m_{108} are permanent cycles, without appealing to the structure of $H^*(B\tilde{E}_7; Z/3)$.

Theorem 4.17. *As a module*

(1) $H^*(BE_7; Z/3) \cong \mathrm{Cotor}_A(Z/3, Z/3)$ *with* $A = H^*(E_7; Z/3)$.

(2) $H^*(BE_8; Z/3) \cong \mathrm{Cotor}_A(Z/3, Z/3)$ *with* $A = H^*(E_8; Z/3)$.

The details of the proof will appear elsewhere ([MS4]).

(V) The case $(E_8, 5)$. [MS]

We choose $L = \{x_3, x_{11}, x_{12}, x_{12}^2, x_{12}^3, x_{12}^4, x_{15}, x_{23}, x_{27}, x_{35}, x_{39}, x_{47}\}$. Then $A \otimes C$ becomes acyclic so that $H(C : d) = \mathrm{Cotor}_A(Z/5, Z/5)$.

Theorem 4.18. *As an algebra $\mathrm{Cotor}_A(Z/5, Z/5)$ with $A = H^*(E_8; Z/5)$ is commutative and generated by the 1040 indecomposable elements.*

For dimensional reasons we can show that all the differentials are trivial in EMSS of E_8 with $Z/5$-coefficient and hence we have

Theorem 4.19. *As a module*

$$H^*(BE_8; Z/5) \cong \mathrm{Cotor}_A(Z/5, Z/5) \quad \text{with} \quad A = H^*(E_8; Z/5).$$

Corollary 4.20. *As a module*

$$H^*(BX_8; Z/5) \cong \mathrm{Cotor}_A(Z/5, Z/5) \quad \text{with} \quad A = H^*(X_8; Z/5)$$

for any associative H-space X_8 such that $H^*(X_8; Z/5) \cong H^*(E_8; Z/5)$ as an algebra over $\mathcal{A}_5$.

Summing up we have

Theorem 4.21. *The Eilenberg-Moore spectral sequence of G with Z/p -coefficient collapses except for $(G, p) = (E_8, 2)$ and $(AdE_6, 3)$.*

As far as the module structure is concerned, the last problems are to determine $H^*(BE_8; Z/2)$ and $H^*(BAdE_6; Z/3)$.

§5. A $Z/2$-resolution over $H^*(E_8; Z/2)$.

As the first step to compute EMSS for $(E_8, 2)$ we will show in this section Shimada's idea how to construct a small $Z/2$-resolution over $A = H^*(E_8; Z/2)$ by making use of the twisted tensor product.

For simplicity let L be a $Z/2$-subspace of A generated by $\{x_3^i$ for $1 \le i \le 8$, x_5^j for $1 \le j \le 4$, x_9^k for $k = 1, 2$, x_{15}^l for $l = 1, 2$, x_{17}, x_{23}, x_{27}, x_{29}, $x_3^m x_9$ for $1 \le m \le 7\}$. Let $s : L \to sL$ be the suspension homomorphism. We express the corresponding elements under the suspension as follows:

$$sL = \{a_i,\ b_j,\ c_k,\ d_l,\ e,\ f,\ g,\ h,\ \alpha_m\}.$$

Let $\iota : L \to A$ be the inclusion and $\theta : A \to L$ a map such that $\theta \circ \iota = 1_L$. Define $\overline{\theta} : A \to sL$ by $\overline{\theta} = s \circ \theta$ and $\overline{\iota} : sL \to A$ by $\overline{\iota} = \iota \circ s^{-1}$. Consider the tensor algebra $T(sL)$. Let I be the ideal generated by $\mathrm{Im}(\psi \circ (\overline{\theta} \otimes \overline{\theta}) \circ \phi) \circ (\ker \overline{\theta})$.

We put $C = T(sL)/I$. Then I is seen to be generated by the following elements where $[x, y] = xy + yx$:

$$
\begin{array}{llll}
[a_8, x] & & \text{for} & \text{any } x \in C, \\
[b_4, x] & & \text{for} & \text{any } x \in C, \\
[c_2, x] & & \text{for} & \text{any } x \in C, \\
[e, x] & & \text{for} & \text{any } x \in C, \\
[a_i, b_j] & & \text{for} & 1 \leq i \leq 7, \quad 1 \leq j \leq 3, \\
[b_j, c_1] & & \text{for} & 1 \leq j \leq 3, \\
[b_j, \alpha_m] & & \text{for} & 1 \leq j \leq 3, \quad 1 \leq m \leq 7,
\end{array}
$$

$$
\begin{array}{llll}
[d_1, d_2], & & & \\
[d_1, f], & [d_2, f], & & \\
[d_1, g], & [d_2, g], & [f, g], & \\
[d_1, h], & [d_2, h], & [f, h], & [g, h];
\end{array}
$$

$[d_1, a_1] + a_1a_5 + a_2a_4 + a_2\alpha_1 + a_3c_1,$

$[d_1, a_2] + a_1a_6 + a_2\alpha_2 + a_3a_4 + a_4c_1,$

$[d_1, a_3] + a_1a_7 + a_2a_6 + a_2\alpha_3 + a_3a_5 + a_3\alpha_2 + a_4^2 + a_4\alpha_1 + a_5c_1,$

$[d_1, a_4] + a_1a_8 + a_2\alpha_4 + a_5a_4 + a_6c_1,$

$[d_1, a_5] + a_2a_8 + a_2\alpha_5 + a_3\alpha_4 + a_5^2 + a_6a_4 + a_6\alpha_1 + a_7c_1,$

$[d_1, a_6] + a_2\alpha_6 + a_3a_8 + a_4\alpha_4 + a_5a_6 + a_6\alpha_2 + a_7a_4 + a_8c_1,$

$[d_1, a_7] + [a_5, a_7] + a_2\alpha_7 + a_3\alpha_6 + a_4\alpha_5 + a_5\alpha_4 + a_6^2 + a_6\alpha_3 + a_7\alpha_2 + a_8\alpha_1,$

$[d_1, b_1] + b_1b_3 + b_2^2,$

$[d_1, b_2] + b_1b_4 + b_3b_2,$

$[d_1, b_3] + b_3^2,$

$[d_1, c_1] + a_1\alpha_4 + a_2c_2 + \alpha_1a_4 + \alpha_2c_1,$

$[d_1, \alpha_1] + a_1\alpha_5 + a_2\alpha_4 + a_3c_2 + \alpha_1a_5 + \alpha_2a_4 + \alpha_2\alpha_1 + \alpha_3c_1,$

$[d_1, \alpha_2] + a_1\alpha_6 + a_3\alpha_4 + a_4c_2 + \alpha_1a_6 + \alpha_2^2 + \alpha_3a_4 + \alpha_4c_1,$

$$[d_1, \alpha_3] + [a_4, \alpha_4] + [\alpha_2, \alpha_3] + a_1\alpha_7 + a_2\alpha_6 + a_3\alpha_5 + a_5c_2 + \alpha_1a_7 \\ + \alpha_2a_6 + \alpha_3a_5 + \alpha_4\alpha_1 + \alpha_5c_1,$$

$[d_1, \alpha_4] + a_5\alpha_4 + a_6c_2 + \alpha_1a_8 + \alpha_2\alpha_4 + \alpha_5a_4 + \alpha_6c_1,$

$[d_1, \alpha_5] + [a_5, \alpha_5] + a_6\alpha_4 + a_7c_2 + \alpha_2a_8 + \alpha_2\alpha_5 + \alpha_3\alpha_4 + \alpha_6a_4 + \alpha_6\alpha_1 + \alpha_7c_1,$

$[d_1, \alpha_6] + [\alpha_2, \alpha_6] + a_5\alpha_6 + a_7\alpha_4 + a_8c_2 + \alpha_3a_8 + \alpha_4^2 + \alpha_5a_6 + \alpha_7a_4,$

$[d_1, \alpha_7] + [a_5, \alpha_7] + [a_6, \alpha_6] + [a_7, \alpha_5] + [\alpha_2, \alpha_7] + [\alpha_3, \alpha_6] + [\alpha_4, \alpha_5],$

$[d_2, a_1] + a_3a_8 + a_5c_2, \qquad [f, a_1] + a_2b_4 + a_3e,$

$[d_2, a_2] + a_4a_8 + a_6c_2, \qquad [f, a_2] + a_3b_4 + a_4e,$

$[d_2, a_3] + a_5a_8 + a_7c_2,$ $[f, a_3] + a_4b_4 + a_5e,$
$[d_2, a_4] + a_6a_8 + a_8c_2,$ $[f, a_4] + a_5b_4 + a_6e,$
$[d_2, a_5] + a_7a_8,$ $[f, a_5] + a_6b_4 + a_7e,$
$[d_2, a_6] + a_8^2,$ $[f, a_6] + a_7b_4 + a_8e,$
$[d_2, a_7],$ $[f, a_7] + a_8b_4,$
$[d_2, b_1] + b_3b_4,$ $[f, b_1] + b_2c_2,$
$[d_2, b_2] + b_4^2,$ $[f, b_2] + b_3c_2,$
$[d_2, b_3],$ $[f, b_3] + b_4c_2,$
$[d_2, c_1] + \alpha_2a_8 + \alpha_4c_2,$ $[f, c_1] + \alpha_1b_4 + \alpha_2e,$
$[d_2, \alpha_1] + \alpha_3a_8 + \alpha_5c_2,$ $[f, \alpha_1] + \alpha_2b_4 + \alpha_3e,$
$[d_2, \alpha_2] + \alpha_4a_8 + \alpha_6c_2,$ $[f, \alpha_2] + \alpha_3b_4 + \alpha_4e,$
$[d_2, \alpha_3] + \alpha_5a_8 + \alpha_7c_2,$ $[f, \alpha_3] + \alpha_4b_4 + \alpha_5e,$
$[d_2, \alpha_4] + \alpha_6a_8,$ $[f, \alpha_4] + \alpha_5b_4 + \alpha_6e,$
$[d_2, \alpha_5] + \alpha_7a_8,$ $[f, \alpha_5] + \alpha_6b_4 + \alpha_7e,$
$[d_2, \alpha_6],$ $[f, \alpha_6] + \alpha_7b_4,$
$[d_2, \alpha_7],$ $[f, \alpha_7],$
$[g, a_1] + a_2a_8 + \alpha_1c_2,$ $[h, a_1] + a_5e + \alpha_1b_4,$
$[g, a_2] + a_3a_8 + \alpha_2c_2,$ $[h, a_2] + a_6e + \alpha_2b_4,$
$[g, a_3] + a_4a_8 + \alpha_3c_2,$ $[h, a_3] + a_7e + \alpha_3b_4,$
$[g, a_4] + a_5a_8 + \alpha_4c_2,$ $[h, a_4] + a_8e + \alpha_4b_4,$
$[g, a_5] + a_6a_8 + \alpha_5c_2,$ $[h, a_5] + \alpha_5b_4,$
$[g, a_6] + a_7a_8 + \alpha_6c_2,$ $[h, a_6] + \alpha_6b_4,$
$[g, a_7] + a_8^2 + \alpha_7c_2,$ $[h, a_7] + \alpha_7b_4,$
$[g, b_1] + b_3e,$ $[h, b_1] + b_2a_8,$
$[g, b_2] + b_4e,$ $[h, b_2] + b_3a_8,$
$[g, b_3],$ $[h, b_3] + b_4a_8,$
$[g, c_1] + c_2^2 + \alpha_1a_8,$ $[h, c_1] + c_2b_4 + \alpha_4e,$
$[g, \alpha_1] + \alpha_2a_8,$ $[h, \alpha_1] + \alpha_5e,$
$[g, \alpha_2] + \alpha_3a_8,$ $[h, \alpha_2] + \alpha_6e,$
$[g, \alpha_3] + \alpha_4a_8,$ $[h, \alpha_3] + \alpha_7e,$
$[g, \alpha_4] + \alpha_5a_8,$ $[h, \alpha_4],$
$[g, \alpha_5] + \alpha_6a_8,$ $[h, \alpha_5],$
$[g, \alpha_6] + \alpha_7a_8,$ $[h, \alpha_6],$
$[g, \alpha_7],$ $[h, \alpha_7].$

Thus we have an additive isomorphism (cf. filtration in the below):

$$C \cong T(a_1, a_2, \ldots, a_7, \alpha_1, \alpha_2, \ldots, \alpha_7, c_1) \otimes T(b_1, b_2, b_3) \otimes Z/2[c_2] \otimes Z/2[d_1] \otimes Z/2[a_8, b_4, d_2, e, f, g, h].$$

Now as in §4 we define a map $d_C = \psi \circ (\overline{\theta} \otimes \overline{\theta}) \circ \phi \circ \overline{\iota} : sL \to T(sL)$ and extend it naturally over $T(sL)$ as derivation. Since $d_C(I) \subset I$ holds, d_C induces a map $C \to C$, which is again denoted by $d_C : C \to C$ by abuse of notation. Furthermore, $d_C \circ d_C = 0$ holds and hence C becomes a differential algebra over $Z/2$. By virtue of the relation

$$d_C \circ \overline{\theta} + \psi \circ (\overline{\theta} \otimes \overline{\theta}) \circ \phi = 0$$

we can construct the twisted tensor product $W = A \otimes C$ with respect to $\overline{\theta}$ in the sense of Brown [Br]. That is, W is an A-comodule with a differential operator

$$d_W = 1 \otimes d_C + (1 \otimes \psi) \circ (1 \otimes \overline{\theta} \otimes 1) \circ (\phi \otimes 1).$$

In the above ϕ is the coproduct of A induced by the multiplication of E_8 and ψ is the product in C. Now we denote d_C and d_W simply by d by abuse of notation.

It is quite easy to write down the image of d for the generators in A; for example

(5.1)
$$\begin{aligned}
dx_3 &= a_1, \\
dx_3^2 &= a_2, \\
dx_3^3 &= a_3 + x_3^2 a_1 + x_3 a_2, \\
dx_3^4 &= a_4, \\
dx_3^5 &= a_5 + x_3^4 a_1 + x_3 a_4, \\
dx_3^6 &= a_6 + x_3^4 a_2 + x_3^3 a_4, \\
dx_3^7 &= a_7 + x_3^6 a_1 + x_3^5 a_2 + x_3^4 a_3 + x_3^3 a_4 + x_3^2 a_5 + x_3 a_6, \\
dx_3^8 &= a_8, \\
dx_5 &= b_1, \\
dx_5^2 &= b_2, \\
dx_5^3 &= b_3 + x_5^2 b_1 + x_5 b_2, \\
dx_9 &= c_1, \\
dx_9^2 &= c_2, \\
dx_{15} &= d_1 + x_3 a_4 + x_5 b_2 + x_3^2 c_1, \\
dx_{15}^2 &= d_2 + x_3^2 a_8 + x_5^2 b_4 + x_3^4 c_2, \\
dx_{17} &= e,
\end{aligned}$$

$$
\begin{aligned}
dx_{23} &= f + x_3b_4 + x_5c_2 + x_3^2e,\\
dx_{27} &= g + x_3a_8 + x_9c_2 + x_5^2e,\\
dx_{29} &= h + x_9b_4 + x_5a_8 + x_3^4e,\\
dx_3x_9 &= \alpha_1 + x_3c_1 + x_9a_1,\\
dx_3^2x_9 &= \alpha_2 + x_3^2c_1 + x_9a_2,\\
dx_3^3x_9 &= \alpha_3 + x_3\alpha_2 + x_3x_9a_2 + x_3^2\alpha_1 + x_3^2x_9a_1 + x_3^3c_1 + x_9a_3,\\
dx_3^4x_9 &= \alpha_4 + x_3^4c_1 + x_9a_4,\\
dx_3^5x_9 &= \alpha_5 + x_3\alpha_4 + x_3x_9a_4 + x_3^4\alpha_1 + x_3^4x_9a_1 + x_3^5c_1 + x_9a_5,\\
dx_3^6x_9 &= \alpha_6 + x_3^2\alpha_4 + x_3^2x_9a_4 + x_3^4\alpha_2 + x_3^4x_9a_2 + x_3^6c_1 + x_9a_6,\\
dx_3^7x_9 &= \alpha_7 + x_3\alpha_6 + x_3x_9a_6 + x_3^2\alpha_5 + x_3^2x_9a_5 + x_3^3\alpha_4 + x_3^3x_9a_4 + x_3^4\alpha_3\\
&\quad + x_3^4x_9a_3 + x_3^5\alpha_2 + x_3^5x_9a_2 + x_3^6\alpha_1 + x_3^6x_9a_1 + x_3^7c_1 + x_9a_7,
\end{aligned}
$$

In the above we omit $\otimes$ where it is obvious.

Here we introduce weight in $A\otimes C$ as follows:

$$
\begin{aligned}
\text{weight } 0: \ A &= \{x_3, x_3^2, x_3^4, x_3^8, x_5, x_5^2, x_5^4, x_9, x_9^2, x_{17}, x_3x_9, x_3^2x_9, x_3^4x_9\},\\
C &= \{a_1, a_2, a_4, a_8, b_1, b_2, b_4, c_1, c_2, e, \alpha_1, \alpha_2, \alpha_4\};\\
\text{weight } 1: \ A &= \{x_3^3, x_3^5, x_3^6, x_5^3, x_3^3x_9, x_3^5x_9, x_3^6x_9\},\\
C &= \{a_3, a_5, a_6, b_3, \alpha_3, \alpha_5, \alpha_6\};\\
\text{weight } 2: \ A &= \{x_3^7, x_{15}, x_{15}^2, x_{23}, x_{27}, x_{29}, x_3^7x_9\},\\
C &= \{a_7, d_1, d_2, f, g, h, \alpha_7\}.
\end{aligned}
$$

The weight of a monomial is the sum of the weights of each element. Define filtration as $F_r = \{x \mid \text{weight} \le r\}$. Consider $E_0(A\otimes C) = \sum(F_i/F_{i-1})$. Then we have an additive isomorphism

$$
\begin{aligned}
E_0(A\otimes C) \cong &(Z/2[x_3, x_4]/(x_3^8, x_9^2)\otimes T(a_1, a_2, \ldots, a_7, \alpha_1, \alpha_2, \ldots, \alpha_7, c_1))\\
&\otimes(\Lambda(x_9^2)\otimes Z/2[c_2])\otimes(Z/2[x_5]/(x_5^4)\otimes T(b_1, b_2, b_3))\\
&\otimes(Z/2[x_{15}]/(x_{15}^2)\otimes Z/2[d_1])\\
&\otimes(\Lambda(x_3^8, x_5^4, x_9^2, x_{17}, x_{23}, x_{27}, x_{29})\otimes Z/2[a_8, b_4, d_2, e, f, g, h]).
\end{aligned}
$$

Then it follows from the formula (5.1) of the d-image of the generators in A that $d(F_i) \subset F_i$ and we can see that $E_0(A\otimes C)$ is acyclic, and so is $A\otimes C$. Thus we have

Theorem 5.2. *$A\otimes C$ is an acyclic injective comodule resolution of $Z/2$ over $H^*(E_8; Z/2)$.*

Corollary 5.3. $H(C : d) = \ker d/\mathrm{Im}\, d \cong \mathrm{Cotor}_A(Z/2, Z/2)$ with $A = H^*(E_8; Z/2)$.

Here, for the generators of C, we have the following d-images:

$$\begin{array}{llll}
d(a_1) = 0, & d(a_2) = 0, & d(a_3) = [a_1, a_2], & d(a_4) = 0, \\
d(a_5) = [a_1, a_4], & d(a_6) = [a_2, a_4], & d(a_7) = [a_1, a_6] + [a_2, a_5] + [a_3, a_4], & \\
d(a_8) = 0, & & & \\
d(b_1) = 0, & d(b_2) = 0, & d(b_3) = [b_1, b_2], & d(b_4) = 0, \\
d(c_1) = 0, & d(c_2) = 0, & & \\
d(\alpha_1) = [a_1, c_1], & d(\alpha_2) = [a_2, c_1], & d(\alpha_3) = [a_1, \alpha_2] + [a_2, \alpha_1] + [a_3, c_1], & \\
d(\alpha_4) = [a_4, c_1], & d(\alpha_5) = [a_1, \alpha_4] + [a_4, \alpha_1] + [a_5, c_1], & &
\end{array}$$

$$d(\alpha_6) = [a_2, \alpha_4] + [a_4, \alpha_2] + [a_6, c_1],$$

$$d(\alpha_7) = [a_1, \alpha_6] + [a_2, \alpha_5] + [a_3, \alpha_4] + [a_4, \alpha_3] + [a_5, \alpha_2] + [a_6, \alpha_1] + [a_7, c_1],$$

$$\begin{array}{ll}
d(d_1) = a_1 a_4 + a_2 c_1 + b_1 b_2, & d(d_2) = a_2 a_8 + b_2 b_4 + a_4 c_2, \\
d(e) = 0, & d(f) = a_1 b_4 + a_2 e + b_1 c_2, \\
d(g) = a_1 a_8 + b_2 e + c_1 c_2, & d(h) = b_1 a_8 + c_1 b_4 + a_4 e.
\end{array}$$

Computation of $\mathrm{Cotor}_A(Z/2, Z/2)$, which is the E_2-term of EMSS, is partially based on the notes left by M. Mori who died several years ago. The details of the calculation will appear elsewhere, together with the proof of the collapsing of EMSS.

§6. A $Z/3$-resolution over $H^*(Ad\,E_6; Z/3)$.

As the first step to compute EMSS for $(Ad\,E_6, 3)$ we will construct a small $Z/3$-resolution over $A = H^*(Ad\,E_6; Z/3)$ by making use of the twisted tensor product.

For simplicity let L be a $Z/3$-subspace of A generated by $\{x_1,\ x_2^i$ for $1 \leq i \leq 8,\ x_3,\ x_7,\ x_8^j$ for $j = 1, 2,\ \ x_9,\ x_{11},\ x_{15}\}$. Let $s : L \to sL$ be the suspension homomorphism. We express the corresponding elements under the suspension as follows:

$$sL = \{a,\ b_i,\ c,\ d,\ e_j,\ f,\ g,\ h\}.$$

As in §5 we construct $C = T(sL)/I$, where I is the ideal generated by $\mathrm{Im}(\psi \circ (\bar\theta \otimes \bar\theta) \circ \phi) \circ (\ker \bar\theta)$. More precisely, I is generated by the following elements where $[x, y] = xy - (-1)^* yx$ with $* = \deg x \cdot \deg y$:

$$\begin{array}{ll}
[a, y] & \text{for any } y \text{ of } C, \\
[x, y] & \text{for } x \neq b_i \text{ and } y = c, d, \\
[x, y] & \text{for } x \neq b_i, e_1 \text{ and } y = f, g, h
\end{array}$$

$$[e_1, f] + e_2 a, \qquad [e_1, g] + e_2 c, \qquad [e_1, h] + e_2 d.$$

$[b_1, c] + b_2a,$ $[b_1, d] + b_4a,$
$[b_2, c] + b_3a,$ $[b_2, d] + b_5a,$
$[b_3, c] + b_4a,$ $[b_3, d] + b_6a,$
$[b_4, c] + b_5a,$ $[b_4, d] + b_7a,$
$[b_5, c] + b_6a,$ $[b_5, d] + b_8a,$
$[b_6, c] + b_7a,$
$[b_7, c] + b_8a,$

$[b_1, e_1] + b_3b_2 + b_4b_1,$
$[b_2, e_1] + b_3^2 - b_4b_2 + b_5b_1,$
$[b_3, e_1] + b_3b_4 + b_6b_1,$
$[b_4, e_1] + b_3b_5 + b_4^2 + b_6b_2 + b_7b_1,$
$[b_5, e_1] + b_3b_6 - b_4b_5 + b_5b_4 + b_6b_3 - b_7b_2 + b_8b_1,$
$[b_6, e_1] + b_3b_7 - b_6b_4,$
$[b_7, e_1] + b_3b_8 + b_4b_7 - b_6b_5 - b_7b_4,$
$[b_8, e_1] - b_4b_8 + b_5b_7 - b_6^2 + b_7b_5 - b_8b_4,$

$[b_1, e_2] + b_6b_3 + b_7b_2,$ $[b_1, f] + b_2d - b_4c + b_5a,$
$[b_2, e_2] + b_6b_4 - b_7b_3 + b_8b_2,$ $[b_2, f] + b_3d - b_5c + b_6a,$
$[b_3, e_2] + b_6b_5,$ $[b_3, f] + b_4d - b_6c + b_7a,$
$[b_4, e_2] + b_6^2 + b_7b_5,$ $[b_4, f] + b_5d - b_7c + b_8a,$
$[b_5, e_2] + b_6b_7 - b_7b_6 + b_8b_5,$ $[b_5, f] + b_6d - b_8c,$
$[b_6, e_2] + b_6b_8,$ $[b_6, f] + b_7d,$
$[b_7, e_2] + b_7b_8,$ $[b_7, f] + b_8d,$
$[b_8, e_2] + b_8^2,$

$[b_1, g] + b_2f - b_3d - b_5c - b_6a,$ $[b_1, h] + b_4f + b_7c,$
$[b_2, g] + b_3f - b_4d - b_6c - b_7a,$ $[b_2, h] + b_5f + b_8c,$
$[b_3, g] + b_4f - b_5d - b_7c - b_8a,$ $[b_3, h] + b_6f,$
$[b_4, g] + b_5f - b_6d - b_8c,$ $[b_4, h] + b_7f,$
$[b_5, g] + b_6f - b_7d,$ $[b_5, h] + b_8f,$
$[b_6, g] + b_7f - b_8d,$
$[b_7, g] + b_8f,$

Thus we have an additive isomorphism (cf. filtration in the below):

$$C \cong T(b_1, b_2, \ldots, b_7) \otimes T(e_1, e_2) \otimes Z/3[a, c, d, f, g, h].$$

As in §5 we define a differential $d_C : C \to C$ which satisfies

$$d_C \circ \overline{\theta} + \psi \circ (\overline{\theta} \otimes \overline{\theta}) \circ \phi = 0$$

so that we can construct the twisted tensor product $W = A \otimes C$ with respect to $\overline{\theta}$. That is, W is an A-comodule with a differential operator

$$d_W = 1 \otimes d_C + (1 \otimes \psi) \circ (1 \otimes \overline{\theta} \otimes 1) \circ (\phi \otimes 1).$$

As before we denote d_C and d_W simply by d by abuse of notation.

Then we have

$$\begin{aligned}
(6.1) \quad & dx_1 = a, \\
& dx_2 = b_1, \qquad dx_2^2 = b_2 - x_2 b_1, \qquad dx_2^3 = b_3, \\
& dx_2^4 = b_4 + x_2 b_3 + x_2^3 b_1, \qquad dx_2^5 = b_5 - x_2 b_4 + x_2^2 b_3 + x_2^3 b_2 - x_2^4 b_1, \\
& dx_2^6 = b_6 - x_2^3 b_3, \qquad dx_2^7 = b_7 + x_2 b_6 - x_2^3 b_4 - x_2^4 b_3 + x_2^6 b_1, \\
& dx_2^8 = b_8 - x_2 b_7 + x_2^2 b_6 - x_2^3 b_5 + x_2^4 b_4 - x_2^5 b_3 + x_2^6 b_2 - x_2^7 b_1, \\
& dx_3 = c + x_2 a, \qquad dx_7 = d + x_2^3 a, \\
& dx_8 = e_1 + x_2^3 b_1, \qquad dx_8^2 = e_2 + x_2^6 b_2 - x_8 e_1 - x_8 x_2^3 b_1, \\
& dx_9 = f + x_2 d - x_2^3 c + x_2^4 a + x_8 a, \\
& dx_{11} = g + x_2 f - x_2^2 d - x_2^4 c - x_2^5 a + x_8 c + x_8 x_2 a, \\
& dx_{15} = h + x_2^3 f + x_2^6 c + x_8 d + x_8 x_2^3 a.
\end{aligned}$$

As before we omit $\otimes$ where it is obvious.

Here we introduce weight in $A \otimes C$ as follows:

weight 0 : $A = \{x_1, x_2^i \text{ for } 1 \le i \le 8\}$, $C = \{a, b_i \text{ for } 1 \le i \le 8\}$,

weight 1 : $A = \{x_3, x_7, x_8\}$, $C = \{c, d, e_1\}$,

weight 2 : $A = \{x_8^2\}$, $C = \{e_2\}$,

weight 3 : $A = \{x_9\}$, $C = \{f\}$,

weight 6 : $A = \{x_{11}, x_{15}\}$, $C = \{g, h\}$.

Define filtration as $F_r = \{x \mid \text{weight} \le r\}$. Consider $E_0(A \otimes C) = \sum_i (F_i / F_{i-1})$. Quite similarly as before, we have an additive isomorphism

$$\begin{aligned}
E_0(A \otimes C) \cong & (Z/3[x_2]/(x_2^9) \otimes T(b_1, b_2, \ldots, b_8)) \\
& \otimes (Z/3[x_8]/(x_8^3) \otimes T(e_1, e_2)) \\
& \otimes (\Lambda(x_1, x_3, x_7, x_9, x_{11}, x_{15}) \otimes Z/3[a, c, d, f, g, h]).
\end{aligned}$$

Then it follows from the formula (6.1) of the d-image of the generators in A that $d(F_i) \subset F_i$ and we can see that $E_0(A \otimes C)$ is acyclic, and so is $A \otimes C$. Thus we have

Theorem 6.2. *$A \otimes C$ is an acyclic injective comodule resolution of $Z/3$ over $H^*(AdE_6; Z/3)$.*

Corollary 6.3. *$H(C:d) = \ker d/\mathrm{Im}\, d \cong Cotor_A(Z/3, Z/3)$, with $A = H^*(AdE_6; Z/3)$.*

Here, for the generators of C, we have the following d-images:

$$da = 0, \quad db_1 = 0, \quad db_2 = b_1^2, \quad db_3 = 0, \quad db_4 = -[b_1, b_3],$$
$$db_5 = [b_1, b_4] - [b_2, b_3], \quad db_6 = b_3^2, \quad db_7 = -[b_1, b_6] + [b_3, b_4],$$
$$db_8 = [b_1, b_7] - [b_2, b_6] + [b_3, b_5] - b_4^2,$$
$$dc = -b_1 a, \quad dd = -b_3 a, \quad de_1 = -b_3 b_1, \quad de_2 = -b_6 b_2 - e_1^2,$$
$$df = -b_1 d + b_3 c - b_4 a - e_1 a, \quad dg = -b_1 f + b_2 d + b_4 c + b_5 a - e_1 c,$$
$$dh = -b_3 f - b_6 c - e_1 d.$$

The details of the computation of $\mathrm{Cotor}_A(Z/3, Z/3)$, which is the E_2-term of EMSS, will appear elsewhere, together with the proof of the collapsing of EMSS.

In concluding let me state the following

Conjecture. *The Eilenberg-Moore spectral sequence with Z/p-coefficient collapses for any exceptional Lie group and any prime p.*

References

[A1] S. Araki, *On the non-commutativity of Pontrjagin rings mod 3 of the compact exceptional groups*, Nagoya Math. J. **17** (1960), 225–260.

[A2] S. Araki, *Differential Hopf algebras and the cohomology mod 3 of the compact exceptional groups E_7 and E_8*, Ann. Math. **73** (1961), 404–436.

[A3] S. Araki, *Cohomology modulo 2 of the compact exceptional groups E_6 and E_7*, J. Math. Osaka City Univ. **12** (1961), 43–65.

[AS] S. Araki - Y. Shikata, *Cohomology mod 2 of the compact exceptional group E_8*, Proc. Japan Acad. **37** (1961), 619–622.

[Bo1] A. Borel, *Topology of Lie groups and characteristic classes*, Bull. AMS **61** (1955), 397–432.

[Bo2] A. Borel, *Sur la cohomologie des espaces fibrés principaux et des espace homogènes de groupes de Lie compacts*, Ann. Math. **57** (1953), 115–207.

[Bo3] A. Borel, *Sur l'homologie et la cohomologie de groupes de Lie compacts connexes*, Amer. J. Math. **76** (1954), 273–342.

[Bo4] A. Borel, *Sous-groupes commutatifs et torsions des groupes de Lie compacts connexes*, Tohoku Math. J. **13** (1961), 216–240.

[Bd] W. Browder, *Homology rings of groups,* Amer. J. Math. **90** (1968), 318–333.

[Br] E. H. Brown Jr., *Twisted tensor product I,* Ann. Math. **69** (1959), 223–246.

[HK] M. Harada - A. Kono, *Cohomology mod p of the 4-connected cover of the classifying space of simple Lie groups,* Advanced Studies in Pure Math. **9** (1986), 109–122.

[IKT] K. Ishitoya - A. Kono - H. Toda, *Hopf algebra structure of mod 2 cohomology of simple Lie groups,* Publ. RIMS Kyoto Univ. **12** (1975), 141–167.

[IS] A. Iwai - N. Shimada, *A remark on resolutions for Hopf algebras,* Publ. RIMS of Kyoto Univ. **1** (1966), 187–198.

[K] R. Kane, *The Homology of Hopf Spaces,* North-Holland Math. Library (1988).

[K1] A. Kono, *Hopf algebra structure of simple Lie groups,* J. Math. Kyoto Univ. **17** (1977), 259–298.

[K2] A. Kono, *Hopf algebra structure and cohomology operations of the mod 2 cohomology of the exceptional Lie groups,* Japanese J. Math. **3** (1977), 49–55.

[KM1] A. Kono - M. Mimura, *Cohomology mod 2 of the classifying space of the compact connected Lie group of type E_6,* J. Pure and Applied Algebra **6** (1975), 61–81.

[KM2] A. Kono - M. Mimura, *Cohomology operations and the Hopf algebra structure of the compact exceptional Lie groups E_7 and E_8,* Proc. London Math. Soc. **(3) 35** (1977), 345–358.

[KM3] A. Kono - M. Mimura, *On the cohomology mod 2 of the classifying space of AdE_7,* J. Math. Kyoto Univ. **18** (1978), 535–541.

[KM4] A. Kono - M. Mimura, *Cohomology mod 3 of the classifying space of the Lie group E_6,* Math. Scand. **46** (1980), 223–235.

[KMS1] A. Kono - M. Mimura - N. Shimada, *Cohomology of classifying space of certain associative H-spaces,* J. Math. Kyoto Univ. **15** (1975), 607–617.

[KMS2] A. Kono - M. Mimura - N. Shimada, *On the cohomology mod 2 of the classifying space of the 1-connected exceptional Lie group E_7,* J. Pure and Applied Algebra **8** (1976), 267–283.

[MM] M. Mimura - M. Mori, *The squaring operations in the Eilenberg-Moore spectral sequence and the classifying space of an associative H-space, I,* Publ. RIMS of Kyoto Univ. **13** (1977), 755–776.

[MS1] M. Mimura - Y. Sambe, *On the cohomology mod p of the classifying spaces of the exceptional Lie groups, I,* J. Math. Kyoto Univ. **19** (1979), 553–581.

[MS2] M. Mimura - Y. Sambe, *On the cohomology mod p of the classifying spaces of the exceptional Lie groups, II,* J. Math. Kyoto Univ. **20** (1980), 327–349.

[MS3] M. Mimura - Y. Sambe, *On the cohomology mod p of the classifying spaces of the exceptional Lie groups, III,* J. Math. Kyoto Univ. **20** (1980), 351–379.

[MS4] M. Mimura - Y. Sambe, *On the cohomology mod p of the classifying spaces of the exceptional Lie groups, IV,* (in preparation).

[MS] M. Mimura - Y. Sambe, *Collapsing of the Eilenberg-Moore spectral sequence mod 5 of the compact exceptional group* E_8, J. Math. Kyoto Univ. **21** (1981), 203–230.

[Mo] J. C. Moore, *Algèbre homologique et homologie des espaces classifiants,* Séminaire H. Cartan, Exposé **7** (1959/1960).

[M] M. Mori, *The Steenrod operations in the Eilenberg-Moore spectral sequence,* Hiroshima Math. J. **9** (1979), 17–34.

[RS1] M. Rothenberg - N. E. Steenrod, *The cohomology of classifying spaces of H-spaces,* Bull. AMS **71** (1965), 872–875.

[RS2] M. Rothenberg - N. E. Steenrod, *The cohomology of classifying spaces of H-spaces,* (mimeographed notes).

[SI] N. Shimada - A. Iwai, *On the cohomology of some Hopf algebra,* Nagoya Math. J. **30** (1971), 103–111.

[S] W. Singer, *Steenrod squares in spectral sequences, I, II,* Trans. AMS **175** (1973), 327–336; 337–353.

[Th] E. Thomas, *Exceptional Lie groups and Steenrod squares,* Michigan J. Math. **11** (1964), 151–156.

[T1] H. Toda, *Cohomology mod 3 of the classifying space* BF_4 *of the exceptional group* F_4, J. Math. Kyoto Univ. **13** (1972), 97–115.

[T2] H. Toda, *Cohomology of the classifying space of the exceptional Lie groups,* Manifolds-Tokyo (1973), 265–271.

[T3] H. Toda, *Cohomology of the classifying spaces,* Advanced Studies in Pure Math. **9** (1986), 75–108.

Mamoru MIMURA
Department of Mathematics
Okayama University
Okayama, JAPAN 700

How can you tell two spaces apart when they have the same n-type for all n ?

By C.A.McGibbon and J.M.Møller*

In memory of J.F.Adams

In 1957, Adams gave the first example of two different homotopy types, say X and Y, whose Postnikov approximations, $X^{(n)}$ and $Y^{(n)}$, are homotopy equivalent for each n. He did this in response to a question posed by J.H.C.Whitehead. Adams gave an explicit description of both spaces and showed they are different, up to homotopy, by noting that one contains a sphere as a retract whereas the other does not, [1]. Recently, in our study of infinite dimensional spaces, we have had to confront the same problem. Often we can prove that for a given space, e.g., $X = S^3 \times K(\mathbf{Z}, 3)$, there are many other spaces, up to homotopy, with the same n-type as X for all n. But when asked to describe one of them, we had to plead ignorance. To correct this situation we began to look for explicit descriptions and for homotopy invariants that are not determined by finite approximations. In doing so, we found some new examples and a new answer, involving automorphism groups, to the question posed in the title. This paper deals with those examples. We find them intriguing, in part, because they indicate an unexpected lack of homogeneity among spaces of the same n -type for all n. Nevertheless, as examples go, Adams's original one remains one of our favorites because of its simplicity and explicit nature. We commend it to the reader.

To describe our results, we first recall Wilkerson's classification theorem. In [13] , he classified, in the following sense, all spaces having the same n-type,

---

* Suported by the Danish Natural Science Research Council

for all n, as a given space X.

Theorem 1. Given a connected CW space X, let $SNT(X)$ denote the set of all homotopy types $[Y]$ such that $Y^{(n)} \simeq X^{(n)}$ for all n. Then there is a bijection of pointed sets,

$$SNT(X) \approx \varprojlim{}^1 AutX^{(n)}$$

where $AutX^{(n)}$ is the group of homotopy classes of homotopy self-equivalences of $X^{(n)}$. □

Recently we used this theorem to show that if X is a nilpotent space with finite type over some subring of the rationals, then the set $SNT(X)$ is either uncountably large or it has just one member; namely $[X]$. For spaces that are rationally equivalent to H-spaces, i.e., H_0-spaces, one can determine which alternative holds by using the following result from [5].

Theorem 2. Let X be a 1-connected, H_0-space with finite type over $\mathbf{Z}_P$ for some set of primes P. Then the following statements are equivalent:
(i) $SNT(X) = *$.
(ii) the canonical map $AutX \longrightarrow AutX^{(n)}$, has a finite cokernel for all n.
(iii)the map $AutX \xrightarrow{f \mapsto f^*} AutH^{\leq n}(X; \mathbf{Z}_P)$ has a finite cokernel for all integers n. □

Using this result we were able to determine, for example, the cardinality of $SNT(X)$ when X is the classifying space of a connected compact Lie group.

Theorem 3. Let G be a connected compact Lie group. Then $SNT(BG) = *$ if and only if $G = T^k$, $SU(n)$ or $PSU(n)$ when $k \geq 0$ and $n = 2$ or 3. □

In other words, except for a few special cases, $SNT(BG)$ is almost always uncountably large. In contrast, we have yet to find a finite complex K for which $SNT(\Omega K)$ is nontrivial! Some partial results on this loop space problem are given in [5]. For example, it is shown there if K is a compact Lie group or a complex Stiefel manifold, then $SNT(\Omega K)$ is the one element set. The Eckmann-Hilton dual problem is considered in [6].

The results just quoted deal only with the cardinality of $SNT(X)$ and do not actually address the question posed in the title. As mentioned earlier, we now seek ways to distinguish different members of $SNT(X)$. What tools should be used in this task ? It might be worth pointing out that mod p cohomology is of little use here. Members of $SNT(X)$ cannot be distinguished, a priori, by their cohomology - even as algebras over the Steenrod algebra - when the spaces involved are nilpotent and of finite type. The reason involves the profinite completion functor, $X \mapsto \widehat{X}$. On the one hand, when X has finite type this map induces an equivalence in mod p cohomology while on the other, for such X, $SNT(\widehat{X})$ has just one element, as Wilkerson showed in [13], Corollary IIc.

The situation for K-theory appears to be different. Indeed, Notbohm claimed at the 1990 Barcelona conference that if X and Y are in $SNT(BG)$ where G is simply connected compact Lie, then X and Y are homotopy equivalent if and only if $K(X)$ and $K(Y)$ are isomorphic as λ-rings. We have not seen the details yet, but the results in [10] make this sound plausible. If this is true, then there may be a computable K-theory invariant (perhaps a power series of some sort) that classifies $SNT(BG)$. On the other hand, obtaining such an invariant may be difficult. As a λ-ring, $K(X)$, for X in $SNT(BG)$, currently seems complicated and intractible. We are certain that Adams and Wilkerson would have loved to have had a K-theory version of their embedding theorem for spaces of this kind. One can safely assume that their failure to achieve such a result was not for lack of trying. For an indication of how complicated this question gets, even in the rank 1 case (how *not* to do it, if you will), see [4].

Having dispensed with something that doesn't work (cohomology), and mentioned something that might work in special cases (K-theory), let us turn to another invariant; the discrete group, $Aut(X)$. Of course, one would not expect $Aut(\ )$ to separate points of $SNT(X)$, in general. Indeed, we find it curious that $Aut(Y)$ is not constant, as Y varies over $SNT(X)$. This is what our examples will illustrate. First recall from [2] that the Postnikov decomposition of X determines the following short exact sequence of groups,

$$0 \longrightarrow \varprojlim{}^1 \pi_1 aut X^{(n)} \longrightarrow Aut(X) \longrightarrow \varprojlim Aut X^{(n)} \longrightarrow 1$$

In our first example, the $\varprojlim{}^1$ term on the left happens to be zero. In this case, $Aut(Y)$ and $Aut(X)$, where Y is in $SNT(X)$, are isomorphic to the inverse limits of "two towers of groups that have the same n-type for all n". We trust that the reader sees the obvious analogy and can supply the proper definition for the term in quotes.

The following example shows that even though two spaces X and Y, have the same n-type for all n, it is possible that one group, $Aut(X)$, is infinite, while the other, $Aut(Y)$, is finite!

Example A Let $X = K(\mathbf{Z}, 3) \times S^3$. Then $Aut(X)$ is isomorphic to the group of upper triangular matrices in $GL(2, \mathbf{Z})$, but there exists a Y in $SNT(X)$ with $Aut(Y) \approx \mathbf{Z}/2$. □

To appreciate this example, it might be helpful to recall a result of Wilkerson; [14], Theorem 2.3. He shows $Aut(Y)$ and $Aut(X)$ are commensurable provided there exists a rational equivalence $f : X \to Y$, and provided both spaces are 1-connected, with finite type, and have only a finite number of nonzero homotopy (or homology) groups. Now, the two spaces in example A are rationally equivalent, although we suspect there are no maps between them, before rationalizing, that induce this equivalence. Therefore, since the automorphism groups in this example are clearly not commensurable, this example shows that Wilkerson's result is, in a sense, best possible.

The proof for example A involves matrix calculations with a lemma that expresses $Aut(Y)$ as a pullback of $Aut(X_o)$ and $Aut(\widehat{X})$. We also use this lemma, together with Mislin's homotopy classification of self-maps of infinite quaternionic projective space to show that $Aut(Y) = \{1\}$ for every Y in the genus of $\mathbf{HP}^\infty$.

Our second example involves the $\varprojlim^1$ term in the short exact sequence above. Following Roitberg, [11], we denote this term by $WI(X)$. Of course, members of this subgroup can be viewed as self maps of X that restrict to the identity map on each finite skeleton. If each of the groups $\pi_1 autX^{(n)}$ is a finitely generated abelian group, (as would be the case when X is nilpotent with finite type over $\mathbf{Z}$), then the possible values of $WI(X)$ are very limited. In fact, it follows from Jensen, [3], Chapter 2, that this subgroup is either zero or else it is a divisible abelian group of the form

$$WI(X) \approx \mathbf{R} \oplus \sum_{all\ p} (\mathbf{Z}/p^\infty)^{n_p}$$

Here $\mathbf{R}$ is regarded as a rational vector space of rank $2^{\aleph_0}$ and each torsion exponent n_p is either finite or $2^{\aleph_0}$.

Assume for the moment that X and Y are 1-connected, with finite type, and that they have the same n-type for all n. It follows at once that the towers $\{\pi_1 autX^{(n)}\}$ and $\{\pi_1 autY^{(n)}\}$, likewise have the same n-type for all n. It is then easy to check that one tower is Mittag-Leffler if and only if the other is. These will then be towers of countable groups and so by Theorem 2 of [5], it follows that $WI(X)$ is zero if and only if the same is true for $WI(Y)$. Consequently, it follows from Jensen's description, above, that if $WI(X)$ and $WI(Y)$ are not isomorphic, then the only way they can differ is in their torsion summands. We were surprised to find that this can actually happen.

Example B Let $X = \mathbf{CP}^\infty \times \Omega S^3 \times S^3$. Then $WI(X) = \mathbf{R}$ and there exists a Y in $SNT(X)$ with $WI(Y) = \mathbf{R} \oplus \mathbf{Q}/\mathbf{Z}$. □

The verification of these examples takes up the rest of the paper. In the first example, the space Y will be constructed as a homotopy pullback of the rationalization, X_o, and the profinite completion, $\widehat{X}$, over the formal completion, $\bar{X}_o$.

$$\begin{array}{ccc} & & \widehat{X} \\ & & \downarrow \\ X_o & \longrightarrow & \bar{X}_o, \end{array}$$

The vertical map in this diagram is fixed. It first rationalizes and then identifies $(\widehat{X})_o$ with $\bar{X}_o$. This identification is valid for 1-connected spaces with finite type, which are the sort of spaces that will be considered here. The horizontal map in the diagram will be altered by composition with certain automorphisms of $\bar{X}_o$. Both maps in the diagram are required to induce isomorphisms on $\pi_*(\ )\otimes\mathbf{Q}\otimes\hat{\mathbf{Z}}$. It is then immediate that the rational homotopy type and the profinite completion of Y are equivalent to those of X. In other words, $[Y] \in \widehat{G}_o(X)$. It is a theorem of Wilkerson that every member of $\widehat{G}_o(X)$ can be obtained this way and that the double coset formula

$$\widehat{G}_o(X) \approx Aut(X_o)\backslash CAut(\bar{X}_o)/Aut(\widehat{X})$$

is valid. The group $CAut(\bar{X}_o)$ in this formula consists of those self maps f for which $\pi_*(f)$ is a $\mathbf{Q} \otimes \hat{\mathbf{Z}}$ module isomorphism. For more details here, see [14].

Lemma 4 Let X be a 1-connected CW-space with finite type such that

$$H_n(X;\pi_{n+1}(\widehat{X}_o)) = 0, \quad \text{for each } n \geq 0.$$

Let φ be in $CAut(\bar{X}_o)$. If Y is the homotopy pullback of

$$X_o \longrightarrow \bar{X}_o \xrightarrow{\varphi} \bar{X}_o \longleftarrow \widehat{X}$$

then $Aut(Y)$ is isomorphic to the group theoretic pullback of

$$Aut(X_o) \longrightarrow CAut(\bar{X}_o) \xrightarrow{\varphi^*} CAut(\bar{X}_o) \longleftarrow Aut(\widehat{X})$$

where in φ^* denotes conjugation by φ. □

The proof of this lemma will be given later. It depends, in part, on W. Meier's results on pullbacks and phantom maps, [8]. Notice that, in this lemma, if $CAut(\bar{X}_o)$ is abelian, then φ^* is the identity. So, under these special conditions, it follows that $Aut(Y)$ is isomorphic to $Aut(X)$ for every $Y \in \widehat{G}_o(X)$. This is the case, for example, when $X = \mathbf{HP}^\infty$. Since Mislin has shown in [9], that $Aut(\mathbf{HP}^\infty) = \{1\}$, we deduce that the same is true for every Y in the genus of this space.

Proof of Example A. Recall that $X = S^3 \times K(\mathbf{Z}, 3)$. Therefore $\bar{X}_o$ is a $K(V, 3)$ where V is a free module of rank 2 over $\mathbf{Q} \otimes \hat{Z}$. We will identify $CAut(\bar{X}_o)$ with the adele group $GL(2, \mathbf{Q} \otimes \hat{\mathbf{Z}})$ and we will represent the self-equivalence φ as follows. First, for each prime p, let B_p be the matrix

$$B_p = \begin{pmatrix} 1 & 0 \\ c_p & 1 \end{pmatrix}$$

where $c_p = 1$ when p is 2, -1 when p is 3 , and 0 when p is 5 or more. Then take φ to be the the map represented by the sequence of matrices $(B_p) = (B_2, B_3, B_5, \ldots)$ in $\Pi_p GL(2, \hat{\mathbf{Z}}_p) \subseteq GL(2, \mathbf{Q} \otimes \hat{\mathbf{Z}})$. Use this φ to construct Y as a pullback of X_o and $\widehat{X}$ over $\bar{X}_o$, as in the lemma. As a pullback,

$$Aut(Y) = \{(\rho, \mu) | \rho \in Aut(X_o), \mu \in Aut(\widehat{X}), \text{and} [\rho] = \varphi^*[\mu] \}.$$

Let R and (M_p) be the matrices that represent $[\rho]$ and $[\mu]$ in $GL(2, \mathbf{Q}\otimes\hat{\mathbf{Z}})$. The matrices M_p can be taken to be upper triangular. This follows from a routine analysis of $[X, X]$ together with some special facts; namely the cohomotopy groups $[K, S]$ and $[\Sigma^3 K, S]$ are trivial, where K denotes a $K(\mathbf{Z}, 3)$ and S denotes the p-completion of a 3-sphere. These facts are due to Zabrodsky, [16].

From the description just given of $Aut(Y)$, it follows that for each prime p there is a matrix equation

$$B_p R = M_p B_p$$

We claim that these equations force R to be $\pm I$. Since B_5 is the identity, it is clear that R is at least upper triangular. Therefore,

$$\text{if } R = \begin{pmatrix} r & s \\ 0 & t \end{pmatrix} \text{ and if } M_2 = \begin{pmatrix} x & y \\ 0 & z \end{pmatrix},$$

then the $p = 2$ equation,

$$\begin{pmatrix} r & s \\ r & s+t \end{pmatrix} = B_2 R = M_2 B_2 = \begin{pmatrix} x+y & y \\ z & z \end{pmatrix}$$

implies that $r = s + t$. Similarly, the $p = 3$ equation,

$$\begin{pmatrix} r & s \\ -r & -s+t \end{pmatrix} = B_3 R = M_3 B_3 = \begin{pmatrix} u-v & v \\ -w & w \end{pmatrix}$$

implies that $r = -s + t$. It follows that $s = 0$ and $r = t = \pm 1$. Hence $R = \pm I = M_p$, for each prime p. Since the inclusions of $Aut(X_o)$ and of each $Aut(\widehat{X}_p)$ in $CAut(\bar{X}_o)$ are injective, it follows that the pullback group $Aut(Y) \approx \mathbf{Z}/2$, as claimed.

It remains to be shown that Y is in $SNT(X)$. To this end, we first establish that $Y \in G(X)$, the localization genus of X. This follows because of the rational entries in each of the matrices B_p. Indeed, it is easy to see that $Y_{(p)}$ is homotopy equivalent to the pullback

$$\begin{array}{ccccc} & & & & \widehat{X}_p \\ & & & & \downarrow \\ X_o & \rightarrow & (\widehat{X}_p)_o & \xrightarrow{\varphi_p} & (\widehat{X}_p)_o \end{array}$$

where φ_p is represented by B_p. Such pullbacks are classified by a double coset formula, as noted earlier. The rational nature of B_p places φ_p in the image of $Aut(X_o)$, which, in turn, is in the double coset of the identity. Thus the induced pullback $Y_{(p)}$, is equivalent to $X_{(p)}$, and so Y and X are in the same genus. Consequently, $Y^{(n)} \in G(X^{(n)})$ for each positive integer n. However, each $G(X^{(n)})$ has only one member; namely $X^{(n)}$. This is an easy consequence of Zabrodsky's genus theorem, [15], and elementary facts about self-maps of X. Therefore $Y^{(n)} \simeq X^{(n)}$ for each n, as was to be shown.

Proof of the lemma The rational homology condition on X, together with the finite type hypothesis, ensures that the group $WI(X) = 0$. Notice that if X satisfies these conditions, then so does every member of $\widehat{G}_o(X)$. Moreover, this ensures that if φ is the identity, then $Aut(X)$ is isomorphic to this pullback, by [8], Theorem 4. Now let Y be in $\widehat{G}_o(X)$ and choose equivalences

$$Y_o \xrightarrow{f} X_o \quad \text{and} \quad \widehat{Y} \xrightarrow{g} \widehat{X}.$$

Then $Y = X_\varphi$, the pullback determined by the equivalence $\varphi = g_o(\bar{f})^{-1}$, by [14], Theorem 3.8. We have isomorphisms

$$Aut(X_o) \xrightarrow{\rho \mapsto f^{-1}\rho f} Aut(Y_o) \quad \text{and} \quad Aut(\widehat{X}) \xrightarrow{\mu \mapsto g^{-1}\mu g} Aut(\widehat{Y})$$

Hence $Aut(Y) \subseteq Aut(Y_o) \times Aut(\widehat{Y}) \approx Aut(X_o) \times Aut(\widehat{X})$, and consists of pairs (ρ, μ) for which $\overline{(f^{-1}\rho f)} = (g^{-1}\mu g)_o$. Equivalently, this is all pairs (ρ, μ) such that $\varphi[\rho]\varphi^{-1} = [\mu]$. □

Example B. Let $X = L \times P \times S$ where L denotes the loop space ΩS^3, P denotes $\mathbf{CP}^\infty$, and S denotes the 3-sphere. Since X is an H-space, there are group isomorphisms,

$$\begin{aligned} WI(X) &\approx \Theta(X,X) && \text{by Roitberg [11], Theorem 3.1} \\ &\approx \varprojlim{}^1[\Sigma X, X^{(n)}] \end{aligned}$$

Here $\Theta(X,X)$ denotes the group of phantom self-maps of X. Notice that

$$\Sigma X \simeq \Sigma(L \vee P \vee S) \vee \Sigma W$$

where ΣW is 4-connected. Of course, for each n, we also have

$$X^{(n)} \simeq L^{(n)} \times P \times S^{(n)}$$

With a bouquet of 4 spaces as a domain, and a product of three spaces for a target, one gets a decomposition of the tower $\{[\Sigma X, X^{(n)}]\}$ as a direct sum of 12 towers. Ten of these are towers of finite groups (this is an easy rational calculation) and consequently their $\varprojlim^1$ terms are zero. The remaining two towers are $\{[\Sigma P, S^{(n)}]\}$ and $\{[\Sigma L, S^{(n)}]\}$. Since ΣL has the homotopy type of a bouquet of spheres, the maps in the last tower are surjections. This forces its $\varprojlim^1$ term to be zero. The other tower, $\{[\Sigma P, S^{(n)}]\}$, is not Mittag-Leffler. Its $\lim^1$ term is shown in [8] to be a rational vector space of rank $2^{\aleph_0}$. Thus we have established

$$WI(X) \approx \Theta(\mathbf{CP}^\infty, S^3) \approx \mathbf{R}$$

We will now describe Y. First, let $\{A, B\}$ be a partition of the set of all prime numbers into two proper subsets, and let Z denote a space that is homotopy equivalent to L at all primes in A and homotopy equivalent to P at all primes in B. In other words, Z is a Zabrodsky mixture of L and P. It can be constructed as a pullback

$$\begin{array}{ccc} & & \mathbf{CP}^\infty_B \\ & & \downarrow \\ \Omega S^3_A & \longrightarrow & K(\mathbf{Q},2) \end{array}$$

wherein the maps are rational equivalences. Let Z' be a second mixture of L and P obtained by reversing the roles of A and B. Then define Y to be $Z \times Z' \times S$. The same argument that was used to calculate $WI(X)$, shows that

$$WI(Y) \approx \Theta(Y,Y) \approx \Theta(Z,S) \oplus \Theta(Z',S)$$

By a $\varprojlim^1$ calculation similar to the one Meier makes in [8], page 480, it follows that

$$\Theta(Z,S) \approx \mathbf{R} \oplus \sum_{p \in A} \mathbf{Z}/p^{\infty}$$

And for Z' in place of Z, the result is almost the same - just replace A with B in the description of the torsion summand. Therefore, since $\mathbf{R} \approx \mathbf{R} \oplus \mathbf{R}$, as rational vector spaces, and the torsion summands add up to $\mathbf{Q}/\mathbf{Z}$, we get

$$WI(Y) \approx \mathbf{R} \oplus \mathbf{Q}/\mathbf{Z},$$

as claimed.

The verification that Y is in $SNT(X)$, is identical to the proof given for Example A. We conclude, as before, that Y is in $G(X) \cap SNT(X)$. Incidently, we refer to such a space as a *clone* of X. In other words, two spaces (or two maps) are said to be clones of each other if they are locally equivalent at each prime and their Postnikov approximations are equivalent. The construction that was used above, $Z \times Z'$, provides, by taking different partitions, uncountably many different clones of $\Omega S^3 \times \mathbf{CP}^{\infty}$. The discovery of nontrivial homotopy clones came as a rude surprise to us, especially after we had expended so much effort trying to rule them out! They are studied in detail in [7].

References

1 J.F. Adams, An example in homotopy theory, Proc. Camb. Phil. Soc. 53, (1957), pp 922-923.

2 A.K. Bousfield and D.M. Kan, Homotopy limits, completions and localizations, Lecture Notes in Math., vol. 304, Springer-Verlag, Berlin Heidelberg New York, 1972.

3 C.U. Jensen, Les Foncteurs Dérivés de $\varprojlim$ et leurs Applications en Théorie des Modules, Lecture Notes in Math., vol. 254, Springer-Verlag, Berlin Heidelberg New York, 1972

4 C.A.McGibbon, K-theory and finite loop spaces of rank one, Math. Proc. Camb. Phil. Soc. (1986), 99, pp. 481 - 487

5 C.A.McGibbon and J.M.Møller, On spaces of the same n-type for all n, to appear in Topology

6 ____________ On infinite dimensional spaces that are rationally equivalent to a bouquet of spheres, to appear in Proc. 1990 Barcelona Conference on Algebraic Topology, Springer LMN

7 ____________ Clones of spaces and maps in homotopy theory, in preparation

8 W.Meier, Pullback theorems and phantom maps, Quart. J. Math. Oxford (2), 29 (1979), pp 469-481

9 G.Mislin, The homotopy classification of self maps of infinite quaternionic projective space, Quart. J. Math. Oxford 38 (1987), pp 245-257

10 D.Notbohm and L. Smith, Fake Lie groups and maximal tori, I,II. and III, Göttingen preprints,1989

11 J. Roitberg, Weak identities, phantom maps, and H-spaces, Israel J. Math. 66, (1989), pp. 319 - 329

12 R.J. Steiner, Localization, completion and infinite complexes, Mathematika, 24, (1977), pp. 1-15.

13 C.W. Wilkerson, Classification of spaces of the same n-type for all n, Proc. AMS. 60 1976, pp. 279-285.

14 ———— Applications of minimal simplical Groups, Topology, Vol. 15 (1976) pp. 111-130.

15 A. Zabrodsky, p-equivalences and homotopy type, Springer LNM. no.418 (1974), pp. 160-171

16 ———— On phantom maps and a theorem of H. Miller, Israel J.Math, Vol. 58, No.2, 1987

C.A.McGibbon
Mathematics Dept.
Wayne State University
Detroit, MI, 48202

J.M.Møller
Matematisk Institut
Universitetsparken 5
DK-2100 København Ø
Danmark

A generalized Grothendieck spectral sequence

David Blanc
Northwestern University

Christopher Stover
University of Chicago

June 5, 1991

Abstract

We construct a generalized Grothendieck spectral sequence for computing the derived functors of a composite functor $T \circ S$, extending the classical version to non-additive functors and non-abelian categories.

1 Introduction

The classical Grothendieck spectral sequence [Gr, Thm 2.4.1] computes the derived functors of a composite functor $T \circ S$ in terms of those of T and S:

That is, suppose $\mathcal{C} \xrightarrow{T} \mathcal{B} \xrightarrow{S} \mathcal{A}$ are additive functors between abelian categories (with enough projectives), and $T(P)$ is S-acyclic for projective $P \in \mathcal{C}$; then for any $C \in \mathcal{C}$ one has a spectral sequence with

$$E^2_{s,t} \cong (L_s S)(L_t T)C \Rightarrow (L_{s+t}(S \circ T))C$$

(see [HS, VIII, Thm 9.3]).

The derived functors $L_\star F$ in question are usually defined as the homology groups of certain chain-complexes; however, they may also be defined as the homotopy groups of suitable simplicial objects (see §2.2.4 below), and this more general definition extends to cases where the functors involved are not additive, or the categories are not abelian.

Now if $\mathcal{B}$ is a category of universal algebras – such as groups, rings, or Lie algebras (§2.1) – then the homotopy groups of a simplicial object $X_\bullet$ over $\mathcal{B}$ support an action of the primary homotopy operations (§3.1.2) associated to $\mathcal{B}$: we then say that $\pi_\star X_\bullet$ is a $\mathcal{B}$-Π-*algebra*. In particular, for any functor $F : \mathcal{C} \to \mathcal{B}$, the derived functors $L_\star F = \{L_0 F, L_1 F, \ldots\}$ together take values in the category of $\mathcal{B}$-Π-algebras. Moreover, if $\mathcal{C}$ is also a category of universal algebras, F induces a functor

$$\bar{F}_* : \mathcal{C}\text{-}\Pi\text{-algebras} \rightarrow \mathcal{B}\text{-}\Pi\text{-algebras}.$$

In this setting we have a generalized Grothendieck spectral sequence:

Theorem 4.4 *Let $\mathcal{C} \xrightarrow{T} \mathcal{B} \xrightarrow{S} \mathcal{A}$ be functors between categories of universal algebras, such that TF is S-acyclic for every free $F \in \mathcal{C}$; then for every $C \in \mathcal{C}$ there is a spectral sequence with*

$$E^2_{s,t} \cong (L_s\bar{S}_t)(L_*T)C \;\Rightarrow\; (L_{s+t}(S \circ T))C.$$

More generally, given a suitable simplicial object $X_\bullet$ over a category $\mathcal{B}$, and a functor $F : \mathcal{B} \rightarrow \mathcal{A}$, one is often interested in determining the homotopy groups of $F(X_\bullet)$ – e.g., in order to calculate derived functors. In particular one may ask how these depend on $\pi_* X_\bullet$. If $\mathcal{B}$ is a category of universal algebras, one has a spectral sequence converging to $\pi_* F X_\bullet$ with E^2-term the derived functors of $\bar{F}_*$ applied to $\pi_* X_\bullet$ (Theorem 4.2).

The generalized Grothendieck spectral sequence is a special case of this; further (degenerate) examples are the Kunneth and Universal Coefficients short exact sequences, and others (§4.2.1).

1.1 notation

For any category $\mathcal{C}$, we denote by $gr\mathcal{C}$ the category of *non-negatively graded objects* over $\mathcal{C}$, and by $s\mathcal{C}$ the category of *simplicial objects* over $\mathcal{C}$. We use the convention that $X_\bullet$ denotes a *simplicial* object (cf. [May, §2]), while $X_* = \{X_i\}_{i=0}^\infty$ denotes a graded object. The category of groups will be denoted $\mathcal{G}p$, that of abelian groups by $\mathcal{A}b\mathcal{G}p$, that of pointed sets by Set_*, and that of pointed simplicial sets by $\mathcal{S}_*$ (rather than $sSet_*$).

We shall assume that all categories are pointed (=have a zero object), and all functors are covariant and pointed (=take zero object to zero object).

1.2 organization

After recalling the needed homotopical algebra in §2, $\mathcal{C}$-Π-algebras and induced functors are defined in §3. The general spectral sequence is set up, and the above theorem is proved, in §4. In §5 we explain why the homotopy group objects over $s\mathcal{C}$ take values in $\mathcal{C}$.

2 universal algebras & homotopical algebra

First, we give some definitions, and remind the reader of some of the (non-abelian) homological algebra needed in our context:

2.1 universal algebras

Recall [McL, I,§7] that a (pointed) *concrete* category $\mathcal{C}$ is one with a faithful functor $U : \mathcal{C} \to Set_*$. In particular, we shall be interested in categories of (possibly graded) *universal algebras* (or varieties of algebras, in the terminology of [McL, V,§6]): that is, categories whose objects, which we shall call simply *algebras*, are (non-negatively graded) sets X ($= \{X_i\}_{i=0}^{\infty}$), together with an action of a fixed set of operators W of the form $\omega : X_{i_1} \times X_{i_2} \times \ldots X_{i_n} \longrightarrow X_k$, satisfying a set of identities E.

For simplicity we assume X (or each X_i) has the underlying structure of a group. (This assumption may be relinquished in certain cases – e.g., for $\mathcal{C} = Set_*$ – but our construction will not work in general for arbitrary universal algebras.)

To avoid confusion with the simplicial dimensions needed later, we shall denote the underlying group in degree i by G_iX. Such a *category of universal graded algebras* will be called a *CUGA*. The ungraded version can be thought of as a CUGA with objects concentrated in degree 0.

If each G_iX is abelian we call X *underlying-abelian*; if this is true of all $X \in \mathcal{C}$, we say the category $\mathcal{C}$ is underlying-abelian (of course, $\mathcal{C}$ need not then be an abelian catgeory).

2.1.1 CUGA's

Examples of CUGA's include *regular graded algebras* in the sense of [B2, §2] (all of which are underlying-abelian), such as:

(i) The category of abelian groups, or more generally of (graded) left R-modules for some ring R.

(ii) The category of graded commutative algebras over a ring k; similarly the category of associative algebras over k.

(iii) $\mathcal{K}_p$, the category of F_p-algebras over the mod-p Steenrod algebra.

(iv) More generally, the category of algebras over E^*E for any ring spectrum E (though these need not be non-negatively graded).

(v) The category of graded Lie rings; similarly the category $\mathcal{L}_p$ of graded restricted Lie algebras over F_p.

CUGA's which are not underlying-abelian include:

(vi) The category of (graded) groups.

(vii) the category of Π-algebras: recall ([B1, §3] or [St, §4]) that a Π-*algebra* is a graded group $X = \{X_i\}_{i=0}^{\infty}$ together with an action of the primary homotopy operations – Whitehead products, compositions, and action of the fundamental group – which satisfies all the universal relations on such operations (cf. [W, XI, §1]). These are modeled on $\pi_* X$, where X is a pointed space.

2.1.2 free algebras and underlying sets

Each CUGA $\mathcal{C}$ is equipped with a pair of adjoint functors $\mathcal{C} \underset{F}{\overset{U}{\rightleftarrows}} grSet_*$ to the category of graded pointed sets, with $U(A)$ the *underlying* graded set of $A \in \mathcal{C}$, and $F(X)$ the *free* algebra on a graded set of *generators* $X = \{X_i\}_{i=1}^{\infty}$ (where in each degree the base point $*$ is identified with the group identity element e). Thus every CUGA $\mathcal{C}$ has enough projectives. It can also be shown to have all limits and colimits, as in [McL, V,§1 & IX,§1].

2.2 homotopical algebra

When $\mathcal{C}$ is a CUGA, Quillen shows in [Q1, II,§4] that $s\mathcal{C}$ can be given a *closed (simplicial) model category* structure as follows:

2.2.1 the model category $s\mathcal{C}$

A map $f : X_\bullet \to Y_\bullet$ in $s\mathcal{C}$ is called a *fibration* if it is a surjection on the basepoint components of the underlying simplicial groups; it is called a *weak equivalence* (w.e.) if it induces an isomorphism on homotopy groups (§3.1.1). A map $i : A_\bullet \to B_\bullet$ is called a *cofibration* if it has the *left lifting property* (LLP) with respect to trivial fibrations – that is, if the dotted arrow exists in the following commutative diagram whenever f is both a fibration and a weak equivalence:

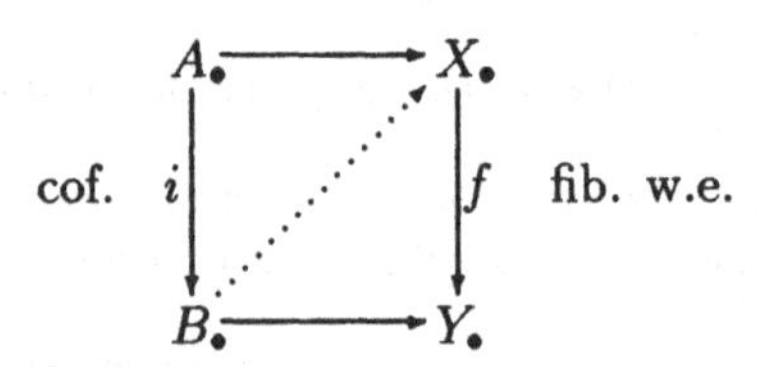

These three classes of maps satisfy certain axioms (cf. [BF, §1]), which allow one to 'do homotopy theory' in $s\mathcal{C}$.

2.2.2 cofibrant and free objects

An object $A_\bullet$ is called *cofibrant* if $* \to A_\bullet$ is a cofibration; the full subcategory of such objects is denoted $s\mathcal{C}_c$. Examples are the *free objects*

$C_\bullet$, where for each $n \geq 0$ there is a (graded) set $T^n \subseteq C_n$ such that $C_n \in \mathcal{C}$ is the free object generated by T^n, and each degeneracy map $s_j : C_n \to C_{n+1}$ takes T^n to T^{n+1}.

2.2.3 resolutions

The *homotopy category* $Ho\mathcal{X}$ of any model category $\mathcal{X}$ is obtained from it by localizing with respect to the weak equivalences, with $\gamma : \mathcal{X} \to Ho\mathcal{X}$ the localization functor. In our case, because all objects in $s\mathcal{C}$ are fibrant, $Ho(s\mathcal{C})$ is equivalent to the category $\pi(s\mathcal{C}_c)$, whose objects are the cofibrant ones of $s\mathcal{C}$, and whose morphisms are homotopy classes of maps (cf. [Q1, I, §1]). Here homotopies between maps in $s\mathcal{C}$ are defined simplicially, as in [Q1, II, §2]; the set of homotopy classes of maps $A_\bullet \to B_\bullet$ is denoted $[A_\bullet, B_\bullet]_{s\mathcal{C}}$.

Under this equivalence of $Ho(s\mathcal{C})$ and $\pi(s\mathcal{C}_c)$, the localization functor is determined by the choice, for each object $X_\bullet \in s\mathcal{C}$, of a cofibrant object $A_\bullet$ with a weak equivalence $A_\bullet \to X_\bullet$. This is called a *resolution* of $X_\bullet$, and all such are homotopy equivalent. We use the embedding $c(-)_\bullet : \mathcal{C} \to s\mathcal{C}$ (which sends $X \in \mathcal{C}$ to the constant simplicial object $c(X)_\bullet$) to define resolutions of objects in $\mathcal{C}$.

2.2.4 derived functors

If $H : \mathcal{X} \to \mathcal{Y}$ is a functor between model categories which preserves weak equivalences between cofibrant objects, the *total left derived functor* of H is the functor $\mathbf{L}H = \tilde{H} \circ \gamma : \mathcal{X} \to Ho(\mathcal{Y})$, where $\tilde{H} : Ho\mathcal{X} \to Ho\mathcal{Y}$ is induced by H on $\mathcal{X}_c$.

Any functor $T : \mathcal{C} \to \mathcal{C}'$ may be *prolonged* to a functor $sT : s\mathcal{C} \to s\mathcal{C}'$, by applying it dimensionwise (by abuse of notation we shall often denote sT simply by T.) In particular, if $T : \mathcal{C} \to \mathcal{C}'$ is a functor from a CUGA into a concrete category, the usual n-th *derived functor* of F, denoted L_nT, assigns to an object $X \in \mathcal{C}$ the object $(L_nT)X = \pi_n((\mathbf{L}sT)c(X)_\bullet) = \pi_n TA_\bullet$, where $A_\bullet \to X$ is any resolution, and $\pi_n(-)$ is defined in §3.1.1 below. The derived functors $L_\star F = \{L_0F, L_1F, \ldots\}$ together take values in the category of $\mathcal{C}'$-Π-algebras (see §3.2).

Note that T need only be given on the full subcategory of free objects of $\mathcal{C}$ in order to define L_nT on $\mathcal{C}$. In this case L_0T is called the *extension* of T to all of $\mathcal{C}$; it agrees with T on the subcategory of free objects, and often agrees with T on all of $\mathcal{C}$ (e.g., if T preserves colimits).

3 homotopy operations over $\mathcal{C}$

Simplicial object $X_\bullet$ over a CUGA $\mathcal{C}$ have *homotopy group objects* $\pi_n X_\bullet \in \mathcal{C}$. These are represented in the homotopy category by certain models:

3.1 models in $s\mathcal{C}$

Let $\mathbf{\Delta}[n]_\bullet$ denote the standard simplicial n-simplex, $\dot{\mathbf{\Delta}}[n]_\bullet = \mathbf{\Delta}[n]_\bullet^{(n-1)}$ its $(n-1)$-skeleton, and $\mathbf{S}^n_\bullet = \mathbf{\Delta}[n]_\bullet/\mathbf{\Delta}[n]_\bullet^{(n-1)}$ the simplicial n-sphere. We denote by $\mathbf{S}^n(k)_\bullet$ the graded simplicial set with $\mathbf{S}^n_\bullet$ in degree k. The simplicial objects $F(\mathbf{S}^n(k)_\bullet) \in s\mathcal{C}$ $(n, k \geq 0)$ should be thought of as the "spheres" of $s\mathcal{C}$. The free simplicial algebras (§2.2.1) which are weakly equivalent to "wedges of spheres" $F(\bigvee_{i \in I} \mathbf{S}^{n_i}(k_i)_\bullet) \cong \coprod_{i \in I} F(\mathbf{S}^{n_i}(k_i)_\bullet)$ will be called *models* for $s\mathcal{C}$; the full subcategory of models in $s\mathcal{C}$ is denoted by $\mathcal{M}$.

3.1.1 homotopy groups

If $\mathcal{C}$ is ungraded, the homotopy groups $\pi_n X_\bullet$ $(n \geq 0)$ of any $X_\bullet \in s\mathcal{C}$ are defined to be those of its underlying simplicial group (§5.2). When $\mathcal{C}$ is graded, $\pi_n X_\bullet$ is the graded group with $G_k(\pi_n X_\bullet) = \pi_n(G_k X_\bullet)$.

The models of $s\mathcal{C}$ then represent the homotopy groups over $s\mathcal{C}$, in the sense that $[F(\mathbf{S}^n(k)_\bullet), X_\bullet]_{s\mathcal{C}} \cong G_k \pi_n X_\bullet$, by the adjointness of U and F (§2.1.2; and compare [Q1, I,§4]).

In fact, $\pi_n X_\bullet$ takes values in $\mathcal{C}$, with the $\mathcal{C}$-structure induced by maps in

$$Hom_{\mathcal{C}}(F(\mathbf{S}^n(k)_n),\ F(\bigvee_{i=1}^{N} \mathbf{S}^n(k_i)_n)\ ) \cong G_k F\{x_1, \ldots x_N\}, \quad \text{where } |x_i| = k_i\ ,$$

which correspond to N-ary operations in W of $\mathcal{C}$. For $n \geq 1$ it turns out that $\pi_n X_\bullet$ is actually an *abelian* object in $\mathcal{C}$ (§5.1.3) – as in the case $\mathcal{C} = \mathcal{G}p$ – so that many of these operations will be trivial (see Lemma 5.2.1).

3.1.2 homotopy operations

In addition, $\pi_* X_\bullet \in gr\mathcal{C}$ has an action of the (primary) $\mathcal{C}$-homotopy operations, which are described as usual by the universal examples – homotopy classes of maps between models. Any such class

$$\alpha \in [F(\mathbf{S}^n(k)_\bullet), F(\bigvee_{i=1}^{N} \mathbf{S}^{n_i}(k_i)_\bullet)]_{s\mathcal{C}} \tag{1}$$

induces a *$\mathcal{C}$-homotopy operation* (natural in $X_\bullet$)

$$\alpha^\# \ :\ \pi_{n_1} G_{k_1}(X_\bullet) \times \pi_{n_2} G_{k_2}(X_\bullet) \times \ldots \pi_{n_N} G_{k_N}(X_\bullet) \longrightarrow \pi_n G_k(X_\bullet)\ .$$

3.2 $\mathcal{C}$-Π-algebras

We define a *$\mathcal{C}$-Π-algebra* A_* to be a sequence $A_0, A_1, \ldots$ of objects in $\mathcal{C}$ (abelian, if $n \geq 1$), together with an action of the $\mathcal{C}$-homotopy operations, subject to the universal relations coming from (1). These form a category, denoted $\mathcal{C}$-Π-Alg, which is itself a CUGA (now *bigraded*, if $\mathcal{C}$ is graded). The *free* $\mathcal{C}$-Π-algebras are those that are isomorphic to $\pi_*(F(\bigvee_i \mathbf{S}^{n_i}(k_i)_\bullet))$ for various n_i, k_i.

Note that the category $s(\mathcal{C}$-Π-$Alg)$ of simplical $\mathcal{C}$-Π-algebras thus has a model category structure, as in §2.2.1.

3.2.1 examples of $\mathcal{C}$-Π-algebras

For some specific CUGA's $\mathcal{C}$, the category of $\mathcal{C}$-Π-algebras has a familiar description:

(I) If the CUGA $\mathcal{C}$ is an abelian category, $s\mathcal{C}$ is equivalent to the category $c\mathcal{C}$ of chain complexes over $\mathcal{C}$ (as in [Do, §1]), with a model $F(\mathbf{S}^n(k)_\bullet)$ in $s\mathcal{C}$ corresponding to a minimal chain complex $\Sigma^n F[x_k]$. Then

$$[F(\mathbf{S}^n(k)_\bullet), F(\bigvee_{i=1}^{N} \mathbf{S}^{n_i}(k_i)_\bullet)]_{s\mathcal{C}} \cong \bigoplus_{n_i=n} G_k F(\mathbf{S}^n(k_i)_\bullet),$$

so there are no homotopy operations in $s\mathcal{C}$, except for the internal ones of $\mathcal{C}$ – and thus the category $\mathcal{C}$-Π-Alg is equivalent to $gr\mathcal{C}$.

(II) If $\mathcal{C} = \mathcal{G}p$, the category of groups, then the homotopy category of $s\mathcal{G}p$ is equivalent to that of connected toplogical spaces (cf. [Kan, §9,11]), so a $\mathcal{G}p$-Π-algebra is just an ordinary Π-algebra (see §2.1.1(vii)), with a shift in dimension.

(III) The homotopy operations for $\mathcal{L}_p$, the category of ungraded Lie algebras over F_p, include a graded Lie bracket which satisfies a Hilton-Milnor theorem (see [Sch], where this is shown to hold over $\mathbf{Z}$). The homotopy groups of the individual models:

$$\pi_* F(\mathbf{S}^n_\bullet) \cong \Lambda(n)$$

are just the "Λ-algebra spheres" $\pi_* LAS_n$ of [6A, 5.4 & 5.4'].

(IV) Similarly, the homotopy operations for commutative algebras over F_2 have been calculated by Bousfield and Dwyer (in [Bo, Dw]): They form a commutative algebra on certain unary "divided power" operations δ_i, of degree i, for each $2 \leq i$, subject to the relations of [Dw, Thm. 2.1].

3.2.2 other categories

We may extend the concept of a $\mathcal{C}$-Π-algebra to categories $\mathcal{C}$ which are not CUGA's, by the following convention:

a) If $\mathcal{C}$ is *any* abelian category, we let $\mathcal{C}\text{-}\Pi\text{-}Alg = gr\mathcal{C}$, as in §3.2.1(I).

b) If $\mathcal{C} = Set_*$, a Set_*-Π-algebra $\{X_i\}_{i=0}^{\infty}$ is an ordinary Π-algebra (§2.1.1(vii)) $\{X_i\}_{i=1}^{\infty}$ in positive degrees, together with a pointed set X_0 in degree 0.

c) If $\mathcal{C}$ is any concrete category which is neither abelian nor a CUGA, we set $\mathcal{C}\text{-}\Pi\text{-}Alg = Set_*\text{-}\Pi\text{-}Alg$, and for each $X_\bullet \in s\mathcal{C}$ we let $\pi_* X_\bullet \in \mathcal{C}\text{-}\Pi\text{-}Alg$ denote $\pi_*(UX_\bullet)$, where $U : \mathcal{C} \to Set_*$ is the faithful "underlying set" functor.

With this convention, for any concrete category $\mathcal{C}$ the functor π_*, and thus the derived functors of any $T : \mathcal{C}' \to \mathcal{C}$, take values in $\mathcal{C}\text{-}\Pi\text{-}Alg$.

3.2.3 Proposition. *Any covariant functor $T : \mathcal{C} \to \mathcal{B}$ from a CUGA $\mathcal{C}$ into a concrete category $\mathcal{B}$ induces a functor $\bar{T}_* : \mathcal{C}\text{-}\Pi\text{-}Alg \to \mathcal{B}\text{-}\Pi\text{-}Alg$, which is the extension* (§2.2.4) *of the functor on free $\mathcal{C}$-Π-algebras defined by $\bar{T}_t(\pi_* A_\bullet) = \pi_t(TA_\bullet)$ for $A_\bullet \in \mathcal{M}$.*

Proof: If $\mathcal{C}$ is a CUGA, it is evident that for any two models $A_\bullet, B_\bullet \in \mathcal{M}$ there is a natural bijection

$$\pi_*(-) : Hom_{Ho(s\mathcal{C})}(A_\bullet, B_\bullet) \to Hom_{\mathcal{C}\text{-}\Pi\text{-}Alg}(\pi_* A_\bullet, \pi_* B_\bullet).$$

In particular, this means that if $\pi_* A_\bullet \cong \pi_* B_\bullet$ *as $\mathcal{C}$-Π-algebras*, then $A_\bullet$ and $B_\bullet$ are actually homotopy equivalent, so $\pi_t(TA_\bullet) \cong \pi_t(TB_\bullet)$. The naturality of the bijection implies that $\bar{T}_t$ is in fact a functor. □

Note that $\bar{T}_*(\pi_* X_\bullet)$ is usually *not* the same as $\pi_* TX_\bullet$ for $X_\bullet \notin \mathcal{M}$ – e.g., for $T = - \otimes \mathbb{Z}/p : \mathcal{A}b\mathcal{G}p \to \mathcal{A}b\mathcal{G}p$.

3.2.4 examples in the abelian case

If $\mathcal{C}$ and $\mathcal{B}$ are abelian categories, then $\mathcal{C}\text{-}\Pi\text{-}Alg$, $\mathcal{B}\text{-}\Pi\text{-}Alg$ are equivalent to $gr\mathcal{C}$, $gr\mathcal{B}$ respectively (§3.2.1(I)), and any $A_\bullet \in \mathcal{M}$ is equivalent to a minimal chain complex $\hat{A}_*$.

a) When T is additive, $\pi_n TA_\bullet \cong T\hat{A}_n = T\pi_n A_\bullet$ and $\bar{T}_*$ is, in each degree, just the 0-th derived functor of T (§2.2.4).

b) When T is not additive, this need not be the case: For example, if $T : \mathcal{A}b\mathcal{G}p \times \mathcal{A}b\mathcal{G}p \to \mathcal{A}b\mathcal{G}p$ is $T(X,Y) = X \otimes Y$, its prolongation to $s\mathcal{A}b\mathcal{G}p \times s\mathcal{A}b\mathcal{G}p$ corresponds (under the equivalence $c\mathcal{A}b\mathcal{G}p \to s\mathcal{A}b\mathcal{G}p$ of §3.2.1(I)) to the chain-complex tensor product, so for any $X_\bullet$, $Y_\bullet$

$$\bar{T}_t(\pi_* X_\bullet, \pi_* Y_\bullet) = \bigoplus_{i+j=t} \pi_i X_\bullet \otimes \pi_j Y_\bullet$$

(compare [Do, §6]).

4 bisimplicial objects

We now consider the category $ss\mathcal{C}$ of *bisimplicial* objects over a CUGA $\mathcal{C}$. We think of an object $A_{\bullet\bullet} \in ss\mathcal{C}$ as having *internal* and *external* simplicial structures, with corresponding homotopy group objects $\pi^i_t A_{\bullet\bullet}$ and $\pi^e_s A_{\bullet\bullet}$ (each in $s\mathcal{C}$). There are two embeddings $c^e, c^i : s\mathcal{C} \hookrightarrow ss\mathcal{C}$, with $c^e(X_\bullet)_{t,s} = X_t$, and $c^i(X_\bullet)_{t,s} = X_s$. We use the convention that for $A_{t,s}$, t is the internal dimension and s is the external one.

The category $ss\mathcal{C}$ can be given a number of different closed model category structures (e.g. [BF, Thm B.6]). We shall need the one defined by Dwyer, Kan and Stover in [DKS]:

4.1 the model category $ss\mathcal{C}$

One can use the models for $s\mathcal{C}$ to provide $ss\mathcal{C}$ with a closed simplicial model category structure, in which a map $f : X_{\bullet\bullet} \to Y_{\bullet\bullet}$ is a weak equivalence if for each $s, t \geq 0$, $f_\star : \pi_s \pi^i_t X_{\bullet\bullet} \to \pi_s \pi^i_t Y_{\bullet\bullet}$ is an isomorphism. Fibrations and cofibrations are defined as in [DKS, §5].

4.1.1 $\mathcal{M}$-free objects

We say that $A_{\bullet\bullet} \in ss\mathcal{C}$ is *$\mathcal{M}$-free* if for each $m \geq 0$ there are (graded) simplicial sets $X[m]_\bullet \simeq \bigvee_i \mathbf{S}^{n_i}(k_i)_\bullet$ such that $A_{\bullet,m} \cong F(X[m]_\bullet)$, and the external degeneracies of $A_{\bullet\bullet}$ are induced under F by maps $X[m]_\bullet \to X[m+1]_\bullet$ which are, up to homotopy, the inclusion of wedge summands. Such an object is cofibrant, and any $X_\bullet \in s\mathcal{C}$ may be resolved (§2.2.3) by such an $\mathcal{M}$-free $A_{\bullet\bullet}$, as follows:

The construction is that of [St, §2]. Let $C\mathbf{S}^n(k)_\bullet$ be the cone on the (graded) simplicial set $\mathbf{S}^n(k)_\bullet$. We obtain the simplicial algebra $W(X_\bullet) \in s\mathcal{C}$ from

$$\coprod_{n,k\geq 0} \Big(\coprod_{f:F(\mathbf{S}^n(k)_\bullet)\to X_\bullet} F(\mathbf{S}^n(k)_\bullet) \amalg \coprod_{H:F(C\mathbf{S}^n(k)_\bullet)\to X_\bullet} F(C\mathbf{S}^n(k)_\bullet) \Big),$$

by identifying the natural sub-object $F(\mathbf{S}^n(k)_\bullet) \overset{i}{\hookrightarrow} F(C\mathbf{S}^n(k)_\bullet)$ of the copy of $F(C\mathbf{S}^n(k)_\bullet)$ indexed by H in the second sum with the copy of $F(\mathbf{S}^n(k)_\bullet)$ indexed by $f = H \circ i$ in the first sum.

This defines a cotriple $W : s\mathcal{C} \to s\mathcal{C}$, with the obvious counit ("evaluation") and comultiplication, and $A_{\bullet\bullet}$ is defined to be the simplicial object over $s\mathcal{C}$ induced by this cotriple (cf. [Go, App., §3]). Clearly each $A_{\bullet,n} = W^{n+1}X_\bullet \in \mathcal{M}$; moreover, the natural augmentation $A_{\bullet\bullet} \to c^e(X_\bullet)_\bullet$ is a weak equivalence in $ss\mathcal{C}$ (see [St, Prop. 2.6]).

4.2 Theorem. *Let $S : \mathcal{B} \to \mathcal{A}$ be a functor from a CUGA to any concrete category; then for any $X_\bullet \in s\mathcal{B}$ there is a first quadrant spectral sequence with*

$$E^2_{s,t} \cong (L_s\bar{S}_t)\pi_*X_\bullet \;\Rightarrow\; (L_{s+t}S)X_\bullet\ .$$

Here $\bar{S}_*$ and $L_s\bar{S}_*$ have values in $\mathcal{A}$-Π-Alg (see §3.2.2 and Proposition 3.2.3). Note that if $X_\bullet$ is cofibrant, $(L_nS)X_\bullet = \pi_n(SX_\bullet)$.

Proof: For any bisimplicial set $Y_{\bullet\bullet}$ there is a spectral sequence with

$$E^2_{s,t} \cong \pi_s\pi^i_tY_{\bullet\bullet} \;\Rightarrow\; \pi_{s+t}\Delta Y_{\bullet\bullet}$$

where the diagonal $\Delta Y_{\bullet\bullet} \in \mathcal{S}_*$ is defined by $(\Delta Y_{\bullet\bullet})_n = Y_{n,n}$, with $d_j = d^e_jd^i_j$, $s_j = s^e_js^i_j$ (cf. [BF, Thm B.5], [Q3]).

Given $X_\bullet \in s\mathcal{B}$, construct an $\mathcal{M}$-free resolution $B_{\bullet\bullet} \to X_\bullet$ as in §4.1.1. Note that $\pi * B_{\bullet\bullet} \to \pi_*X_\bullet$ is a free simplicial resolution in the category of $\mathcal{C}$-Π-algebras, by definition of the weak equivalences for $ss\mathcal{B}$. Therefore, by Proposition 4.2 $\pi^i_tSB_{\bullet\bullet} = \bar{S}_t\pi^i_*B_{\bullet\bullet}$ (since each $B_{\bullet,n} \in \mathcal{M}$), and in the spectral sequence for $SB_{\bullet\bullet} \in ss\mathcal{A}$ we have

$$E^2_{s,t} \cong \pi_s\pi^i_tSB_{\bullet\bullet} \cong \pi_s\bar{S}_t\pi^i_*B_{\bullet\bullet} \cong (L_s\bar{S}_t)\pi_*X_\bullet \tag{2}$$

Since $B_{\bullet\bullet}$ is weakly equivalent to $c^e(X_\bullet)_\bullet$, by definition 4.1 $\pi_s\pi^i_tB_{\bullet\bullet} \cong \pi_s\pi^i_tc^e(X_\bullet)_\bullet$, so the spectral sequence for $B_{\bullet\bullet}$ collapses and $\pi_*(\Delta B_{\bullet\bullet}) \cong \pi_*X_\bullet$. Moreover, by the construction of §4.1.1 $\Delta B_{\bullet\bullet}$ is a free object in $s\mathcal{B}$, so in fact $\Delta B_{\bullet\bullet} \to X_\bullet$ is a resolution in $s\mathcal{C}$, and thus $\Delta SB_{\bullet\bullet} = S\Delta B_{\bullet\bullet} \cong (\mathbf{L}S)X_\bullet$ and (2) converges to $(L_{s+t}S)X_\bullet$. □

4.2.1 examples

(i) If $\mathcal{B} = \mathcal{A} = R\text{-}Mod$, M is an R-module, and $S = Hom_{R-Mod}(-, M)$ or $S = -\otimes M$, the spectral sequence reduces to the Universal Coefficients short exact sequences for (co)homology.

(ii) For $S = \otimes : R\text{-}Mod \times R\text{-}Mod \to R\text{-}Mod$ (§3.2.4) one similarly obtains the Kunneth short exact sequence for homology.

(iii) If $\mathcal{B} = \mathcal{G}p$, $\mathcal{A} = R\text{-}Mod$, and $S = M \otimes_{\mathbb{Z}} Ab(-)$ for some $M \in \mathcal{A}$, one obtains the Hurewicz spectral sequence of [B1, §2]. This generalizes to the homology functor in any CUGA $\mathcal{C}$ (in particular, the André-Quillen homology of a supplemented algebra over a ground ring k – see [Q2, §1]), by taking $S = Ab : \mathcal{C} \to \mathcal{C}_{ab}$ to be the abelianization functor (cf. §5.1.4).

4.3 Theorem. *Let $\mathcal{Y}$ be an arbitrary model category, and $H : \mathcal{Y} \to s\mathcal{B}$ a functor which preserves cofibrancy and weak equivalences between cofibrants. Let $S : \mathcal{B} \to \mathcal{A}$ be a functor from a CUGA to any concrete category. For every $Y \in \mathcal{Y}$ there is a* Grothendieck spectral sequence *with*

$$E^2_{s,t} \cong (L_s\bar{S}_t)(L_*H)Y \;\Rightarrow\; (L_{s+t}(S \circ H))Y.$$

Proof: We may assume $Y \in \mathcal{Y}$ is cofibrant, and let $W_\bullet = HY \in s\mathcal{B}$, so $\pi_t W_\bullet = (L_tH)Y$ and the spectral sequence of Theorem 4.2 has $E^2_{s,t} \cong (L_s\bar{S}_t)(L_*H)Y$. For $B_{\bullet\bullet}$ as in the Theorem, we have $\Delta B_{\bullet\bullet} \simeq W_\bullet = HY$, and since H preserves cofibrancy, $\Delta S B_{\bullet\bullet} \cong S\Delta B_{\bullet\bullet} \simeq SW_\bullet \simeq (S \circ H)Y$ so the spectral sequence converges to $(L_{s+t}(S \circ H))Y$. □

It should be pointed out that we have different assumptions on the three categories in question: $\mathcal{Y}$ may be any model category - e.g., $s\mathcal{C}$, for a wide range of allowable $\mathcal{C}$'s (cf. [Q1, II,§4]; $\mathcal{B}$ must actually be a CUGA, as defined in §2.1; while $\mathcal{A}$ may be any concrete category.

If in fact $\mathcal{Y} = s\mathcal{C}$ for some CUGA $\mathcal{C}$, and H is the prolongation of $T : \mathcal{C} \to \mathcal{B}$, then $H = sT$ will preserve cofibrancy if T takes free objects in $\mathcal{C}$ to free objects in $\mathcal{B}$. However, in this case the requirement may be weakened using the following

Definition: An object $B \in \mathcal{B}$ is called *S-acyclic* if $(L_nS)B = 0$ for $n > 0$, and $(L_0S)B \cong SB$ (with the obvious map).

4.4 Theorem. *Let $\mathcal{C} \xrightarrow{T} \mathcal{B} \xrightarrow{S} \mathcal{A}$ be covariant functors, $\mathcal{C}$, $\mathcal{B}$ CUGA's, and $\mathcal{A}$ any concrete category. Suppose that TF is S-acyclic for every* free *$F \in \mathcal{C}$. Then for every $C \in \mathcal{C}$ there is a Grothendieck spectral sequence with*

$$E^2_{s,t} \cong (L_s\bar{S}_t)(L_*T)C \;\Rightarrow\; (L_{s+t}(S \circ T))C.$$

Proof: In fact we show that the theorem holds for any $C_\bullet \in s\mathcal{C}$ (rather than just $C_\bullet = c(C)_\bullet$). We may assume $C_\bullet \in s\mathcal{C}$ is free; then it suffices to produce an object $W_\bullet$ and a map $f : W_\bullet \to TC_\bullet$ in $s\mathcal{B}$ such that

(a) $W_\bullet$ is cofibrant,

(b) $f : W_\bullet \to TC_\bullet$ is a weak equivalence,

(c) $Sf : SW_\bullet \to (ST)C_\bullet$ is also a weak equivalence,

since then we can proceed as in the proof of Theorem 4.3. Such a $W_\bullet$ may be obtained by a diagonal construction, as follows.

(I) Recall that if $f : X_\bullet \to Y_\bullet$ is a map of simplicial sets such that for each $l \geq 0$, the inclusion $\dot{\Delta}[l]_\bullet \overset{i_l}{\hookrightarrow} \Delta[l]_\bullet$ has the LLP (see §2.2.1) with respect to f, then f is a weak equivalence (in fact, a trivial fibration – cf. [Q1, II,§1]).

(II) Let $\Delta[m]_\bullet \tilde{\times} \Delta[l]_\bullet$ denote the bisimplicial set with $(\Delta[m]_\bullet \tilde{\times} \Delta[l]_\bullet)_{s,t} = \Delta[m]_s \times \Delta[l]_t$, and similarly $(\Delta[m]_\bullet \tilde{\times} \dot{\Delta}[l]_\bullet)_{s,t} = \Delta[m]_s \times \dot{\Delta}[l]_t$. Assume that a map $f : X_{\bullet\bullet} \to Y_{\bullet\bullet}$ of bisimplicial sets has the property that, for all $m, l \geq 0$, the map $i_{m,l} : \Delta[m]_\bullet \tilde{\times} \dot{\Delta}[l]_\bullet \hookrightarrow \Delta[m]_\bullet \tilde{\times} \Delta[l]_\bullet$ has the LLP with respect to f. Then $f_{m,\bullet} : X_{m,\bullet} \to Y_{m,\bullet}$ is a weak equivalence (of simplicial sets) for each $m \geq 0$, by (I).

(III) Now for any $Y_{\bullet\bullet} \in ss\mathcal{B}$ we may use the "small object" construction of [Q1, II,§3] to obtain an object $Z_{\bullet\bullet}$ with a map $f : Z_{\bullet\bullet} \to Y_{\bullet\bullet}$ as follows:

Define $Z_{\bullet\bullet}$ to be the direct limit of a sequence $0 = Z^0_{\bullet\bullet} \hookrightarrow \dots Z^{n-1}_{\bullet\bullet} \hookrightarrow Z^n_{\bullet\bullet} \dots$ of cofibrant objects in $ss\mathcal{B}$ with compatible maps $f^n : Z^n_{\bullet\bullet} \to Y_{\bullet\bullet}$, which are defined inductively by the pushout diagrams

$$\begin{array}{ccc} \coprod_{\mathcal{D}} F(\Delta[m]_\bullet \tilde{\times} \dot{\Delta}[l]_\bullet) & \xrightarrow{\coprod_{\mathcal{D}} g_d} & Z^{n-1}_{\bullet\bullet} \\ {\scriptstyle \coprod_{\mathcal{D}} F(i_{m,l})} \downarrow & \boxed{\text{PO}} & \downarrow \\ \coprod_{\mathcal{D}} F(\Delta[m]_\bullet \tilde{\times} \Delta[l]_\bullet) & \longrightarrow & Z^n_{\bullet\bullet} \end{array}$$

where the coproducts are taken over the set $\mathcal{D} = \mathcal{D}_n$ of all commutative diagrams of the form:

$$\begin{array}{ccc} F(\Delta[m]_\bullet \tilde{\times} \dot{\Delta}[l]_\bullet) & \xrightarrow{g_d} & Z^{n-1}_{\bullet\bullet} \\ {\scriptstyle F(i_{m,l})} \downarrow & & \downarrow {\scriptstyle f^{n-1}} \\ F(\Delta[m]_\bullet \tilde{\times} \Delta[l]_\bullet) & \xrightarrow{h_d} & Y_{\bullet\bullet} \end{array} \qquad (d)$$

for all m, l (in all degrees k, in the graded case).

Using the adjointness of F and U (§2.1.2), one sees that each $i_{m,l}$, for all m, l (and k), has the LLP with respect to the underlying map of $f : Z_{\bullet\bullet} \to Y_{\bullet\bullet}$, so that $Z_{m,\bullet} \overset{w.e.}{\simeq} Y_{m,\bullet}$ for each $m \geq 0$, by (II).

Applying the construction of (III) to $Y_{\bullet\bullet} = c^e(TC_\bullet)_\bullet \in s\mathcal{B}$, we obtain such a $Z_{\bullet\bullet}$. We claim that the diagonal $W_\bullet \overset{Def}{=} \Delta Z_{\bullet\bullet}$ has properties (a)-(c) above:

First, note that $W_\bullet = \Delta Z_{\bullet\bullet}$ is free (and thus cofibrant) in $s\mathcal{B}$, by construction – so it satisfies (a). Next, we have $Z_{m,\bullet} \overset{w.e.}{\simeq} Y_{m,\bullet} = c(TC_m)_\bullet$ for each $m \geq 0$, and thus

$$\pi_s \pi_t^e Z_{\bullet\bullet} \cong \pi_s \pi_t^e Y_{\bullet\bullet} = \begin{cases} \pi_s TC_\bullet & \text{if } t = 0 \\ 0 & \text{if } t > 0 \end{cases},$$

so the Quillen spectral sequence (cf. [Q3]) shows $W_\bullet = \Delta Z_{\bullet\bullet} \overset{w.e.}{\simeq} \Delta Y_{\bullet\bullet} = TC_\bullet$ – i.e., (b) holds.

Moreover, each $Z_{m,\bullet} \in s\mathcal{B}$ is also free, by construction, so it is a resolution of TC_m, and thus (§2.2.4) $\pi_s S Z_{m,\bullet} = (L_s S)(TC_m)$ ($= 0$, for $s > 0$ by the assumption of S-acyclicity). Thus the Quillen spectral sequence for $SZ_{\bullet\bullet} \in ss\mathcal{A}$ shows that $SW_\bullet = S\Delta Z_{\bullet\bullet} \cong \Delta S Z_{\bullet\bullet} \overset{w.e.}{\simeq} STC_\bullet$ – so (c) holds, too. Therefore, $W_\bullet \to TC_\bullet$ has the required properties, which completes the proof of the Theorem. □

4.4.1 remark

When B and $\mathcal{A}$ are abelian and S is additive, then (§3.2.4) $\bar{S}_*$ may be identified with S (applied dimensionwise), and one has the classical Grothendieck spectral sequence of [Gr, Thm. 2.4.1].

Note however that while $\mathcal{C}$ can be any abelian category with enough projectives (in which case we require that TP be S-acyclic for any projective $P \in \mathcal{C}$), $\mathcal{B}$ must actually be a CUGA, since the construction of §4.1.1 depends on the existence of free objects. Of course, if $\mathcal{B}$ is any *abelian* category with enough projectives, the standard construction of the resolution of $TC_\bullet$ in $cc\mathcal{B} \simeq ss\mathcal{B}$ (cf. [HS, VIII, §9]) can be used instead.

5 appendix: homotopy group objects for $s\mathcal{C}$

To illustrate some of the structure of $\mathcal{C}$-Π-algebras, we here show that the homotopy groups of a simplicial object $X_\bullet \in s\mathcal{C}$ actually take value in $\mathcal{C}$, and are abelian in dimensions≥ 1.

5.1 abelian objects in $\mathcal{C}$

In the case of CUGA's, abelian objects have a convenient explicit description:

5.1.1 operations

For any CUGA $\mathcal{C}$, we fix a subset Ω of the semi-groupoid of all operations W, not including the group operation of the G_k's, which (together with these group operations) generates W. Let Ω' denote the subset of W consisting of the n-ary operations $\omega(x, y, \ldots) \in \Omega$ with $n \geq 2$, together with the "operation commutators" $[\omega, i](x_1, \ldots x_{i-1}, x, y, x_{i+1}, \ldots, x_n) \stackrel{Def}{=}$

$$\omega(x_1,...,x_{i-1}, x, x_{i+1},...,x_n)\omega(x_1,...,x_{i-1}, y, x_{i+1},...,x_n)\omega(x_1,...,x_{i-1}, xy, x_{i+1},...,x_n)^{-1}$$

for all $\omega \in \Omega$, $1 \leq i \leq n$.

(For $\omega(x) = x^{-1}$, the inverse operation of the underlying group structure, this is just the group commutator: $[\omega, 1](x, y) = x^{-1}y^{-1}xy$).

We shall assume that the operations in Ω are *pointed* in the sense that they vanish if at least one operand is 0. This is then true of Ω', too.

5.1.2 ideals

In this situation we call a sub-algebra $X \subseteq Y$ an *ideal* of Y if for each $\theta \in \Omega'$ and $y_1, \ldots, y_n \in Y$, we have $\theta(y_1, \ldots, y_n) \in X$ if $y_j \in X$ for some $1 \leq j \leq n$. Thus in particular, $G_kX \lhd G_kY$ for every $k \geq 0$, and the quotient graded group Y/X is an object of $\mathcal{C}$. For example, $Ker(f)$ is an ideal of X for any $f : X \to Y$.

5.1.3 abelian objects

An object A in a CUGA $\mathcal{C}$ is called *abelian* if $Hom_{\mathcal{C}}(X, A)$ has a *natural* abelian group structure for any $X \in \mathcal{C}$. We denote by $\mathcal{C}_{ab}$ the full subcategory of abelian objects in $\mathcal{C}$. Equivalently, A is abelian iff there are in $\mathcal{C}$ "abelian group object structure maps": $\mu : A \times A \to A$ (*group operation*), $\nu : A \to A$ (*inverse*), and $\eta : \star \to A$ (*identity*), fitting into the obvious commutative diagrams.

Since $\mathcal{C}$ is a CUGA, *any* $A \in \mathcal{C}$ has an underlying group structure, which takes the form of "group structure maps" on the underlying (graded) set of A, viz.: $\bar{\mu} : A \times A \to A$, $\bar{\nu} : A \to A$, and $\bar{\eta} : \star \to A$ – again fitting into suitable commutative diagrams.

It is straightforward to verify that if A has "abelian group object structure maps" μ, ν, and η as above, then $\bar{\mu}$, $\bar{\nu}$, and $\bar{\eta}$ must equal μ, ν, and η respectively, as maps of graded sets. Conversely, if the given group structure maps $\bar{\mu}$, $\bar{\nu}$, and $\bar{\eta}$ can be lifted to $\mathcal{C}$ (i.e., are in the image of the faithful functor U of §2.1.2), then the liftings will be "abelian group object structure maps".

It is also readily verified that if the operations Ω' act trivially on an object A in $\mathcal{C}$, then $\bar{\mu}$, $\bar{\nu}$, and $\bar{\eta}$ in fact lift to $\mathcal{C}$, so A is abelian – and conversely, Ω' must act trivially on any abelian $A \in \mathcal{C}$.

5.1.4 abelianization

For any $X \in \mathcal{C}$, let $P(X) \subset X$ be the sub-algebra generated by the image of X under Ω'. This an ideal of X, and the graded group $Ab(X) = X/P(X)$ thus lies in $\mathcal{C}$.

Note that $Ab(X)$ lies in $\mathcal{C}_{ab}$, since Ω' acts trivially on it. In fact, $Ab(X)$ is the *abelianization* of X – i.e., any map from X into $B \in \mathcal{C}_{ab}$ factors uniquely through the natural map $X \to Ab(X)$.

5.1.5 examples

In most of the examples of §2.1.1, $\mathcal{C}_{ab}$ is just a category of graded R-modules, for suitable rings R, with other operations trivial.

When $\mathcal{C} = \mathcal{K}_p$, then $\mathcal{C}_{ab}$ is a suitable category of unstable modules over the mod-p Steenrod algebra $\mathcal{A}_p$, with trivial multiplication: For $p = 2$, $(\mathcal{K}_2)_{ab} = \Sigma\mathcal{U} =$ the category of $\mathcal{A}_2$-modules with $Sq^i x = 0$ for $|x| \leq i$. For $p > 2$, $(\mathcal{K}_p)_{ab} = \mathcal{V} =$ the category of $\mathcal{A}_p$-modules with $\mathcal{P}^i x = 0$ for $|x| \leq 2i$ (cf. [Mi, §1]).

Abelian Π-algebras, and the category $(\Pi\text{-}Alg)_{ab}$, are discussed in [B3].

5.2 homotopy group objects over $\mathcal{C}$

A simplicial object $X_\bullet$ over any concrete category $\mathcal{C}$ has *homotopy groups* $\pi_* X_\bullet$ – namely, those of the underlying simplicial set. In our case, since $X_\bullet$ has the underlying structure of a simplical group, these are defined to be the homology of the (not necessarily abelian) chain complex $\{N_* X_\bullet, \partial\}$ (cf. [May, §17]), where

$$N_n X_\bullet = \bigcap_{1 \leq j \leq n} ker\{d_j : X_n \to X_{n-1}\} \subset X_n, \text{ and } \partial_n = d_0|_{N_n X_\bullet} \text{ for each } n \geq 0.$$

5.2.1 Lemma. *For any CUGA $\mathcal{C}$ and $X_\bullet \in s\mathcal{C}$, $\pi_k X_\bullet \in \mathcal{C}_{ab} \subset \mathcal{C}$ for all $k \geq 1$, and $\pi_0 X_\bullet \in \mathcal{C}$.*

Proof: The image of $d_0 : N_{k+1} X_\bullet \to N_k X_\bullet$ is an ideal (§5.1.2) in $Z_k X_\bullet = Ker(\partial_n)$, since if $x = d_0 x'$ for some $x' \in N_{k+1} X_\bullet$, then for any $\omega \in \Omega$ and $x, y, \ldots \in Z_k X_\bullet$ we have $\omega(x, y, \ldots) = d_0(\omega(x', s_0 y, s_0 \ldots))$ and

$d_i(\omega(x', s_0 y, s_0 \ldots)) = 0$ for $i \geq 1$ (because $\mathcal{C}$ is pointed). Thus $\pi_k X_\bullet \in \mathcal{C}$ for all $k \geq 0$.

Now let $k \geq 1$: for any operation $\omega \in \Omega'$ and $x, y, \ldots \in Z_k X_\bullet$, let $c = \omega(s_0 x, s_0 y, s_0 \ldots) \cdot \omega(s_0 x, s_1 y, s_0 \ldots)^{-1}$; then $d_0 c = \omega(x, y, \ldots)$, and $d_j c = 0$ for $j \geq 1$. Thus any element of $P(Z_k X_\bullet)$ is a boundary in $N_k X_\bullet$, so $\pi_k X_\bullet = Ab(\pi_k X_\bullet)$ is an abelian object in $\mathcal{C}$ (§5.1.3). □

5.2.2 Corollary. *For any functor $T : \mathcal{C}' \to \mathcal{C}$, the 0-th derived functor of T takes values in $\mathcal{C}$, and higher derived functors take values in $\mathcal{C}_{ab}$.*

References

[B1] D. Blanc, "A Hurewicz spectral sequence for homology", *Trans. AMS* **318** (1990) No. 1, pp. 335-354.

[B2] D. Blanc, "Derived functors of graded algebras", *J. Pure Appl. Alg* **64** (1990) No. 3, pp. 239-262.

[B3] D. Blanc, "Abelian Π-algebras and their projective dimension", *Preprint* 1991.

[Bo] A.K. Bousfield, "Operations on Derived Functors of Non-additive Functors", (Brandeis preprint, 1967).

[6A] A.K. Bousfield, E.B. Curtis, D.M. Kan, D.G. Quillen, D.L. Rector, & J.W. Schlesinger, "The mod-p lower central series and the Adams spectral sequence", *Topology* **5** (1966), pp. 331-342.

[BF] A.K. Bousfield & E.M. Friedlander, "Homotopy theory of Γ-spaces, spectra, and bisimplicial sets", in *Geometric Applications of Homotopy Theory, II*, ed. M.G. Barratt & M.E. Mahowald, Springer-Verlag *Lec. Notes Math.* **658**, Berlin-New York 1978, pp. 80-130.

[Do] A. Dold, "Homology of symmetric products and other functors of complexes", *Ann. of Math.* **68** (1958) No. 1, pp. 54-80.

[Dw] W.G. Dwyer, "Homotopy operations for simplicial commutative algebras", *Trans. AMS* **260** (1980) No. 2, pp. 421-435.

[DKS] W.G. Dwyer, D.M. Kan, & C.R. Stover, "An E^2 model category structure for pointed simplicial spaces", *J. Pure & Appl. Alg.*, (to appear).

[Go] R. Godement, *Topologie algébrique et théorie des faisceaux*, Act. Sci. & Ind. No. 1252, Publ. Inst. Math. Univ. Strasbourg **XIII**, Hermann, Paris 1964.

[Gr] A. Grothendieck, "Sur quelques points d'algèbre homologique", *Tôhoku Math. J.* (Ser. 2) **9** (1957) Nos. 2-3, pp. 119-221.

[HS] P.J. Hilton & U. Stammbach, *A Course in Homological Algebra* Springer-Verlag, Berlin-New York 1971.

[Kan] D.M. Kan, "On homotopy theory and c.s.s. groups", *Ann. of Math.*(2) **68** (1958), pp. 38-53.

[McL] S. Mac Lane, *Categories for the working mathematician*, Grad. Texts in Math. No. 5, Springer-Verlag, Berlin-New York 1971.

[May] J.P. May, *Simplicial Objects in Algebraic Topology*, Univ. of Chicago Press, Chicago-London 1967.

[Mi] H. Miller, "Correction to 'The Sullivan conjecture on maps from classifying spaces' " *Ann. of Math.* **121** (1985), pp. 605-609.

[Q1] D.G. Quillen, *Homotopical Algebra*, Springer-Verlag *Lec. Notes Math.* 20, Berlin-New York 1963.

[Q2] D.G. Quillen, "On the (co-)homology of commutative rings", in: *Applications of categorical algebra*, Proc. Symp. Pure Math. **XVII** American Mathematical Society, Providence, RI 1970 pp. 65-87.

[Q3] D.G. Quillen, "Spectral sequences of a double semi-simplicial group", *Topology* **5** (1966), pp. 155-156.

[Sch] J.W. Schlesinger, "The semi-simplicial free Lie ring", *Trans. AMS* **122** (1966) No. 2, pp. 436-442.

[St] C.R. Stover, "A Van Kampen spectral sequence for higher homotopy groups", *Topology*, **29** (1990) No. 1, pp. 9-26.

[W] G.W. Whitehead, *Elements of homotopy theory*, Grad. Texts in Math. No. 61, Springer-Verlag, Berlin-New York 1971.

Localization of the Homotopy Set of the Axes of Pairings

Nobuyuki Oda

Department of Applied Mathematics, Faculty of Science, Fukuoka University, 8-19-1, Nanakuma, Jonanku, Fukuoka, 814-01, Japan

Introduction

Let $f : X \to Z$ and $g : Y \to Z$ be maps. We write $f \perp g$ when there exists a continuous map $\mu : X \times Y \to Z$ with $\mu|X \times \{*\} \simeq f$ and $\mu|\{*\} \times Y \simeq g$. We call such a map $\mu : X \times Y \to Z$ a *pairing* with *axes* f and g [13]. For a fixed map $v : X \to Z$, we define the homotopy set of the axes of pairings [14] by

$$v^{\perp}(Y, Z) = \{ \ [g] : Y \to Z \ \mid \ v \perp g \ \}.$$

The set $v^{\perp}(Y, Z)$ is a generalization of Gottlieb group $G(X)$ and $G_n(X)$ of [3, 4], Varadarajan set $G(Y, Z)$ of [17] and Woo Kim group $G_n^f(X, A, *)$ of [18]. Some fundamental properties of the set $v^{\perp}(Y, Z)$ were studied in [14]. The set $v^{\perp}(Y, Z)$ is a homotopy invariant subset of the homotopy set $[Y, Z]$. Thus we obtain a family of homotopy invariant subsets

$$\{ \ v^{\perp}(Y, Z) \mid v : X \to Z \ \textit{for any space} \ \ X \ \}$$

of $[Y, Z]$.

Some authors applied the localizations of [2, 6, 12, 16] to the study of Gottlieb group and related topics; Mataga [11] characterized Gottlieb space by localizations. Lang [9] obtained fundamental results on the localization of evaluation subgroups $G_n(X)$. Kim [8] generalized the results to the generalized evaluation subgroups of Woo and Kim [18].

The purpose of this paper is to show that the group $v^{\perp}(\Sigma^n Y, Z)$ is natural with respect to localization of space, where $\Sigma^n Y$ is the n-th suspension space of Y and $n \geq 2$.

In §1 we consider spaces of continuous maps and show that $v^{\perp}(Y, Z)$ is a certain *evaluation subset* (Theorem 1.4).

In §2 we consider localizations of $v^{\perp}(\Sigma^n Y, Z)$ for $n \geq 2$. Let $l_p : Z \to Z_p$ denote the p-localization map for any prime number p. If $p = 0$, we denote by $l_p : Z \to Z_p$ the rationalization map $l_Q : Z \to Z_Q$. Then we obtain the following results.

Theorem 2.2. *Let $n \geq 2$ and $p = 0$ or a prime number. Let X be a connected finite CW-complex and Y a finite CW-complex and Z a connected nilpotent space. Let $v : X \to Z$ and $\alpha : \Sigma^n Y \to Z$ be maps. If $(l_p \circ v) \perp (l_p \circ \alpha)$, then there exists a positive integer m such that $v \perp (m\alpha)$, where $(m, p) = 1$ when p is a prime number.*

Theorem 2.4. *Let $n \geq 2$ and $p = 0$ or a prime number. Let X be a connected finite CW-complex and Y a finite CW-complex and Z a connected nilpotent space. Let $v : X \to Z$ be a map. Then*

$$(l_p \circ v)^{\perp}(\Sigma^n Y, Z_p) \cong \{v^{\perp}(\Sigma^n Y, Z)\}_p \,,$$

where the right hand side of the above formula is the p-localization of abelian group.

In §3 we consider some applications of the results of the previous sections. Especially we consider localizations of G-spaces and W-spaces; we characterize them by localization. Finally we prove the following *pull-back theorem.*

Theorem 3.8. (Pull-back Theorem) *Let $n \geq 2$. Let X be a connected finite CW-complex and Y a finite CW-complex and Z a connected nilpotent space of finite type. Let $v : X \to Z$ be a map. Then $v^{\perp}(\Sigma^n Y, Z)$ is the pull-back of the diagram*

$$\{(l_p \circ v)^{\perp}(\Sigma^n Y, Z_p) \to (l_Q \circ v)^{\perp}(\Sigma^n Y, Z_Q) \mid p \text{ is a prime number} \},$$

where $l_Q : Z \to Z_Q$ is the rationalization map.

We work in the category of topological spaces with base point. We denote by $*$ the base point for any space X unless otherwise stated. So the map $f : X \to Y$ is a base point preserving map and the symbol $* : X \to Y$ also denotes the constant map. For two maps $f, g : X \to Y$ the symbol $f \simeq g$ means that f is homotopic to g with base point preserving homotopy. We denote by $[f] : X \to Y$ the homotopy class of a map $f : X \to Y$ and $[X, Y]$ the set of homotopy classes $[f] : X \to Y$.

The map $1_X : X \to X$ is the *identity map* defined by $1_X(x) = x$ for any element $x \in X$. Let $X \vee Y$ denote the *wedge sum* (or *one point union*) of X and Y. We consider $X \vee Y$ as a subspace of the product space $X \times Y$ through the inclusion map $j : X \vee Y \to X \times Y$.

The map $\nabla_X : X \vee X \to X$ denotes the *folding map* defined by $\nabla_X(x, *) = x = \nabla_X(*, x)$ for any element x of X. The map $\Delta_X : X \to X \times X$ denotes the *diagonal map* defined by $\Delta_X(x) = (x, x)$ for any element x of X.

For integers m and n, the symbol (m, n) denotes the *greatest (positive) common divisor.* Let Q_0 denote the field of rational numbers. For a prime number p, we define $Q_p = \{ n/q \mid n \text{ and } q \text{ are integers and } (q,p) = 1 \}$.

1. Evaluation subsets

Let X and Y be spaces with base point. We denote by $f : X \xrightarrow{\circ} Y$ and $f \overset{\circ}{\simeq} g$ base point free map and base point free homotopy respectively. We denote by $X^{\circ}Y$ the space of all *base point free* continuous maps $f : X \xrightarrow{\circ} Y$ with the *compact-open topology.* The space $X^{\circ}Y$ contains a subspace $X^{*}Y$ which consists of *base point preserving* maps $f : X \to Y$. We see that $X^{\circ}Y = (X^{+})^{*}Y$, where $X^{+} = X \cup \{+\}$ (disjoint union of X and a new base point $+$). When $v : X \to Y$ is a base point of $X^{\circ}Y$ or $X^{*}Y$, we write $(X^{\circ}Y, v)$ or $(X^{*}Y, v)$.

Let $A \subset X$ and $B \subset Y$ be subspaces. Then we call a map $f : X \xrightarrow{\circ} Y$ a *map of pairs* $f : (X, A) \xrightarrow{\circ} (Y, B)$ when $f(A) \subset B$. We define

$$(X,A)^{\circ}(Y,B) = \{f : (X,A) \xrightarrow{\circ} (Y,B)\},$$

$$\begin{aligned}(X,A)^{*}(Y,B) &= \{(X,A)^{\circ}(Y,B)\} \cap \{X^{*}Y\} \\ &= \{f : (X,A) \to (Y,B)\}.\end{aligned}$$

Let us consider the adjoint map

$$\tau : (X \times Y)^{\circ}Z \to Y^{\circ}(X^{\circ}Z) \tag{1.1}$$

which is defined by $\{\tau(F)(y)\}(x) = F(x,y)$ for any elements $F \in (X \times Y)^{\circ}Z$, $x \in X$ and $y \in Y$.

If X is a locally compact space, then the adjoint map τ in (1.1) is a bijection of sets, and moreover, $F \overset{\circ}{\simeq} F' : X \times Y \xrightarrow{\circ} Z$ if and only if $\tau(F) \overset{\circ}{\simeq} \tau(F') : Y \xrightarrow{\circ} X^{\circ}Z$.

Proposition 1.2. *Let X be a locally compact space. Then the following results hold through the bijection of* (1.1).

(1) $(X \times Y)^{*}Z \cong (Y, *)^{\circ}(X^{\circ}Z, X^{*}Z)$ *(bijection as sets).*

(2) $\mu \simeq \mu' : X \times Y \to Z$ *if and only if*
$\tau(\mu) \overset{\circ}{\simeq} \tau(\mu') : (Y, *) \xrightarrow{\circ} (X^{\circ}Z, X^{*}Z)$

Proof. (1) Let $\mu : X \times Y \to Z$ be an element of $(X \times Y)^{*}Z$. Then it defines the adjoint $\tau(\mu) : Y \xrightarrow{\circ} X^{\circ}Z$ such that $\tau(\mu)(*) \in X^{*}Z$. Conversely any such map $h : (Y, *) \xrightarrow{\circ} (X^{\circ}Z, X^{*}Z)$ defines the adjoint $\tau^{-1}(h) : X \times Y \to Z$. (2) is proved similarly. *q.e.d.*

We call a map $\mu : X \times Y \to Z$ a *pairing* [13] with the *axes* $f : X \to Z$ and $g : Y \to Z$, when it satisfies

$$\mu \mid X \vee Y \simeq \nabla_Z \circ (f \vee g) : X \vee Y \to Z.$$

We write $f \perp g$ when there exists a pairing $\mu : X \times Y \to Z$ with the axes $f : X \to Z$ and $g : Y \to Z$.

Let $v : X \to Z$ be a map. We call a map $g : Y \to Z$ *v-cyclic* if $v \perp g$. Then we define the following set of the homotopy classes of the v-cyclic maps $g : Y \to Z$ [14];

$$v^{\perp}(Y, Z) = \{ [g] : Y \to Z \mid v \perp g \} \subset [Y, Z].$$

Examples 1.3. (1) $(1_X)^{\perp}(S^1, X) = G(X)$ (Gottlieb [3]).
(2) $(1_X)^{\perp}(S^n, X) = G_n(X)$ (Gottlieb [4]).
(3) $(1_X)^{\perp}(Y, X) = G(Y, X)$ (Varadarajan [17]).
(4) Let $f : A \to X$ be a map. Then $f^{\perp}(S^n, X) = G_n^f(X, A, *)$ (Woo and Kim [18]).

The *evaluation map*

$$\omega : X^{\circ}Z \overset{\circ}{\to} Z$$

is defined by $\omega(f) = f(*)$ for any map $f : X \overset{\circ}{\to} Z$ of $X^{\circ}Z$.

Let $v : X \to Z$ be a map. Then the evaluation map $\omega : (X^{\circ}Z, v) \to Z$ induces a map

$$\omega_* : [Y, (X^{\circ}Z, v)] \to [Y, Z]$$

in homotopy sets.

The following result is a generalization of Theorem III.1 of Gottlieb [3], Proposition 1.1 of Gottlieb [4], Theorem 3.2 of Lim [10] and Theorem 2.1 of Woo and Kim [18]. In the following theorem we assume that a source space has *nondegenerate* base point, namely, $(Y, *)$ is an NDR (*neighborhood deformation retract*) pair, so that we can use *homotopy extension property.*

Theorem 1.4. *Let X be a locally compact space and $v : X \to Z$ a map. If $(Y, *)$ is an* NDR *pair, then the following relation holds.*

$$\omega_*[Y, (X^{\circ}Z, v)] = v^{\perp}(Y, Z).$$

Proof. We consider the adjoint map τ in (1.1). If $[h] \in [Y, (X^{\circ}Z, v)]$, then $\tau^{-1}(h) : X \times Y \to Z$ is a pairing with axes $v : X \to Z$ and some map

$g : Y \to Z$ such that $\omega_*(h) = g$ by Proposition 1.2(1). So we have $\omega_*([h]) = [g] \in v^{\perp}(Y, Z)$.

Conversely, for any element $[g] \in v^{\perp}(Y, Z)$, we have a pairing $\mu : X \times Y \to Z$ with the axes v and g, namely, $\mu|X \times \{*\} \simeq v$ and $\mu|\{*\} \times Y \simeq g$. Set $\mu|X \times \{*\} = \bar{v}$ and $\mu|\{*\} \times Y = \bar{g}$. Then we have a map

$$\tau(\mu) : Y \to (X^{\circ}Z, \bar{v}).$$

Since $\bar{v} \simeq v$ and $(Y, *)$ is an NDR pair and hence has homotopy extension property, we have a map $h : Y \to (X^{\circ}Z, v)$ such that

$$\tau(\mu) \overset{\circ}{\simeq} h : (Y, *) \overset{\circ}{\to} (X^{\circ}Z, X^{*}Z).$$

Then through the adjoint map τ in (1.1), we have a pairing $\mu' : X \times Y \to Z$ such that

$$\tau(\mu') = h \text{ and } \mu \simeq \mu' : X \times Y \to Z$$

by Proposition 1.2(2). We see that $\mu'|X \times \{*\} = v$ as maps and $\mu'|\{*\} \times Y \simeq \bar{g} \simeq g$. Then we have $\omega_*([h]) = [\bar{g}] = [g]$. *q.e.d.*

We call a map $\theta : A \to B \vee C$ a *copairing* [13] with the *coaxes* $h : A \to B$ and $r : A \to C$ if it satisfies the condition that

$$j \circ \theta \simeq (h \times r) \circ \Delta_A \ : \ A \to B \times C$$

for the inclusion map $j : B \vee C \to B \times C$.

Given a copairing $\theta : A \to B \vee C$, we define a map $\alpha \dot{+} \beta : A \to X$ for any maps $\alpha : B \to X$ and $\beta : C \to X$ by

$$\alpha \dot{+} \beta = \nabla_X \circ (\alpha \vee \beta) \circ \theta .$$

This defines a pairing $\dot{+} : [B, X] \times [C, X] \to [A, X]$ of sets.

We have the following results (Theorems 1.4, 1.5 and 2.5 of [14]).

Theorem 1.5. (1) *Let* $f_1 : X_1 \to Z$, $f_2 : X_2 \to X_1$, $g_1 : Y_1 \to Z$ *and* $g_2 : Y_2 \to Y_1$ *be maps. Then* $f_1 \perp g_1$ *implies* $(f_1 \circ f_2) \perp (g_1 \circ g_2)$.
(2) *Let* $f : X \to Z$, $g : Y \to Z$ *and* $w : Z \to W$ *be maps. Then* $f \perp g$ *implies* $(w \circ f) \perp (w \circ g)$.

Theorem 1.6. *Let* $f : X \to Z$, $g_1 : Y_1 \to Z$ *and* $g_2 : Y_2 \to Z$ *be maps from CW-complexes. Then the following results hold.*
(1) $f \perp g_1$ *and* $f \perp g_2$ *implies* $f \perp \{\nabla_Z \circ (g_1 \vee g_2)\}$, *where* $g_1 \vee g_2 : Y_1 \vee Y_2 \to Z \vee Z$.
(2) *Let* $\theta : Y \to Y_1 \vee Y_2$ *be a copairing. Then* $f \perp g_1$ *and* $f \perp g_2$ *implies*

$f \perp (g_1 \dot{+} g_2)$.

By Theorems 1.5 and 1.6, we see that the set $v^{\perp}(\Sigma^n Y, Z)$ has a group structure for all $n \geq 1$. Let $f : \Sigma^{n-1}A \to \Sigma^{n-1}Y$ and $\omega : Z \to W$ be maps. Then we see that the induced maps

$$(\Sigma f)^* : v^{\perp}(\Sigma^n Y, Z) \to v^{\perp}(\Sigma^n A, Z), \quad \text{and}$$

$$\omega_* : v^{\perp}(\Sigma^n Y, Z) \to (\omega \circ v)^{\perp}(\Sigma^n Y, W)$$

are homomorphisms of groups for all $n \geq 1$ by Theorems 1.5 and 1.6.

2. Localizations

We now consider localizations of the homotopy set of the axes of pairings $v^{\perp}(\Sigma^n Y, Z)$ for $n \geq 2$. Let $p = 0$ or a prime number. We use the symbol

$$l_p : Z \to Z_p$$

for the p-localization map of the space Z for any prime number p. If $p = 0$, then $l_p : Z \to Z_p$ denotes the rationalization map $l_Q : Z \to Z_Q$. The space Z is assumed to be a connected nilpotent space for the existence of p-localization [2, 6, 16]. Let $v : X \to Z$ be a map. Then the map

$$l_{p*} : (X^\circ Z, v) \to (X^\circ(Z_p), l_p \circ v)$$

is defined by $l_{p*}(f) = l_p \circ f$ for any map $f \in X^\circ Z$. We now show that this map l_{p*} is a p-localization map.

Proposition 2.1. *Let $n \geq 2$ and $p = 0$ or a prime number. Let X be a connected finite CW-complex and Y a finite CW-complex and Z a connected nilpotent space. Then*

$$\psi_p \ : \ [\Sigma^n Y, (X^\circ Z, v)] \otimes Q_p \to [\Sigma^n Y, (X^\circ(Z_p), l_p \circ v)]$$

is an isomorphism, where ψ_p is defined by $\psi_p(\alpha \otimes r) = r l_{p*}(\alpha)$ for any elements $\alpha \in [\Sigma^n Y, (X^\circ Z, v)]$ and $r \in Q_p$.

Proof. Consider the following diagram.

$$\begin{array}{ccccc} [\Sigma^n Y, (X^* Z, v)] & \xrightarrow{i_*} & [\Sigma^n Y, (X^\circ Z, v)] & \xrightarrow{\omega_*} & [\Sigma^n Y, Z] \\ \downarrow l''_{p*} & & \downarrow l_{p*} & & \downarrow l'_{p*} \\ [\Sigma^n Y, (X^*(Z_p), \bar{v})] & \xrightarrow{i'_*} & [\Sigma^n Y, (X^\circ(Z_p), \bar{v})] & \xrightarrow{\omega'_*} & [\Sigma^n Y, Z_p] \end{array}$$

where the base points of $X^*(Z_p)$ and $X^\circ(Z_p)$ are $\bar{v} = l_p \circ v$. The horizontal sequences can be extended to long homotopy exact sequences. We see that the maps l''_{p*} and l'_{p*} are p-localizations by Theorem 3.11 (p.77) of [6], Proposition 5.1 (p.141) of [2] and Proposition 5.3 (p.142) of [2]. The lower long homotopy exact sequence defines the Q_p-module structure of $[\Sigma^n Y, (X^\circ(Z_p), \bar{v})]$. Then we have the result by the long homotopy exact sequences and the 5-lemma.

Theorem 2.2. *Let $n \geq 2$ and $p = 0$ or a prime number. Let X be a connected finite CW-complex and Y a finite CW-complex and Z a connected nilpotent space. Let $v : X \to Z$ and $\alpha : \Sigma^n Y \to Z$ be maps. If $(l_p \circ v) \perp (l_p \circ \alpha)$, then there exists a positive integer m such that $v \perp (m\alpha)$, where $(m, p) = 1$ when p is a prime number.*

Proof. (We follow the method of the proof of Theorem 2.3 of Lang [9].) Consider the following commutative diagram.

$$\begin{array}{ccc} [\Sigma^n Y, (X^\circ Z, v)] \otimes Q_p & \xrightarrow{\omega_* \otimes 1} & [\Sigma^n Y, Z] \otimes Q_p \\ \psi_{p*} \downarrow \cong & & \cong \downarrow \phi_{p*} \\ [\Sigma^n Y, (X^\circ(Z_p), l_p \circ v)] & \xrightarrow{\omega'_*} & [\Sigma^n Y, Z_p] \end{array}$$

The left vertical map is an isomorphism by Proposition 2.1. The right vertical map is given by

$$\phi_{p*}(\alpha \otimes r) = r l_{p*}(\alpha) = r(l_p \circ \alpha)$$

for any elements $\alpha \in [\Sigma^n Y, Z]$ and $r \in Q_p$. This map is also an isomorphism by Theorem 3.11 (p.77) of [6] or Proposition 5.3 (p.142) of [2].

Let us suppose that $(l_p \circ v) \perp (l_p \circ \alpha)$. Then there exists an element δ of $[\Sigma^n Y, (X^\circ(Z_p), l_p \circ v)]$ such that $\omega'_*(\delta) = l_p \circ \alpha$ by Theorem 1.4. Since ψ_{p*} is an isomorphism by Proposition 2.1, we have an element $\Sigma a_i \otimes (k_i/q_i)$ of $[\Sigma^n Y, (X^\circ Z, v)] \otimes Q_p$ such that $\psi_{p*}(\Sigma\{a_i \otimes (k_i/q_i)\}) = \delta$, where $(q_i, p) = 1$ for all i when p is a prime number. Now, set $q = \Pi q_i$ and $\bar{q}_i = q/q_i$, then we have $\Sigma(a_i \otimes (k_i/q_i)) = (\Sigma \bar{q}_i k_i a_i) \otimes (1/q)$.

Thus there exists an element $\beta (= \Sigma \bar{q}_i k_i a_i)$ of $[\Sigma^n Y, (X^\circ Z, v)]$ and an integer q such that $\psi_{p*}(\beta \otimes (1/q)) = \delta$, where $(q, p) = 1$ when p is a prime number. Now consider the homomorphism

$$\omega_* : [\Sigma^n Y, (X^\circ Z, v)] \to [\Sigma^n Y, Z].$$

We see that $\mathrm{Im}(\omega_*) = v^\perp(\Sigma^n Y, Z)$ by Theorem 1.4. We have

$$\phi_{p*}(\omega_*(\beta) \otimes 1) = \omega'_*(\psi_{p*}(\beta \otimes 1)) = \omega'_*(q\delta) = q(l_p \circ \alpha) = \phi_{p*}(q\alpha \otimes 1),$$

and hence, setting $\omega_*(\beta) - q\alpha = \gamma$, we see that the order of γ is finite, say, s and $(s, p) = 1$ when p is a prime number. Then we have $\omega_*(s\beta) - sq\alpha = s\gamma = 0$

or $sq\alpha = \omega_*(s\beta) \in v^\perp(\Sigma^n Y, Z)$, where $(sq, p) = 1$ when p is a prime number. Set $m = sq$, then we have shown that $v \perp (m\alpha)$ by Theorem 1.4. We can choose $m > 0$, since $v \perp (m\alpha)$ also implies $v \perp (-m\alpha)$ by Theorem 1.5(1).

q.e.d.

The following result is a generalization of Theorem 2.3 of Lang [9] and Theorem 2.1 of Kim [8].

Theorem 2.3. *Let $n \geq 2$. Let X be a connected finite CW-complex and Y a finite CW-complex and Z a connceted nilpotent space. Let $v : X \to Z$ and $\alpha : \Sigma^n Y \to Z$ be maps. Then $v \perp \alpha$ if and only if $(l_p \circ v) \perp (l_p \circ \alpha)$ for every prime number p.*

Proof. Necessity is a consequence of Theorem 1.5(2). We now prove sufficiency.

Localizing at 2, we see that $v \perp (q_0 \alpha)$ for a positive integer q_0 such that $(q_0, 2) = 1$ by Theorem 2.2. If $q_0 \neq 1$, then let p be a prime number which divides q_0. Then localizing at p, we have a positive integer s_0 such that $(s_0, p) = 1$ and $v \perp (s_0 \alpha)$. Let $(q_0, s_0) = q_1$. Then there exist integers r_0 and t_0 such that $q_0 r_0 + s_0 t_0 = q_1$. Then $v \perp q_0 \alpha$ and $v \perp s_0 \alpha$ implies $v \perp (q_0 r_0 \alpha + s_0 t_0 \alpha)$ by Theorem 1.6(2) or $v \perp (q_1 \alpha)$. Then by induction we have a sequence of positive integers $q_0 > q_1 > q_2 > \cdots$ such that $v \perp (q_i \alpha)$ for all i until we get $q_k = 1$ for some k. This completes the proof.

The following result is a generalization of Corollary 2.5 of Lang [9] and Corollary 2.3 of Kim [8].

Theorem 2.4. *Let $n \geq 2$ and $p = 0$ or a prime number. Let X be a connected finite CW-complex and Y a finite CW-complex and Z a connected nilpotent space. Let $v : X \to Z$ be a map. Then*

$$(l_p \circ v)^\perp(\Sigma^n Y, Z_p) \cong \{v^\perp(\Sigma^n Y, Z)\}_p ,$$

where the right hand side of the above formula is the p-localization of abelian group.

Proof. (We follow the method of the proof of Corollary 2.5 of Lang [9]). Consider the isomorphism of localization

$$\phi_p : [\Sigma^n Y, Z] \otimes Q_p \to [\Sigma^n Y, Z_p].$$

We show that the restriction map

$$\phi_p \ : \ v^{\perp}(\Sigma^n Y, Z) \otimes Q_p \to (l_p \circ v)^{\perp}(\Sigma^n Y, Z_p) \tag{2.5}$$

is well-defined and bijective.

Let $[g] \in v^{\perp}(\Sigma^n Y, Z)$ be any element. Then there exists a pairing $\mu : X \times \Sigma^n Y \to Z$ for $v \perp g$. Let $h = \tau(\mu) : \Sigma^n Y \to (X^\circ Z, v)$ be the adjoint of μ. Since the induced map $l_{p*} : (X^\circ Z, v) \to (X^\circ(Z_p), l_p \circ v)$ is the p-localization map by Proposition 2.1, the composition map

$$l_{p*} \circ h : \Sigma^n Y \to (X^\circ Z, v) \to (X^\circ(Z_p), l_{p*} \circ v)$$

can be factored through the localization map $l_Y : \Sigma^n Y \to \Sigma^n Y_p$, namely, $l_{p*} \circ h = h_p \circ l_Y$ for a map $h_p : \Sigma^n Y_p \to (X^\circ(Z_p), l_p \circ v)$. If $r \in Q_p$, then $\tau^{-1}(h_p \circ r l_Y) : X \times \Sigma^n Y \to Z$ is a pairing for $(l_p \circ v) \perp \{r(l_p \circ g)\}$. This shows that the map ϕ_p in (2.5) is well-defined.

Let α be any element of $(l_p \circ v)^{\perp}(\Sigma^n Y, Z_p)$. We remark that any element of $[\Sigma^n Y, Z] \otimes Q_p$ can be written as $\beta \otimes (1/q)$ for an element $\beta \in [\Sigma^n Y, Z]$ and an integer q such that $(q, p) = 1$ when p is a prime number (cf. Proof of Theorem 2.2 or Corollary 2.5 of [9]). Suppose that $\phi_p(\beta \otimes (1/q)) = \alpha$ for an element $\beta \in [\Sigma^n Y, Z]$ and an integer q such that $(q, p) = 1$ when p is a prime number. We see $l_{p*}(\beta) = q\alpha \in (l_p \circ v)^{\perp}(\Sigma^n Y, Z_p)$ and hence by Theorem 2.2, there exists an integer s such that $s\beta \in v^{\perp}(\Sigma^n Y, Z)$ and $(s, p) = 1$ when p is a prime number. Then $s\beta \otimes 1/(sq)$ is an element of $v^{\perp}(\Sigma^n Y, Z) \otimes Q_p$ and we have

$$\phi_p\{s\beta \otimes 1/(sq)\} = \phi_p(\beta \otimes (1/q)) = \alpha.$$

This completes the proof.

3. Some applications

We denote by $f_p : A_p \to Z_p$ the p-localization of a map $f : A \to Z$ and $l_A : A \to A_p$ the p-localization map. Then the following diagram is homotopy commutative by Proposition 3.5 (p.134) of [2] or Theorem 3.A of [6].

$$\begin{array}{ccc} A & \xrightarrow{f} & Z \\ {\scriptstyle l_A} \downarrow & & \downarrow {\scriptstyle l_p} \\ A_p & \xrightarrow{f_p} & Z_p \end{array}$$

Proposition 3.1. *Let $n \geq 2$. Let X be a connected nilpotent finite CW-complex and Y a finite CW-complex and Z a connected nilpotent space. Let $v : X \to Z$ and $\alpha : \Sigma^n Y \to Z$ be maps. Then $v \perp \alpha$ if and only if $v_p \perp \alpha_p$ for* *every prime number p.*

Proof. Suppose that $v \perp \alpha$. Then there exists a pairing $\mu : X \times \Sigma^n Y \to Z$ with axes $v : X \to Z$ and $\alpha : \Sigma^n Y \to Z$. Then the localization $\mu_p : X_p \times \Sigma^n Y_p \to Z_p$ is a pairing for $v_p \perp \alpha_p$.

Conversely, suppose that $v_p \perp \alpha_p$ for every prime number p. Let $l_X : X \to X_p$ and $l_Y : \Sigma^n Y \to \Sigma^n Y_p$ be p-localization maps. Then we have $(v_p \circ l_X) \perp (\alpha_p \circ l_Y)$ by Theorem 1.5(1). Since $v_p \circ l_X = l_p \circ v$ and $\alpha_p \circ l_Y = l_p \circ \alpha$ by naturality of localization, we have $(l_p \circ v) \perp (l_p \circ \alpha)$ for every prime number p. Then by Theorem 2.3 we have $v \perp \alpha$. *q.e.d.*

We call a space X a *Gottlieb space* (*G-space*) when $G_m(X) = \pi_m(X)$ for all m (Gottlieb [4], cf. Examples 1.3(2)).

We call a space X a *Varadarajan space* when $G(\Sigma Y, X) = [\Sigma Y, X]$ for all spaces Y (cf. Examples 1.3(3)).

We call a space X a *Whitehead space* (*W-space*) when all Whitehead products vanish, namely, $[\alpha, \beta] = 0$ for all $\alpha \in \pi_n(X)$, $\beta \in \pi_m(X)$ and all integers $m, n \geq 1$. (Siegel [15]).

We call a space X a *Generalized Whitehead space* (*GW-space*) when all generalized Whitehead products [1] vanish, namely, $[\alpha, \beta] = 0$ for all $\alpha \in [\Sigma A, X]$ and $\beta \in [\Sigma B, X]$ and all spaces A and B (Iwase and Mimura [7]).

Let $S \subset [X, Z]$ and $T \subset [Y, Z]$ be subsets of the homotopy sets. Then we write $S \perp T$ when $s \perp t$ for all $s \in S$ and $t \in T$.

Proposition 3.2. (1) *A space X is a Gottlieb space if and only if $1_X \perp \pi_n(X)$ for all $n \geq 1$.*
(2) *A space X is a Varadarajan space if and only if $1_X \perp [\Sigma A, X]$ for all spaces A.*
(3) *A space X is a Whitehead space if and only if $\pi_n(X) \perp \pi_m(X)$ for all n and $m \geq 1$.*
(4) *A space X is a Generalized Whitehead space if and only if $[\Sigma A, X] \perp [\Sigma B, X]$ for all spaces A and B.*

Proof. (1) and (2) are direct consequences of the definitions. (3) and (4) are obtained by Proposition 5.1 of Arkowitz [1]. *q.e.d.*

Proposition 3.3. *Let $n \geq 2$ and $p = 0$ or a prime number. Let X be a connected nilpotent finite CW-complex and Y a finite CW-complex and Z a connected nilpotent space. Let $v : X \to Z$ be a map. Then the following equality holds.*

$$(v_p)^{\perp}(\Sigma^n Y, Z_p) = (l_p \circ v)^{\perp}(\Sigma^n Y, Z_p).$$

Proof. Let $l_X : X \to X_p$ and $l_Y : \Sigma^n Y \to \Sigma^n Y_p$ be the p-localization maps. Let $\alpha \in (v_p)^{\perp}(\Sigma^n Y, Z_p)$ be any element. Then $v_p \perp \alpha$ and hence $(v_p \circ l_X)\perp\alpha$ by Theorem 1.5(1). Since $v_p \circ l_X = l_p \circ v$, we have $(l_p \circ v)\perp\alpha$ or $\alpha \in (l_p \circ v)^{\perp}(\Sigma^n Y, Z_p)$.

Conversely, for any element $\alpha \in (l_p \circ v)^{\perp}(\Sigma^n Y, Z_p)$, we have $(l_p \circ v)\perp\alpha$ and hence $(l_p \circ v)_p \perp \alpha_p$ (cf. Proof of Proposition 3.1) or $v_p \perp \alpha_p$. Then we have $v_p \perp (\alpha_p \circ l_Y)$ by Theorem 1.5(1). Since $\alpha = \alpha_p \circ l_Y$, we have $v_p \perp \alpha$ and hence $\alpha \in (v_p)^{\perp}(\Sigma^n Y, Z_p)$.

Theorem 3.4. *Let X be a 1-connected finite CW-complex. Then the following results hold.*
(1) X is a Gottlieb space if and only if X_p is a Gottlieb space for every prime number p.
(2) X is a Whitehead space if and only if X_p is a Whitehead space for every prime number p.

Proof. (1) We have to show that $1_X \perp \pi_n(X)$ if and only if $1_{X_p} \perp \pi_n(X_p)$ for every prime number p.

Let $l_p : X \to X_p$ be the p-localization map. We remark that

$$(1_{X_p})^{\perp}(S^n, X_p) = (l_p)^{\perp}(S^n, X_p)$$

by Proposition 3.3 (Set $v = 1_X$). Now, let us suppose that $1_X \perp \pi_n(X)$, namely, $(1_X)^{\perp}(S^n, X) = \pi_n(X)$. Then by Theorem 2.4, we have

$$(1_{X_p})^{\perp}(S^n, X_p) = (l_p)^{\perp}(S^n, X_p) = \{(1_X)^{\perp}(S^n, X)\}_p = \{\pi_n(X)\}_p = \pi_n(X_p).$$

Conversely, suppose that $1_{X_p} \perp \pi_n(X_p)$ for every prime number p and all $n \geq 2$. Let $\alpha \in \pi_n(X)$ be any element and $n \geq 2$. Then $1_{X_p} \perp (l_p \circ \alpha)$ for every prime number p by our assumption. It follows that $(1_{X_p} \circ l_p)\perp(l_p \circ \alpha)$ by Theorem 1.5(1), and hence $(l_p \circ 1_X)\perp(l_p \circ \alpha)$ for every prime number p. Then we have $1_X \perp \alpha$ by Theorem 2.3.
(2) We show that $\pi_n(X)\perp\pi_m(X)$ if and only if $\pi_n(X_p)\perp\pi_m(X_p)$ for every prime number p.

First suppose that $\pi_n(X_p)\perp\pi_m(X_p)$ for every prime number p. Let $\alpha \in \pi_n(X)$ and $\beta \in \pi_m(X)$ be any elements. Then $(l_p \circ \alpha)\perp(l_p \circ \beta)$ by our assumption. Then we have $\alpha\perp\beta$ by Theorem 2.3. It follows then that $\pi_n(X)\perp\pi_m(X)$.

Conversely, suppose that $\pi_n(X)\perp\pi_m(X)$. Let $\gamma \in \pi_n(X_p)$ and $\delta \in \pi_m(X_p)$ be any elements. Then we can write $\gamma = \alpha_p \circ q$ and $\delta = \beta_p \circ s$ for some $q \in \pi_n(S^n_p)$, $s \in \pi_m(S^m_p)$, and $\alpha \in \pi_n(X)$, $\beta \in \pi_m(X)$. Since $\alpha\perp\beta$ by our assumption, we have $\alpha_p \perp \beta_p$ by Proposition 3.1 and hence $(\alpha_p \circ q)\perp(\beta_p \circ s)$ by Theorem 1.5(1) or $\gamma\perp\delta$. *q.e.d.*

Remark. Theorem 3.4(1) is the result of Theorem 1 of Mataga [11] and Corollary 2.6 of Lang [9]. It is not known whether the results of Theorem 3.4 hold for Varadarajan space and generalized Whitehead space.

If p is a prime number, then we denote by C_p the *Serre class* of all abelian torsion groups whose elements's orders are prime to p. We denote by C_0 the Serre class of all abelian torsion groups.

Theorem 3.5. *Let $n \geq 2$ and $p = 0$ or a prime number. Let X be a connected finite CW-complex and Y a finite CW-complex and Z a connected nilpotent space. Let $v : X \to Z$ be a map. Then $[\Sigma^n Y, Z]/v^{\perp}(\Sigma^n Y, Z) \in C_p$ if and only if $(l_p \circ v)^{\perp}(\Sigma^n Y, Z_p) = [\Sigma^n Y, Z_p]$.*

Proof. Let us suppose that $[\Sigma^n Y, Z]/v^{\perp}(\Sigma^n Y, Z) \in C_p$. Let $\alpha \in [\Sigma^n Y, Z_p]$. By the isomorphism of localization

$$\phi_p \; : \; [\Sigma^n Y, Z] \otimes Q_p \cong [\Sigma^n Y, Z_p],$$

we have an element $\beta \in [\Sigma^n Y, Z]$ and an integer q such that $(q, p) = 1$ when p is a prime number and $\phi_p(\beta \otimes (1/q)) = \alpha$ or $l_{p*}(\beta) = q\alpha$. Since by our assumption $r\beta \in v^{\perp}(\Sigma^n Y, Z)$ for some positive integer r such that $(r, p) = 1$ when p is a prime number, we have

$$rq\alpha = l_{p*}(r\beta) \in l_{p*}\{v^{\perp}(\Sigma^n Y, Z)\} \subset (l_p \circ v)^{\perp}(\Sigma^n Y, Z_p)$$

and $(rq, p) = 1$ when p is a prime number. Since $(rq, p) = 1$ and $(l_p \circ v)^{\perp}(\Sigma^n Y, Z_p)$ is a Q_p-module by Theorem 2.4, we have $\alpha \in (l_p \circ v)^{\perp}(\Sigma^n Y, Z_p)$.

Conversely, assume that $(l_p \circ v)^{\perp}(\Sigma^n Y, Z_p) = [\Sigma^n Y, Z_p]$. Let $\alpha \in [\Sigma^n Y, Z]$, then

$$l_{p*}(\alpha) \in [\Sigma^n Y, Z_p] = (l_p \circ v)^{\perp}(\Sigma^n Y, Z_p).$$

Hence by Theorem 2.2 there is a positive integer m such that $(m, p) = 1$ when p is a prime number and $m\alpha \in v^{\perp}(\Sigma^n Y, Z)$. *q.e.d.*

We call a space X a *Gottlieb space* mod p when $\pi_m(X)/G_m(X) \in C_p$ for all m [5, 11].

The following result is a special case of Theorem 3 of Mataga [11].

Theorem 3.6. *Let $p = 0$ or a prime number. Let X be a 1-connected finite CW-complex and p a prime number. Then X is a Gottlieb space* mod *p if and only if X_p is a Gottlieb space.*

Proof. Set $X = Z$, $Y = S^0$ and $v = 1_X$ in Theorem 3.5. Then we have

$$[S^n, X]/(1_X)^{\perp}(S^n, X) \in C_p \text{ if and only if } (l_p)^{\perp}(S^n, X_p) = [S^n, X_p],$$

where $l_p : X \to X_p$ is the p-localization map. Since $(1_X)^\perp(S^n, X) = G_n(X)$ and $(l_p)^\perp(S^n, X_p) = (1_{X_p})^\perp(S^n, X_p) = G_n(X_p)$ by Proposition 3.3, we have the result. *q.e.d.*

A space X is called a *Hopf space* mod p if there exists a map $\mu : X \times X \to X$ such that $\mu|\{*\} \times X$ and $\mu|X \times \{*\}$ induce p-isomorphisms in homotopy groups.

The following theorem is the result of Thorem 2.8 (Haslam) of Lang [9] (cf. Theorem 1 of Haslam [5]).

Theorem 3.7. *Let $p = 0$ or a prime number. Let X be a 1-connected finite CW-complex. If X is a Hopf space* mod p, *then X is a Gottlieb space* mod p.

Proof. If X is an Hopf space mod p, then there exists a pairing $\mu : X \times X \to X$ such that $\mu|\{*\} \times X$ and $\mu|X \times \{*\}$ induce p-isomorphisms in homotopy groups. Then the p-localization $\mu_p : X_p \times X_p \to X_p$ is a pairing for $1_{X_p} \perp 1_{X_p}$, that is, X_p is a Hopf space and hence X_p is a Gottlieb space. It follows from Theorem 3.6 that X is a Gottlieb space mod p. *q.e.d.*

Since $v^\perp(\Sigma^n Y, Z)$ $(n \geq 2)$ is a subgroup of an abelian group $[\Sigma^n Y, Z]$, we see that $v^\perp(\Sigma^n Y, Z)$ is a finitely generated abelian group when Y is a finite CW-complex and Z is a connected space of *finite type*, that is, $\pi_n(Z)$ is a finitely generated group for all $n \geq 0$. In the following theorem we assume that the target space Z is a connected nilpotent space of finite type.

Theorem 3.8. (Pull-back Theorem) *Let $n \geq 2$. Let X be a connected finite CW-complex and Y a finite CW-complex and Z a connected nilpotent space of finite type. Let $v : X \to Z$ be a map. Then $v^\perp(\Sigma^n Y, Z)$ is the pull-back of the diagram*

$$\{(l_p \circ v)^\perp(\Sigma^n Y, Z_p) \to (l_Q \circ v)^\perp(\Sigma^n Y, Z_Q) \mid p \text{ is a prime number }\},$$

where $l_Q : Z \to Z_Q$ is the rationalization map.

Proof. By Theorem 5.1 of Chapter II of [6] or 6.2 (Prime fracture lemma) (p.146) of [2], we see that the homotopy set $[\Sigma^n Y, Z]$ is a pull-back of the diagram

$$\{[\Sigma^n Y, Z_p] \to [\Sigma^n Y, Z_Q] \mid p \text{ is a prime number } \}.$$

Suppose we are given an element β_p of $(l_p \circ v)^\perp(\Sigma^n Y, Z_p)$ for each prime number p such that

$$l_{Q*}(\beta_p) = l_{Q*}(\beta_q)$$

for any p and q. Let $\alpha \in [\Sigma^n Y, Z]$ be the pull-back of $\{\beta_p\}$. Then we have

$$l_p \circ \alpha = \beta_p \in (l_p \circ v)^\perp(\Sigma^n Y, Z_p)$$

for every prime number p. Then we have $v \perp \alpha$ by Theorem 2.3, namely, $\alpha \in v^\perp(\Sigma^n Y, Z)$. This completes the proof.

References

[1] M. Arkowitz, *The generalized Whitehead product*, Pacific J. Math. **12** (1962), 7-23.

[2] A. Bousfield and D. M. Kan, *Homotopy limits, completions and localizations* (Lecture Notes in Math. **304**), Berlin Heidelberg New York: Springer-Verlag, 1972.

[3] D. H. Gottlieb, *A certain subgroup of the fundamental group*, Amer. J. Math. **87** (1965), 840-856.

[4] D. H. Gottlieb, *Evaluation subgroups of homotopy groups*, Amer. J. Math. **91** (1969), 729-756.

[5] H. B. Haslam, *G-spaces* mod *F and H-spaces* mod *F*, Duke Math. J. **38** (1971), 671-679.

[6] P. Hilton, G. Mislin and J. Roitberg, *Localization of nilpotent groups and spaces* (Mathematics Studies **15**), Amsterdam Oxford : North-Holland / New York : American Elsevier, 1974.

[7] N. Iwase and M. Mimura, *Generalized Whitehead spaces with few cells*, Preprint.

[8] J.-R. Kim, *Localizations and generalized evaluation subgroups of homotopy groups*, J. Korean Math. Soc. **22** (1985), 9-18.

[9] G. E. Lang Jr, *Localizations and evaluation subgroups*, Proc. Amer. Math. Soc. **50** (1975), 489-494.

[10] K. L. Lim, *On evaluation subgroups of generalized homotopy groups*, Canad. Math. Bull. **27** (1984), 78-86.

[11] Y. Mataga, *Localization of G-spaces*, Proc. Japan Acad. **50** (1974), 448-452.

[12] M. Mimura, G. Nishida and H. Toda, *Localization of CW-complexes and its applications*, J. Math. Soc. Japan **23** (1971), 593-624.

[13] N. Oda, *Pairings and copairings in the category of topological spaces*, preprint.

[14] N. Oda, *The homotopy set of the axes of pairings*, Canad. J. Math. **42** (1990), 856-868.

[15] J. Siegel, *G-spaces, H-spaces and W-spaces*, Pacific J. of Math. **31** (1969), 209-214.

[16] D. Sullivan, *Genetics of homotopy theory and the Adams conjecture*, Ann. of Math. **100** (1974), 1-79.

[17] K. Varadarajan, *Generalized Gottlieb groups*, J. Indian Math. Soc. **33** (1969), 141-164.

[18] M. H. Woo and J.-R. Kim, *Certain subgroups of homotopy groups*, J. Korean Math. Soc. **21** (1984), 109-120.

Fibrewise reduced product spaces

I.M. James

The purpose of this article is to establish, under appropriate conditions, a fibrewise version of the well-known theorem that the reduced product has the same homotopy type as the loop-space on the suspension. The precise theorem we generalize is that proved by tom Dieck, Kamps and Puppe in [1], an improvement on the original theorem given in [3]. Eggar [2] has given a fibrewise version of the same theorem but under more restrictive conditions which enable a different method of proof to be used.

We work over an arbitrary base space B. Let X be a fibrewise pointed space over B with section s and projection p. The *fibrewise reduced product space* $J_B X$ of X is a fibrewise identification space of the disjoint union of the fibrewise topological products

$$X, X \times_B X, X \times_B X \times_B X, \ldots .$$

In each fibre X_b ($b \in B$) the identification is that which produces the reduced product space $J(X_b)$, i.e. the free monoid on the points of X_b with neutral element $s(b)$. Similarly with fibrewise pointed maps and fibrewise pointed homotopies.

With this topology it is easy to check that the restriction of $J_B X$ to a subset B' of B is equivalent, as a fibrewise pointed space over B', to the fibrewise reduced product space $J_{B'} X_{B'}$. More generally $\lambda^* J_B X$ is equivalent to $J_{B'} \lambda^* X$ for each space B' and map $\lambda : B' \to B$, where λ^* means the induced fibrewise pointed space. In particular J_B $(B \times T)$ is equivalent, as a fibrewise pointed space over B, to $B \times JT$ for each pointed space T, where JT denotes the ordinary reduced product space of T.

Disregarding the topology, for a moment, one can view $J_B X$ as a fibrewise monoid. However it is important to note that the fibrewise multiplication

$$J_B X \times_B J_B X \to J_B X$$

is not, in general continuous. In fact even the action of $X \subset J_B X$ on $J_B X$ is not continuous (an example is given on p.230 of [1], with X the rational line and B the point-space). This trouble can be avoided if we exclude the section. For example, let s_0 be another section of X such that $s_0 B$ does not meet sB. Then the action of $s_0 B$ on $J_B X$ is continuous.

The reduced fibrewise cone of X is denoted by $\Gamma_B X$ and the reduced fibrewise suspension by $\sum_B X$. We regard X as a subspace of $\Gamma_B X$, in the usual way, so that $\sum_B X$ can be obtained from $\Gamma_B X$ by fibrewise collapsing X.

In the proof of our main theorem an essential role is played by the fibrewise pointed space $K_B X$ obtained from the fibrewise topological product $J_B X \times_B \Gamma_B X$ by identifying (z, x) with $(z.x, b)$ for all $b \in B$, $z \in JX_b$ and $x \in X_b$. We regard $J_B X$ as a subspace of $K_B X$, in the obvious way, and note that $K_B X$ is fibrewise contractible, by a routine fibrewise generalization of the argument given in (17.7) of [1].

For any fibrewise pointed space Y over B the fibrewise path-space and the fibrewise loop-space are defined as in [4], except that for present purposes we need to modify the definitions so that paths are parametrized over intervals of arbitrary length rather than unit length. So $\Lambda_B Y$ denotes the modified fibrewise path-space, in this note, and $\Omega_B Y$ the fibrewise loop-space. Of course $\Omega_B Y$ is a fibrewise topological monoid, using juxtaposition of paths to define the fibrewise multiplication. Similarly a fibrewise action of $\Omega_B Y$ on $\Lambda_B Y$ is defined. Note that $\Lambda_B Y$ is fibrewise contractible, by the usual path- shrinking construction (see (17.4) of [1]). Furthermore the projection $\rho : \Lambda_B Y \to Y$, given by the termination of each path, is a fibrewise fibration, in the sense of [4], and $\Omega_B Y$ is the fibrewise fibre.

By a *fibrewise distance function* on the given fibrewise pointed space X we mean a map $u : X \to I = [0, 1]$ such that $u^{-1}(0) = B$, the section of X. Given u a fibrewise pointed map

$$g : X \to \Omega_B \Sigma_B X$$

is defined, as in (17.2) of [1], and extended to a fibrewise pointed map

$$h : J_B X \to \Omega_B \Sigma_B X$$

using the fibrewise monoid structure. In fact h in turn can be extended to a fibrewise pointed map

$$k : K_B X \to \Lambda_B \Sigma_B X$$

similarly. Of course these fibrewise pointed maps depend on the choice of fibrewise distance function u. However since the set of such functions is convex the fibrewise pointed homotopy class is independent of that choice, and so such fibrewise pointed maps may be described as *canonical.*

In fact these fibrewise pointed maps can be defined whenever X is fibrewise non-degenerate, in the sense of [4]. For then X has the fibrewise pointed homotopy type of the fibrewise mapping cylinder $\check{X}_B$ of the section, as explained in §22 of [4], and since $\check{X}_B$ admits a fibrewise distance function, by direct construction, we can transfer the associated canonical fibrewise map from $\check{X}_B$ to X. Thus the canonical fibrewise map is defined, up to fibrewise pointed homotopy, for all fibrewise non-degenerate spaces and in particular for all fibrewise well-pointed spaces, in the sense of [4].

In our main theorem, to be stated in a moment, we shall assert that the canonical fibrewise pointed map

$$h : J_B X \to \Omega_B \Sigma_B X$$

is a fibrewise pointed homotopy equivalence under certain conditions. One of these conditions is that X is fibrewise well-pointed. This implies that $J_B X$ is also fibrewise well-pointed, by a routine fibrewise generalization of Lemma 5 in (17.1) of [1]. Moreover $\Sigma_B X$ is fibrewise well-pointed, as shown in (20.20) of [4], and so $\Omega_B \Sigma_B X$ is fibrewise well-pointed. If we can show that the canonical fibrewise map h is a fibrewise homotopy equivalence, therefore, then we can at once conclude that h is a fibrewise pointed homotopy equivalence, by a routine fibrewise generalisation of the argument given in (5.13) of [6].

Theorem. Let X be a vertically connected fibrewise well-pointed space over B. Suppose that X admits a numerable fibrewise categorical covering. Then the canonical fibrewise pointed maps

$$J_B X \to \Omega_B \Sigma_B X$$

are fibrewise pointed homotopy equivalences.

Here, as in [5], we describe a subset of X as fibrewise categorical if it is fibrewise contractible in X.

In some applications it is more convenient to use the unreduced fibrewise suspension rather than the reduced fibrewise suspension. In fact the former comes equipped with a family of sections, running from one polar section to the other along the track corresponding to the section of X. When X is fibrewise well-pointed, as in our theorem, the natural projection from the unreduced fibrewise suspension to the reduced fibrewise suspension is a fibrewise pointed homotopy equivalence, by (20.8) and (20.20) of [4]. Consequently we can replace the reduced fibrewise suspension by the unreduced fibrewise suspension provided the latter is regarded as a fibrewise pointed space using one of the polar sections (or indeed any section belonging to the above family).

The condition that X is vertically connected can easily be weakened. For suppose that B admits a numerable covering by subsets V such that X_V is vertically connected over V, and that X satisfies the other assumptions of the theorem. The theorem shows that the restriction

$$h_V : J_V X_V \to \Omega_V \Sigma_V X_V$$

of the canonical fibrewise map is a fibrewise homotopy equivalence for each V. Therefore h itself is a fibrewise homotopy equivalence by Dold's theorem (see (9.1) of [1]). It follows that the theorem holds on the assumption that X is locally trivial, in the sense of fibrewise homotopy type, that B is locally contractible, and that the fibres of X are well-pointed and pathwise-connected. But then, as Eggar [2] has shown, the theorem follows easily by Dold's theorem from the special case when B is a point. The argument which follows is a generalization of the argument used in the special case but does not depend on the special case.

The proof of our theorem uses the following lemma, which is a straightforward generalization of the corresponding lemma 14 in (17.8) of [1]. We choose our notation to correspond to the application we are going to make of it.

Lemma. Let E be a fibrewise pointed space over B. Let $q : K \to E$ be a fibrewise pointed map, where K is a fibrewise pointed space, such that the restrictions $q \mid_B: K \mid_B \to E \mid_B$ and $q \mid_{B'}: K \mid_{B'} \to E \mid_{B'}$ are weak fibrewise fibrations, where $B \subset E$ is the section and $B' = E - B$. Suppose that there exists a halo V of B in E such that $V - B$ admits a numerable categorical covering. Also suppose that there exists a strong fibrewise deformation retraction $r : V \to B$ and a strong fibrewise deformation retraction $R : K \mid_V \to K \mid_B$ over r. Finally suppose that for each section s_0 of $V - B$ the restriction $K \mid_{B_0} \to K \mid_B$ of R is a fibrewise homotopy equivalence, where $B_0 = s_0 B$. Then there exists a fibrewise pointed space L, containing K as a strong fibrewise deformation retract, and a fibrewise extension $\pi : L \to E$ of q which is a weak fibrewise fibration. Furthermore $K \mid_B$ is a strong fibrewise deformation retract of $L \mid_B$ and $K \mid_{B'}$ is a strong fibrewise deformation retract of $L \mid_{B'}$.

Recall that $\Sigma_B X$ is obtained from the disjoint union of $X \times I$ and B by identifying (x, t) with $p(x)$ whenever $t = 0$ or 1 or $x = sp(x)$. Without real loss of generality we can replace X here by the fibrewise mapping cylinder of the section s_0 of a fibrewise pointed space X_0 satisfying the conditions of the theorem. Then $\Sigma_B X$ may be regarded as a fibrewise identification space of the disjoint union

$$X_0 \times I \amalg B \times I \times I \amalg B \times I.$$

Note that the section s_0 of X_0 is vertically homotopic in X to the section s of X, and that $B_0 = s_0 B$ is disjoint from $B = sB$. Moreover every section of X_0 is vertically homotopic to s_0 in X_0 and hence to s in X, since X_0 is vertically-connected.

Now consider the neighbourhood V of B in $\Sigma_B X$ formed by the image of

$$X_0 \times ([0, \frac{1}{4}) \cup (\frac{3}{4}, 1]) \amalg B \times [0, \frac{1}{2}) \times I.$$

In fact V is a halo of B with respect to the function $v : \Sigma_B X \to I$ given by

$$v(x,t) = min(2, 4t, 4(1-t)) \qquad (x \in X_0)$$

$$v(b,s,t) = min(1, 2s, 4t, 4(1-t)) \qquad (b \in B).$$

Consider the numerable covering $\{V_0, V_1\}$ of $V - B$, where V_0 is formed by the images of $B \times (0, \frac{1}{2}) \times I$ and $X \times (0, \frac{1}{4})$ while V_1 is formed by the images of $B \times (0, \frac{1}{2}) \times I$ and $X \times (\frac{3}{4}, 1)$. Clearly $(X - B) \times \frac{1}{8}$ is a fibrewise deformation retract of V_0 and $(X - B) \times \frac{7}{8}$ is a fibrewise deformation retract of V_1. Hence V_0 and V_1 admit numerable fibrewise categorical coverings, since $X - B$ admits such a covering, and so their union $V - B$ admits a numerable fibrewise categorical covering.

We apply the lemma to the fibrewise pointed map $q : K_B X \to \Sigma_B X$ which is induced by the composition

$$J_B X \times_B \Gamma_B X \to \Gamma_B X \to \Sigma_B X$$

of the second projection and the fibrewise quotient map. The strong fibrewise deformation retraction ϕ_t of V to B, constructed as in (17.9) of [1], can be lifted explicitly to a strong fibrewise deformation retraction Φ_t of $K_B X \mid_V$ to $K_B X \mid_B = J_B X$. Also the fibrewise map of $J_B X$ into itself given by translating with respect to s_0, is fibrewise homotopic to the identity, since s_0 is vertically homotopic to s. Therefore the restriction of Φ_1 to $K_B X \mid_{B_0}$ is a fibrewise homotopy equivalence.

The conditions of the lemma are therefore satisfied in our situation. We write $L = L_B X$, for consistency, so that $\pi : L_B X \to \Sigma_B X$ is the weak fibrewise fibration extending q. We compare π with the fibrewise fibration $\rho : \Lambda_B \Sigma_B X \to \Sigma_B X$, defined by the termination of each path. Let $f : L_B X \to K_B X$ be a fibrewise retraction so that $jf \simeq \mathrm{id}\ rel\ K_B X$ by a fibrewise homotopy, where $j : K_B X \to L_B X$ denotes the inclusion. Then $\rho k f = qf = \pi j f \simeq \pi$, $rel\ K_B X$, by a fibrewise homotopy ψ_t. Since ρ is a fibrewise fibration we can lift ψ_t to a fibrewise deformation

$$\Psi_t : L_B X \to \Lambda_B \Sigma_B X$$

of kf. Now $L_B X$ is fibrewise contractible since $K_B X$ is fibrewise contractible and f is a fibrewise homotopy equivalence. Therefore Ψ_1 is a fibrewise homotopy equivalence, over B, and hence a fibrewise homotopy equivalence over $\Sigma_B X$, by the fibrewise version of Dold's theorem given in (23.11) of [4]. By restricting Ψ_1 to the fibrewise fibres we obtain a fibrewise homotopy equivalence

$$\ell : L_B X \mid_B \to \Omega_B \Sigma_B X$$

over B. However $\rho\Psi_T = \psi_t$ is a fibrewise homotopy *rel* $K_B X$, hence fibrewise constant on B, and so Ψ_t determines a fibrewise homotopy

$$h_t : J_B X \to \Omega_B \Sigma_B X.$$

Here $h_0 = h$, the canonical fibrewise map, and $h_1 = \ell \mid J_B X$. But ℓ is a fibrewise homotopy equivalence, as we have just seen, and $J_B X$ is a fibrewise deformation retract of $L_B X \mid_B$, by the lemma. Therefore h_1 is a fibrewise homotopy equivalence and so h_0 is a fibrewise homotopy equivalence, as asserted.

This proves the theorem, which can be used to extend the standard results about reduced product spaces to the fibrewise case. For example, suppose that X is a fibrewise Hopf space in the strict sense, i.e. with a two-sided neutral element provided by the section in each fibre. Then X is a fibrewise retract of $J_B X$, by direct construction. When the conditions of the theorem are satisfied, therefore, we deduce that X is a fibrewise retract of $\Omega_B \Sigma_B X$.

For an example of quite a different type we show how our theorem can be used to obtain a model of the space of free loops on a sphere of odd dimension. Consider the Stiefel manifold $V_{2n,2}$ $(n = 1, 2, ...)$ of orthonormal 2- frames in $\mathbf{R}^{2n}$. Thus $V_{2n,2}$ consists of pairs (x, y) of points of S^{2n-1} such that $x \perp y$. We fibre $V_{2n,2}$ over S^{2n-1}, in the usual way, by sending (x, y) to x; the fibre over x is the orthogonal $(2n-1)$-sphere S_x^{2n-2}. Identifying $\mathbf{R}^{2n}$ with $\mathbf{C}^n$, in the usual way, we equip $V_{2n,2}$ with the section $s : S^{2n-1} \to V_{2n,2}$, where $s(x) = (x, ix)$, and so regard $V_{2n,2}$ as a fibrewise pointed space over S^{2n-1}. The fibrewise reduced product space $J_{S^{2n-1}}(V_{2n,2})$ can therefore be constructed, so that the fibre over x is the free monoid generated by the points of S_x^{2n-2}, the neutral element being ix. I assert that $J_{S^{2n-1}}(V_{2n,2})$, as a space, has the same homotopy type as the free loop-space on S^{2n-1}.

To see this first observe that the unreduced fibrewise suspension $\Sigma_{S^{2n-1}} V_{2n,2}$ of $V_{2n,2}$ is equivalent, as a fibrewise space, to $S^{2n-1} \times S^{2n-1}$, regarded as a fibrewise space over S^{2n-1} using the first projection. In fact if t is the suspension parameter, running from O to 1, a fibrewise topological equivalence

$$\phi : \Sigma_{S^{2n-1}} V_{2n,2} \to S^{2n-1} \times S^{2n-1}$$

is given by

$$\phi(t, (x, y)) = (x, x\, cos\, \pi t + y\, sin\, \pi t).$$

Now ϕ transforms σ_0, the section of $\Sigma_{S^{2n-1}} V_{2n,2}$, into the diagonal section of $S^{2n-1} \times S^{2n-1}$. Hence the fibrewise loop-space of $\Sigma_{S^{2n-1}} V_{2n,2}$, with σ_0 as section, is equivalent to the fibrewise loop-space of $S^{2n-1} \times S^{2n-1}$, with the diagonal as section, and this in turn is equivalent, as a space, to the free loop-space on S^{2n-1}. So the assertion follows from the second version of our theorem, in which vertical connectedness is only required locally.

The properties of $V_{2n,2}$ being used here are first that it has a section and secondly that the fibrewise suspension is trivial. A similar result can be obtained for any sphere-bundle with these properties, in particular for the tangent sphere-bundle to any almost-parallelizable manifold with Euler characteristic zero.

References.

1. T. tom Dieck, K.H. Kamps and D. Puppe, Homotopietheorie (Lecture Notes in Mathematics no. 157). Springer (1970).

2. M.G. Eggar, Ex-homotopy theory, Compositio Math. 27 (1973), 185-195.

3. I.M. James, Reduced product spaces, Ann. of Math. 62 (1955), 170-197.

4. I.M. James, Fibrewise Topology. Cambridge Univ. Press (1989).

5. I.M. James and J.R. Morris, Fibrewise category, Proc. Roy. Soc. Edinb. (to appear).

6. D. Puppe, Homotopiemengen und ihre induzierten Abbildingen, I, Math. Zeit. 69 (1967), 299-344.

MATHEMATICAL INSTITUTE, UNIVERSITY OF OXFORD.

COMPUTING HOMOTOPY TYPES USING CROSSED N-CUBES OF GROUPS

by Ronald Brown

Dedicated to the memory of Frank Adams

Introduction.

The aim of this paper is to explain how, through the work of a number of people, some algebraic structures related to groupoids have yielded algebraic descriptions of homotopy n-types. Further, these descriptions are explicit, and in some cases completely computable, in a way not possible with the traditional Postnikov systems, or with other models, such as simplicial groups.

These algebraic structures take into account the action of the fundamental group. It follows that the algebra has to be at least as complicated as that of groups, and the basic facts on the use of the fundamental group in 1-dimensional homotopy theory are recalled in Section 1. It is reasonable to suppose that it is these facts that a higher dimensional theory should imitate.

However, modern methods in homotopy theory have generally concentrated on methods as far away from those for the fundamental group as possible. Such a concentration has its limitations, since many problems in the applications of homotopy theory require a non-trivial fundamental group (low dimensional topology, homology of groups, algebraic K-theory, group actions, ...). We believe that the methods outlined here continue a classical tradition of algebraic topology. Certainly, in this theory non-Abelian groups have a clear role, and the structures which appear arise directly from the geometry, as algebraic structures on sets of homotopy classes.

It is interesting that this higher dimensional theory emerges not directly from groups, but from groupoids. In Sections 2 and 3 we state some of the main facts about the use of multiple groupoids in homotopy theory, including two notions of *higher homotopy groupoid*, and the related notions of *crossed complex* and of *crossed n-cube of groups.* Theorem 2.4 shows how to calculate standard homotopy invariants of 3-types for the classifying space of a crossed square. We also show in Section 3 how crossed n-cubes of groups and the notion of n-cube of fibrations may, with the use of a Generalized Seifert-Van Kampen Theorem due to Brown and Loday, 1987a, be used for the computation of homotopy types in some practical cases (Proposition 3.3).

An interesting methodological point is that the description of the whole n-type has, by these methods, better algebraic properties than do the individual invariants (homotopy groups, Whitehead products, etc.). As an application, we give some explicit results on 3-types, including computations of Whitehead products at this level. In Section 4 we prove a result from Section 2. In Section 5, we show that all simply connected 3-types arise from a crossed square of Abelian groups (Theorem 5.1).

Baues, 1989, 1991, also considers algebraic models of homotopy types involving non-Abelian groups, and in the second reference considers *quadratic modules* and *quadratic chain complexes.* It seems that the sets of results of the two techniques have a non-trivial intersection, but neither is contained in the other. Thus a further comparison, and possibly integration, of the two types of methods would be of interest.

Joyal and Tierney have also announced a model of 3-types using *braided 2-groupoids.* These models are equivalent to the *braided crossed modules* of Brown and Gilbert, 1989, which are there related to simplicial groups and used to discuss automorphism structures of crossed modules.

1. Groups and homotopy 1-types.

The utility of groups in homotopy theory arises from the standard functors

$$\pi_1 : \text{(spaces with base point)} \longrightarrow \text{(groups)}$$
$$B : \text{(groups)} \longrightarrow \text{(spaces with base point)}$$

known as the *fundamental group* and *classifying space* functors respectively. The classifying space functor is the composite of geometric realisation and the nerve functor N from groups to simplicial sets. These functors have the properties:

1.1 *There is a natural equivalence of functors* $\pi_1 \circ B \simeq 1$.

1.2 *If* G *is a group, then* $\pi_i BG$ *is* 0 *if* $i \neq 1$.

The fundamental group of many spaces may be calculated using the Seifert-Van Kampen theorem, or using fibrations of spaces. Further, if X is a connected CW-complex and G is a group, then there is a natural bijection

$$[X,BG] \cong Hom(\pi_1 X,G) ,$$

where the square brackets denote pointed homotopy classes of maps. It follows that there is a map $X \longrightarrow B\pi_1 X$ inducing an isomorphism of fundamental groups. It is in this sense that groups are said to model homotopy 1-types, and a computation of a group G is also regarded as a computation of the 1-type of the classifying space BG .

A standard block against generalising this theory to higher dimensions has been that higher homotopy groups are Abelian. The algebraic reason for this is that group objects in the category of groups are Abelian groups. This seems to kill the case for a 'higher dimensional group theory', and in 1932 was the reason for an initial dissatisfaction with Čech's definition of higher homotopy groups (Čech, 1932). Incidentally, Čech also suggested the idea of higher homotopy groups went back to Dehn, who never published it (Dieudonné, 1989). The difficulties of basic homotopy theory are shown by the fact that Hurewicz never published the proofs of the results announced in his four notes on homotopy groups (Hurewicz, 1935, 1936); that a proof of the Homotopy Addition Theorem did not appear in print till Hu, 1953; and that even current proofs of this basic theorem are not easy (e.g. G.W.Whitehead, 1978).

It has for some time been established that most of the theory of the fundamental group is better expressed in terms of the fundamental groupoid (Brown, 1968, 1988) in that theorems:

have more natural and convenient expression;

have more powerful statements;

and have simpler proofs.

As an example, we mention the description in Brown, 1988, of the fundamental groupoid of the orbit space of a discontinuous action. Thus it is natural to ask:

Can a 'better' higher homotopy theory be obtained using some notion of 'higher homotopy groupoid'?

Expectations in this direction were expressed in Brown, 1967.

By now, some of the answers to this question seem to be of the 'best possible situation' kind, and suggest that homotopy theory is in principle coincident with a 'higher dimensional group(oid) theory'. Such a theory is a significant generalisation of group theory. In view of the many applications of group theory in mathematics and science, the wider uses of this generalisation, and the principles underlying it, need considerable further study. For example, some possibilities are sketched in Brown, 1989b, 1990, Brown, Gilbert, Hoehnke, and Shrimpton, 1991. Also, the known applications in homotopy theory have so far used what seems only a small part of the algebra.

2. Multiple groupoids.

The simplest object to consider as a candidate for 'higher dimensional groupoid' is an *n-fold groupoid.* This is defined inductively as a *groupoid object in the category of* $(n - 1)$*-fold groupoids*, or alternatively, as a *set with* n *compatible groupoid structures*. The compatibility condition is that if $\circ_i$ and $\circ_j$ are two distinct groupoid structures, then the *interchange law* holds, namely that

$$(a \circ_i b) \circ_j (c \circ_i d) = (a \circ_j c) \circ_i (b \circ_i d)$$

whenever both sides are defined. This is often expressed in terms of the diagram

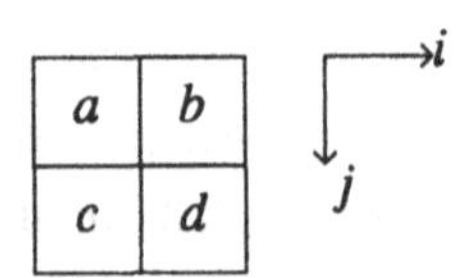

Note that Ehresmann, 1963, defines double categories, and the

above definition is a simple extension of that concept. The argument that a group object in the category of groups is an Abelian group now yields that a double groupoid contains a family of Abelian groups, one for each vertex of the double groupoid. More generally, a basic result is that a double groupoid contains also two *crossed modules over groupoids* (Brown and Spencer, 1976). For example, the *horizontal crossed module* is defined analogously to the second relative homotopy group. It consists in dimension 2 of the elements of the form

$$\begin{array}{ccc} & 1_H & \\ \partial\alpha & \boxed{\alpha} & 1_V \\ & 1_H & \end{array}$$

where 1_H and 1_V denote identities for the horizontal and vertical structures respectively. In dimension 1 it consists of the vertical 'edge' groupoid. The boundary ∂ is as shown, and the action is not hard to define, as suggested by the following diagram:

$$\alpha^b = \begin{array}{ccc} b & \boxed{\square} & b \\ \partial\alpha & \boxed{\alpha} & \\ b^{-1} & \boxed{\square} & b^{-1} \end{array}$$

where $\square$ denotes a horizontal identity. The existence of these crossed modules in any double groupoid, and the fact that double groupoids can be constructed from any given crossed module (Brown and Spencer, 1976), together illustrate that double groupoids are in some sense *more non-Abelian than groups.* This in principle makes them more satisfactory as models for two dimensional homotopy theory than are the second homotopy groups. In fact it is known that crossed modules over groupoids, and hence also double groupoids, model all homotopy 2-types (see Mac Lane and Whitehead, 1950, but note that they use 3-type for what is now called 2-type).

One of the features of the use of multiple groupoids is that they are most naturally considered as cubical objects of some kind, since they have structure in different directions. Analogous simplicial objects may in some cases be defined, but their properties are often difficult to establish, and are sometimes obtained by referring to the cubical analogue. For a general background to problems on algebraic models of homotopy types, see Grothendieck, 1983, although this work does not take into account the use of multiple groupoids.

The first example of which we know of a 'higher homotopy groupoid' was found in 1974 (see Brown and Higgins, 1978), 42 years after Čech's definition of homotopy groups, namely the *fundamental double groupoid of a pair of spaces.* It is conveniently expressed in the more general situation of filtered spaces as follows (Brown and Higgins, 1981b, as modified in Brown and Higgins, 1991, Section 8). Let

$$X_* : X_0 \subseteq X_1 \subseteq \ldots \subseteq X_n \subseteq \ldots \subseteq X$$

be a filtered space. Let RX_* be the *singular cubical complex* of X_*, consisting for all $n \geq 0$ of the filtered maps $I^n_* \to X_*$, where the standard n-cube I^n is filtered by its skeleta, and with the standard face and degeneracy maps. Let ρX_* consist in dimension n of the homotopy classes, through filtered maps and rel vertices, of such filtered maps. (The modification in the 1991 paper is to take the homotopies rel vertices.) The standard gluing of cubes in each direction imposes an extra structure of n compositions on $(RX_*)_n$ for each $n \geq 1$. It is a subtle fact (Brown and Higgins, 1981b) that this structure is inherited by ρX_* to give the latter the structure of n-fold groupoid in each dimension n. There is also an extra, easily verified structure, on both RX_* and ρX_*, namely that of *connections*: these are extra degeneracy operations which arise in the cubical context from the monoid structure *max* on the unit interval I. The total structure on ρX_* is called that of *ω-groupoid* (Brown and Higgins, 1981a,b). This gives our first example of a *higher homotopy groupoid.*

The aim of the introduction of this functor

$$\rho : (\text{filtered spaces}) \to (\omega\text{-groupoids}),$$

was that the proof of the usual Seifert-Van Kampen Theorem for the fundamental group generalised to a corresponding theorem for ρ (Brown and Higgins, 1981b). One main feature of the proof is that ω-groupoids, being cubical objects, are appropriate for encoding subdivision methods, since they easily allow an 'algebraic inverse to subdivision'. It is not easy to formulate a corresponding simplicial method. (See Jones, 1984, for a possible approach.) Another feature crucial in the proof is the use of the connections to express facts related to the Homotopy Addition Lemma. It seems that connections are an important new part of the cubical theory, since they allow for 'degenerate' elements in which adjacent faces are identical, as in the simplicial theory.

The *classifying space* BG of an ω-groupoid G is the geometric realisation of its underlying cubical set. These classifying spaces model only a restricted range of homotopy types, namely those which fibre over a $K(\pi,1)$ with fibre a topological Abelian group (Brown and Higgins, 1991). Nonetheless, these restricted models have useful applications. A principal reason for this is the equivalence proved in Brown and Higgins, 1981a, between ω-groupoids and the classical tool in homotopy theory of crossed complex.

A *crossed complex* is a structure which encapsulates the properties of the fundamental groupoid $\pi_1(X_1,X_0)$ and the relative homotopy groups $\pi_n(X_n,X_{n-1},p)$, $p \in X_0$, $n \geq 2$, for a filtered space X_* , together with the actions of the fundamental groupoid and the boundary maps. The notion was first considered in the reduced case (i.e. when X_0 is a singleton) by Blakers, 1948, under the name *group system*. It was studied in the free case, and under the name *homotopy system*, by Whitehead, 1949. The term *crossed complex* is due to Huebschmann, 1980, who used crossed n-fold extensions to represent the elements of the $(n + 1)$st cohomology group of a group (see also Holt, 1979, Mac Lane, 1979, Lue, 1981), and to determine differentials in the Lyndon-Hochschild-Serre spectral sequence (Huebschmann, 1981). Lue, 1981, gives a good background to the general algebraic setting of crossed complexes. Crossed complexes have the advantage of being able to include information on chain complexes with a group G of operators and on presentations of the group G. The category of crossed complexes also has a monoidal closed structure (Brown and Higgins, 1987), which is convenient for expressing homotopies and higher homotopies.

The Generalized Seifert-Van Kampen Theorem for the fundamental ω-groupoid of a filtered space (Brown and Higgins, 1981b) implies immediately a similar theorem for the fundamental crossed complex, and this theorem has a number of applications, including the Relative Hurewicz Theorem. The latter theorem is thus seen in a wider context, related to excision, and in a formulation dealing initially with the natural map $\pi_n(X,A) \to \pi_n(X \cup CA)$. This formulation was a model for the $(n + 1)$-ad Hurewicz theorem (Brown and Loday, 1987b). Other recent applications of crossed complexes are given in Baues, 1988, 1991, Brown and Higgins, 1987, 1989, 1991, Baues and Brown, 1990, Baues and Conduché, 1991.

More general algebraic models related to groupoids are associated not with filtered objects but with n-cubes of objects. Let $<n>$ denote the set $\{1,2,\ldots,n\}$. An *n-cube* C in a category $\mathcal{C}$ is a commutative diagram with vertices C_A for $A \subseteq <n>$ and morphisms $C_A \to C_{A\cup\{i\}}$ for $A \subseteq <n>$, $i \in <n>$, and $i \notin A$. In particular, a 1-cube is a morphism, and a 2-cube is a commutative square.

Let $\mathbf{X}$ be an n-cube of pointed spaces. Loday, 1982, defines the *fundamental cat^n-group* $\Pi\mathbf{X}$. (We are following the terminology and notation of Brown and Loday, 1987a.) Here, a *cat^n-group* may be defined to be an n-fold groupoid in the category of groups. Alternatively, it is an $(n + 1)$-fold groupoid in which one of the structures is a group. (This is one of several equivalent definitions considered in Loday, 1982.)

For simplicity, we describe $\Pi\mathbf{X}$ in a special case, namely when $\mathbf{X}$ arises from a pointed $(n + 1)$-ad $\mathfrak{X} = (X;X_1,\ldots,X_n)$ by the rule: $\mathbf{X}^{<n>} = X$ and for A properly contained in $<n>$, $\mathbf{X}^A = \bigcap_{i\notin A} X_i$, with maps the inclusions. Let Φ be the space of maps $I^n \to X$ which take the faces of I^n in the ith direction into X_i. Notice that Φ has the structure of n compositions derived from the gluing of cubes in each direction. Let $* \in \Phi$ be the constant map at the base point. Then $G = \pi_1(\Phi,*)$ is certainly a group. Gilbert, 1988, identifies G with Loday's ΠX, so that Loday's results, obtained by methods of simplicial spaces, show that G becomes a cat^n-group. It may also be shown that the extra groupoid structures are inherited from the compositions on Φ. It is this cat^n-group which is written $\Pi \mathfrak{X}$ and is called the *fundamental cat^n-group of the* $(n + 1)$*-ad* $\mathfrak{X}$. This construction of Loday is our second example of a *higher homotopy groupoid.* We emphasise that the existence of this structure is itself a non-trivial fact, containing homotopy theoretic information. Also the results of Gilbert, 1988, are for the case of n-cubes of spaces.

The nerve NG mentioned in Section 1 may be defined, not only for a group but also for a groupoid G, to be in dimension i the set of groupoid maps $\pi_1(\Delta^i,\Delta^i_0) \to G$, where Δ^i_0 is the set of vertices of the i-simplex Δ^i. It follows by iteration that N defines also a functor

$$((n+1)\text{-fold groupoids}) \to ((n+1)\text{-simplicial sets}).$$

Hence there is a *classifying space functor*

$$B : (\text{cat}^n\text{-groups}) \to (\text{pointed spaces})$$

defined as the composite of geometric realisation and the nerve functor to $(n + 1)$-simplicial sets. Loday, 1982, proves that if G is a cat^n-group, then BG is $(n + 1)$*-coconnected,* i.e. $\pi_i BG = 0$ for $i > n + 1$. He also shows, with a correction due to Steiner, 1986, that if X is a connected, $(n + 1)$-coconnected CW-complex, then there is a cat^n-group G such that X is of the homotopy type of BG. In fact, Loday gives an equivalence between a localisation of the category of cat^n-groups and the pointed homotopy category of connected, $(n + 1)$-coconnected CW-complexes. This can be put provocatively as

$(n + 1)$-fold groupoids model all homotopy $(n + 1)$-types.

That is, the generalisation from groups or groupoids to

$(n + 1)$-fold groupoids is as good for modelling homotopy types as might be expected. This result also shows the surprising richness of the algebraic structure of $(n + 1)$-fold groupoids.

There is an important structure equivalent to that of cat^n-groups, namely that of *crossed n-cubes of groups* (Ellis and Steiner, 1987). The main intuitive idea is that a crossed n-cube of groups is a crossed module in the category of crossed $(n - 1)$-cubes of groups. This leads to the following definition (*loc. cit.*).

Definition 2.1. Let $\langle n\rangle$ denote the set $\{1,2,\ldots,n\}$. A *crossed n-cube of groups* is a family of groups, M_A, $A \subseteq \langle n\rangle$, together with morphisms

$$\mu_i : M_A \to M_{A\setminus\{i\}}, \quad (i \in \langle n\rangle, A \subseteq \langle n\rangle),$$

and functions

$$h : M_A \times M_B \to M_{A\cup B}, \quad (A,B \subseteq \langle n\rangle),$$

such that if ${}^a b$ denotes $h(a,b)b$ for $a \in M_A$ and $b \in M_B$ with $A \subseteq B$, then for $a,a' \in M_A$, $b,b' \in M_B$, $c \in M_C$ and $i,j \in \langle n\rangle$, the following hold:

(1) $\mu_i a = a$ if $i \notin A$

(2) $\mu_i\mu_j a = \mu_j\mu_i a$

(3) $\mu_i h(a,b) = h(\mu_i a,\mu_i b)$

(4) $h(a,b) = h(\mu_i a,b) = h(a,\mu_i b)$ if $i \in A$ and $i \in B$

(5) $h(a,a') = [a,a']$

(6) $h(a,b) = h(b,a)^{-1}$

(7) $h(a,b) = 1$ if $a = 1$ or $b = 1$

(8) $h(aa',b) = {}^a h(a',b)h(a,b)$

(9) $h(a,bb') = h(a,b)\,{}^b h(a,b')$

(10) ${}^a h(h(a^{-1},b),c)\;{}^c h(h(c^{-1},a),b)\;{}^b h(h(b^{-1},c),a) = 1$

(11) ${}^a h(b,c) = h({}^a b,{}^a c)$ if $A \subseteq B \cap C$.

A morphism of crossed n-cubes $(M_A) \to (N_A)$ is a family of morphisms of groups $f_A : M_A \to N_A$ $(A \subseteq \langle n\rangle)$ which commute with the maps μ_i and the functions h. This gives us a category $\mathcal{C}rs^n\mathcal{G}p$. Ellis and Steiner, 1987, show that this category is equivalent to that of cat^n-groups, and this is the reason for the choice of structure and axioms in Definition 2.1. This equivalence shows that there is a *classifying space functor*

$$B : \mathcal{C}rs^n gp \longrightarrow \mathcal{T}op \ .$$

This functor would be difficult to describe directly. (See Porter, 1990, for a different account of such a functor.) The results for cat^n-groups imply that a localisation of the category $\mathcal{C}rs^n gp$ is equivalent to the homotopy category of pointed, connected, $(n + 1)$-coconnected CW-complexes.

The *fundamental crossed n-cube of groups functor* Π' is defined from n-cubes of pointed spaces to crossed n-cubes of groups: $\Pi'\mathbf{X}$ is simply the crossed n-cube of groups equivalent to the cat^n-group $\Pi\mathbf{X}$. It is easier to identify Π' in classical terms in the case $\mathbf{X}$ is the n-cube of spaces arising from a pointed $(n + 1)$-ad $\mathcal{X} = (X;X_1,\ldots,X_n)$. Let $X^{<n>} = X$ and for A properly contained in $<n>$ let $X^A = \bigcap_{i \notin A} X_i$. Then $M = \Pi'\mathcal{X}$ is given as follows (Ellis and Steiner, 1987): $M_\emptyset = \pi_1(X^\emptyset)$; if $A = \{i_1,\ldots,i_r\}$, then M_A is the homotopy $(r + 1)$-ad group

$$\pi_{r+1}(X^A;X^A \cap X_{i_1},\ldots,X^A \cap X_{i_r}) \ ;$$

the maps μ_i are given by the usual boundary maps; the h-functions are given by the generalised Whitehead product. Note that whereas these separate elements of structure had all been considered previously, the aim of this theory is to consider the whole structure, despite its apparent complications. This global approach is necessary for the Generalized Seifert-Van Kampen Theorem, stated below. That $\Pi'\mathcal{X}$ satisfies the laws for a crossed n-cube of groups follows immediately since $\Pi'\mathcal{X}$ is the crossed n-cube of groups derived from the cat^n-group $\Pi\mathbf{X}$. From now on, we abbreviate Π' to Π , the meaning being clear from the context.

A crossed n-cube of groups M gives rise to an *n-cube of crossed n-cubes of groups* $\square M$ where

$$((\square M)(A))_B = \begin{cases} M_B & \text{if } A' \subseteq B \\ 1 & \text{otherwise} \end{cases}$$

Then $B\square M$ is an n-cube of spaces. The generalisation to this context of the result on the fundamental group of the classifying space of a group is that there is a natural isomorphism of crossed n-cubes of groups

$$\Pi B \square M \cong M \ .$$

(See Loday, 1982, for the cat^n-group case, and Brown and Higgins, 1981b, 1991, for the analogous crossed complex case.) This result confirms the appropriate nature of the axioms (1)-(11) of Definition 2.1.

A description of the homotopy groups of BG for a cat^n-group G has been given in Loday, 1982, in terms of the homology groups of a non-Abelian chain complex. This, with some extra work, yields a result on the homotopy invariants of the classifying space of a *crossed square* (i.e. a crossed 2-cube of groups). It is useful first to make the axioms for these explicit.

A *crossed square* (Loday, 1982) consists of a commutative square of morphisms of groups

$$\begin{array}{ccc} L & \xrightarrow{\lambda} & M \\ {\scriptstyle\lambda'}\downarrow & & \downarrow{\scriptstyle\mu} \\ N & \xrightarrow{\nu} & P \end{array} \qquad (2.2)$$

together with actions of P on the groups $L,M,N,$ and a function $h : M \times N \to L$. This structure shall satisfy the following axioms, in which we assume that M and N act on L,M,N via P :

(2.3)(i) the morphisms $\lambda, \lambda', \mu, \nu$ and $\mu\lambda = \nu\lambda'$ are crossed modules and λ and λ' are P-equivariant;

(ii) $h(mm',n) = h({}^{m}m', {}^{m}n)h(m,n)$, $h(m,nn') = h(m,n)h({}^{n}m, {}^{n}n')$;

(iii) $\lambda h(m,n) = m\,{}^{n}m^{-1}$, $\lambda' h(m,n) = {}^{m}n n^{-1}$;

(iv) $h(\lambda l,n) = l\,{}^{n}l^{-1}$, $h(m,\lambda' l) = {}^{m}l\,l^{-1}$;

(v) $h({}^{p}m, {}^{p}n) = {}^{p}h(m,n)$;

for all $l \in L$, $m,m' \in M$, $n,n' \in N$, $p \in P$.

We now describe the homotopy groups of BG for a crossed square G as above. The first part of the following result is a special case of results in Loday, 1982.

Theorem 2.4. *Let G be the crossed square (2.2). Then the homotopy groups of BG may be computed as the homology groups of the non-Abelian chain complex*

$$L \xrightarrow{(\lambda^{-1},\lambda')} M \rtimes N \xrightarrow{\mu\cdot\nu} P \qquad (2.5)$$

where $\mu\cdot\nu : (m,n) \mapsto (\mu m)(\nu n)$. This implies that

$$\pi_i BG \cong \begin{cases} P/(\mu M)(\nu N) & \text{if } i = 1, \\ (M \times_P N)/\{(\lambda l, \lambda' l) : l \in L\} & \text{if } i = 2, \\ (Ker\ \lambda) \cap (Ker\ \lambda') & \text{if } i = 3, \\ 0 & \text{if } i \geq 4. \end{cases} \qquad (2.6)$$

Further, under these isomorphisms, the composition $\eta^* : \pi_2 BG \to \pi_3 BG$ with the Hopf map $\eta : S^3 \to S^2$ is induced by the function $M \times_P N \to L$, $(m,n) \mapsto h(m,n)$, and the Whitehead product $\pi_2 \times \pi_2 \to \pi_3$ on BG is induced by the function $((m,n),(m',n')) \mapsto h(m',n)h(m,n')$. The first Postnikov invariant of BG is the cohomology class determined by the crossed module

$$(M \rtimes N)/Im(\lambda^{-1},\lambda) \xrightarrow{\mu\cdot\nu} P. \qquad \square$$

We will explain the proof of this result in Section 4.

3. n-cubes of fibrations.

As in Brown and Loday, 1987a, an n-cube of maps $\mathbf{X}$ yields an n-cube of fibrations $\bar{\mathbf{X}}$. (See Edwards and Hastings, 1976, Cordier and Porter, 1990.) Following Steiner, 1986, we parametrize this as a commutative diagram consisting of spaces $\bar{\mathbf{X}}_{A,B}$ (A,B disjoint subsets of $\langle n\rangle$) and fibration sequences

$$\bar{\mathbf{X}}_{A\cup\{i\},B} \to \bar{\mathbf{X}}_{A,B} \to \bar{\mathbf{X}}_{A,B\cup\{i\}}, \quad A \cap B = \emptyset, \ i \in \langle n\rangle \setminus (A \cup B) \qquad (3.1)$$

The n-cube of fibrations $(\bar{\mathbf{X}}_{A,B})$ contains an n-cube of spaces $\bar{\mathbf{X}}_{\emptyset,*}$ homotopy equivalent to $\mathbf{X}$ (i.e. there is a morphism $\mathbf{X} \to \bar{\mathbf{X}}_{\emptyset,*}$ consisting of homotopy equivalences $\mathbf{X}_B \to \bar{\mathbf{X}}_{\emptyset,B}$). The n-cube of maps $\mathbf{X}$ is called *connected* if all the spaces $\bar{\mathbf{X}}_{A,B}$ are path-connected.

Just as the Seifert-Van Kampen Theorem enables one to compute the fundamental group of a union of connected spaces, so the Generalised Seifert-Van Kampen Theorem (GSVKT) enables one to compute the fundamental crossed n-cube of a union of connected n-cubes. This result is Theorem 5.4 of Brown and Loday, 1987a, where it is proved by induction on n starting with the usual SVKT. It may be restated in terms of crossed n-cubes of groups, rather than cat^n-groups, as follows.

Theorem 3.2. *Let* $\mathbf{X}$ *be an* n*-cube of spaces, and suppose that* $\mathcal{U} = \{U^\lambda\}$ *is an open cover of the space* $\mathbf{X}_{<n>}$, *such that* $\mathcal{U}$ *is closed under finite intersections. Let* $\mathbf{U}^\lambda$ *be the* n*-cube of spaces obtained from* $\mathbf{X}$ *by inverse images of the* U^λ. *Suppose that each* $\mathbf{U}^\lambda$ *is a connected* n*-cube of spaces. Then:*

(C): *the* n*-cube* $\mathbf{X}$ *is connected, and*

(I): *the natural morphism of crossed* n*-cubes of groups*

$$\mathrm{colim}^\lambda\ \Pi\mathbf{U}^\lambda \longrightarrow \Pi\mathbf{X}$$

is an isomorphism.

The colimit in this theorem is taken in the category of crossed n-cubes of groups, and so the validity of (I) confirms again that the axioms for crossed n-cubes of groups are well chosen.

The connectivity statement (C) of this theorem generalises the famous $(n + 1)$*-ad connectivity theorem*, which is usually regarded as a difficult result (at the time of writing, no recent proof is in print except that referred to here). Of course, the connectivity result is related to the fact that a colimit of zero objects is zero.

The isomorphism statement (I) implies the characterisation by a universal property of the critical group of certain $(n + 1)$-ads. (See Brown and Loday, 1987b, for the general procedure and explicit results on the triad case, using a non-Abelian tensor product, and Ellis and Steiner, 1987, for the general case.) The earlier result in this area is in Barratt and Whitehead, 1952, but there the assumption is made of simply connected intersection, and the result is proved by homological methods, so that it has no possibility for dealing with the occurrence of a non-Abelian $(r + 1)$-ad homotopy group. It is clearly advantageous to see the Barratt and Whitehead result, including the $(n + 1)$-ad connectivity theorem, as a special case of a theorem which has other consequences, for example an $(n + 1)$-ad Hurewicz theorem (Brown and Loday, 1987b).

These results, with Theorem 2.4, illustrate how situations in homotopy theory may require constructions on non-Abelian groups for the convenient statement of a theorem, let alone its proof. The methods of crossed n-cubes of groups give a (largely unstudied) range of new constructions in group theory.

Theorem 3.2 allows in some cases for the computation of the fundamental crossed n-cube of groups $\Pi\mathbf{X}$ of an n-cube of spaces $\mathbf{X}$. We now consider to what extent it also allows computation of the $(n + 1)$-type of the space $\mathbf{X}_{<n>}$.

Let $\mathbf{X}$ be a connected n-cube of spaces, and let $X = \mathbf{X}_{<n>}$. It is proved in Loday, 1982, that there is an n-cube of fibrations $\mathbf{Z}$ and maps of n-cubes of fibrations

$$\bar{\mathbf{X}} \xleftarrow{f} \mathbf{Z} \xrightarrow{g} \overline{B\square(\Pi\mathbf{X})}$$

such that f is a level weak homotopy equivalence and g induces an isomorphism of π_1 at each level. Assume now that X is of the homotopy type of a CW-complex. Then from f and g we obtain a map

$$\phi : X \longrightarrow B\Pi\mathbf{X}$$

inducing an isomorphism of π_1, namely the composite, *at this level*, of g with a homotopy inverse of f, and with the map $\mathbf{X}_{<n>} \longrightarrow \bar{\mathbf{X}}_{<n>}$. We do not expect ϕ to be a homotopy equivalence in general, since the n-cube of fibrations $\overline{B\square(\Pi\mathbf{X})}$ has special properties not necessarily satisfied by $\bar{\mathbf{X}}$.

We say an n-cube of spaces $\mathbf{X}$ is an *Eilenberg-Mac Lane n-cube of spaces* if it is connected and all the spaces $\bar{\mathbf{X}}_{A,\emptyset}$ are spaces of type $K(\pi,1)$. A chief example of this is the n-cube of spaces $B\square M$ derived from a crossed n-cube of groups. In fact, $\overline{(B\square M)}_{A,B}$ is not only path-connected but also $(|B| + 1)$-coconnected. This n-cube of fibrations may also be constructed directly in terms of the structure of M, using the techniques of Loday, 1982.

Proposition 3.3. *Let* $\mathbf{X}$ *be a connected* n*-cube of spaces such that* $\mathbf{X}_{<n>}$ *is of the homotopy type of a CW-complex. Suppose that for* $A,B \subseteq <n>$, *such that* $A \cap B = \emptyset$, $i \in <n>\setminus(A \cup B)$, *and* $r = |B|$, *the induced morphism* $\pi_{r+2}\bar{\mathbf{X}}_{A,B} \to \pi_{r+2}\bar{\mathbf{X}}_{A,B\cup\{i\}}$ *is zero. Then the canonical (up to homotopy) map* $\phi : \mathbf{X}_{<n>} \to B\Pi\mathbf{X}$ *is an* $(n+1)$*-equivalence.*

Proof. This is a simple consequence of the five lemma applied by induction on $|B|$ to the maps of homotopy exact sequences of the fibration sequences (3.1) of the n-cubes of fibrations $\bar{\mathbf{X}}$ and $\overline{B\square(\Pi\mathfrak{X})}$. □

Example 3.4. Let M and N be normal subgroups of a group P , and let the space X be given as the homotopy pushout

$$\begin{array}{ccc} K(P,1) & \longrightarrow & K(P/M,1) \\ \downarrow & & \downarrow \\ K(P/N,1) & \longrightarrow & X \end{array}$$

Brown and Loday, 1987a, apply the case $n = 2$ of Theorem 3.1 to show that the above square of spaces has fundamental crossed square given by the 'universal' crossed square

$$\begin{array}{ccc} M \otimes N & \to & M \\ \downarrow & & \downarrow \\ N & \to & P , \end{array} \tag{3.5}$$

where $M \otimes N$ is the non-Abelian tensor product (*loc. cit.*), with generators $m \otimes n$ for $m \in M$ and $n \in N$ and relations

$$mm' \otimes n = ({}^{m}m' \otimes {}^{m}n)(m \otimes n) , \quad m \otimes nn' = (m \otimes n)({}^{n}m \otimes {}^{n}n')$$

for all $m,m' \in M$, $n,n' \in N$. The h-map of this crossed square is $(m,n) \mapsto m \otimes n$. It follows from Proposition 3.3 that the 3-type of X is also given by this crossed square. This result has been stated in Brown, 1989b, 1990, and we have now given the proof. Note that Theorem 2.4 allows one to compute $\eta^* : \pi_2 \to \pi_3$ and the Whitehead product map $\pi_2 \times \pi_2 \to \pi_3$.

By contrast, the Postnikov description of the 3-type of X requires the description of the first k-invariant

$$k^{(3)} \in H^3(P/MN,(M \cap N)/[M,N]) ,$$

which in this case is represented by the crossed module

$M \circ N \longrightarrow P$, where $M \circ N$ is the coproduct of the crossed P-modules M and N (see Brown, 1984, and also Gilbert and Higgins, 1989). This k-invariant determines (up to homotopy) a space $X^{(2)}$, which could be taken to be the classifying space of the above crossed module, constructed either by regarding the crossed module as a crossed 1-cube of groups, or as in Brown and Higgins, 1991. One then needs a second Postnikov invariant

$$k^{(4)} \in H^4(X^{(2)}, Ker(M \otimes N \longrightarrow P)) .$$

This description of the 3-type of X is less explicit than that given by the crossed square (3.5), from which we obtained the homotopy groups and the action of π_1 in the first place. Note also that if M, N, P are finite, then so also is $M \otimes N$ (Ellis, 1987), so that in this case the crossed square (3.5) is finite.

As an example, in this way one finds that if $P = M = N$ is the dihedral group D_n of order $2n$, with generators x and y and relations $x^2 = y^n = xyxy = 1$, where n is even, then the suspension $SK(D_n,1)$ of $K(D_n,1)$ has π_3 isomorphic to $(\mathbb{Z}_2)^4$ generated by the elements of $D_n \otimes D_n$

$$x \otimes x \ , \ (x \otimes y)^{n/2} \ , \ y \otimes y \ , \ (x \otimes y)(y \otimes x) \ .$$

Further, $\eta^*(\bar{x}) = x \otimes x$, $\eta^*(\bar{y}) = y \otimes y$, where $\bar{x}$ and $\bar{y}$ denote the corresponding generators of $\pi_2 SK(D_n,1) = (D_n)^{ab}$ (if n is odd, only the $x \otimes x$ term appears in π_3). The element $(x \otimes y)(y \otimes x)$ is the Whitehead product $[\bar{x},\bar{y}]$. Other computations of η^* and of Whitehead products at this level in spaces $SK(P,1)$ may be deduced from the calculations of non-Abelian tensor products given in Brown, Johnson and Robertson, 1987. (This paper covers the case of dihedral, quaternionic, metacyclic and symmetric groups, and all groups of order ≤ 31.)

Problems in this area are given in Brown, 1990.

4. Proof of Theorem 2.4.

We now explain the results on η^* and Whitehead products in the second part of Theorem 2.4.

Let G be the crossed square (2.2). Then the square of crossed squares $\square G$ may be written in abbreviated form as follows:

$$\begin{array}{ccc} \begin{bmatrix} 1 & 1 \\ 1 & P \end{bmatrix} & \rightarrow & \begin{bmatrix} 1 & 1 \\ N & P \end{bmatrix} \\ \downarrow & & \downarrow \\ \begin{bmatrix} 1 & M \\ 1 & P \end{bmatrix} & \rightarrow & \begin{bmatrix} L & M \\ N & P \end{bmatrix} \end{array} \tag{4.1}$$

Let us write $\mathbf{Y}$ for the square of spaces $B\square G$. Then $\Pi\mathbf{Y}$ is isomorphic to the original crossed square G. Further the 2-cube of fibrations $\bar{\mathbf{Y}}$ associated to $\mathbf{Y}$ is homotopy equivalent to the following diagram:

$$\begin{array}{ccccc} BL & \longrightarrow & BM & \rightarrow & B(L \rightarrow M) \\ \downarrow & & \downarrow & & \downarrow \\ BN & \longrightarrow & BP & \rightarrow & B(N \rightarrow P) \\ \downarrow & & \downarrow & & \downarrow \\ B(L \rightarrow N) & \rightarrow & B(M \rightarrow P) & \rightarrow & B(G) \end{array} \tag{4.3}$$

For a general square of spaces $\mathbf{X}$ as follows

$$\begin{array}{ccc} C & \xrightarrow{f} & A \\ {\scriptstyle g}\downarrow & & \downarrow{\scriptstyle a} \\ B & \xrightarrow{b} & X \end{array} \quad . \tag{4.4}$$

the associated 2-cube of fibrations is equivalent to the following diagram

$$\begin{array}{ccccc} F(\mathbf{X}) & \rightarrow & F(g) & \rightarrow & F(a) \\ \downarrow & & \downarrow & & \downarrow \\ F(f) & \longrightarrow & C & \longrightarrow & A \\ \downarrow & & \downarrow & & \downarrow \\ F(b) & \longrightarrow & B & \longrightarrow & X \end{array} \tag{4.5}$$

where each row and column is a homotopy fibration sequence. So we deduce the second part of Theorem 2.4 from the following more general result.

Proposition 4.6. *Let* $\mathbf{X}$ *be the square of pointed spaces as in (4.4). Suppose that the induced morphism* $\pi_2 C \rightarrow \pi_2 X$ *is zero. Then there is a commutative diagram*

$$\begin{array}{ccccc} \pi_2 X & \xleftarrow{\quad \delta' \quad} & & & \pi_1 F(f) \times_{\pi_1 C} \pi_1 F(g) \\ {\scriptstyle \eta^*}\downarrow & & & & \downarrow{\scriptstyle h'} \\ \pi_3 X & \xrightarrow{\partial} & \pi_2 F(a) & \xrightarrow{\partial'} & \pi_1 F(\mathbf{X}) \end{array} \tag{4.7}$$

in which δ' *is defined by a difference construction,* ∂ , ∂' *are boundaries in homotopy exact sequences of fibrations,* η^* *is induced by composition with the Hopf map* η , *and* h' *is the restriction of the* h*-map of the crossed square* $\Pi\mathbf{X}$.

Proof. This result is a refinement of Lemma 4.2 of Brown and Loday, 1987a. It is proved by similar methods. One first considers the suspension square of S^1 :

$$\begin{array}{ccc} S^1 & \longrightarrow & E^2_+ \\ \downarrow & & \downarrow \\ E^2_- & \longrightarrow & S^2 \end{array} .$$

The fundamental crossed square of this suspension square is determined by Theorem 3.2, compare Example 3.4, as in Brown and Loday, 1987a, and is

$$\begin{array}{ccc} \mathbb{Z} & \xrightarrow{0} & \mathbb{Z} \\ {\scriptstyle 0}\downarrow & & \downarrow{\scriptstyle 1} \\ \mathbb{Z} & \xrightarrow{1} & \mathbb{Z} \end{array}$$

with h-map $\mathbb{Z} \times \mathbb{Z} \to \mathbb{Z}$ given by $(m,n) \mapsto mn$, so that $h(1,1)$ represents the Hopf map η . But the diagram (4.7) for the suspension square of S^1 may now be completely determined, and is the universal example for Proposition 4.6. This completes the proof of the proposition. □

For the proof of the final part of Theorem 2.4 we have to explain how the particular crossed module given in the theorem determines the homotopy 2-type. This is proved by considering the Moore complex of the diagonal simplicial group of the bisimplicial group arising as the nerve of the associated cat^2-group.

5. Simply connected 3-types and crossed squares of Abelian groups.

It is known that the 3-type of a simply connected space X is determined by the homotopy groups $\pi_2 X$, $\pi_3 X$ and the quadratic function $\eta^* : \pi_2 X \to \pi_3 X$ induced by composition with the Hopf map $\eta : S^3 \to S^2$. This essentially results from the fact that for abelian groups A and B the cohomology group $H^4(K(A,2),B)$ is isomorphic to the group of quadratic functions

$A \longrightarrow B$ (Eilenberg and Mac Lane, 1954). The aim of this section is to show that all simply connected 3-types can be modelled by a crossed square of Abelian groups. It is not known if simply connected $(n+1)$-types can be modelled by crossed n-cubes of Abelian groups.

Theorem 5.1. *Let C and D be Abelian groups such that C is finitely generated, and let $t : C \longrightarrow D$ be a quadratic function. Then there is a crossed square*

$$G \qquad \begin{array}{ccc} L & \xrightarrow{\lambda} & M \\ {\scriptstyle \lambda'}\downarrow & & \downarrow{\scriptstyle 1} \\ M & \xrightarrow{-1} & M \end{array}$$

of abelian groups whose classifying space $X = BG$ satisfies $\pi_2 X \cong C$, $\pi_3 X \cong D$ and such that these isomorphisms map η^* to the quadratic map t.

Proof. The quadratic function t has first to be extended to a biadditive map. We use a slight modification of a definition of Eilenberg and Mac Lane, 1954, §18.

Let $t : C \longrightarrow D$ be a quadratic function on Abelian groups C,D. A *biadditive extension* of t is an abelian group M and an epimorphism $\alpha : M \longrightarrow C$ of Abelian groups together with a biadditive map $\phi : M \times M \longrightarrow D$ such that for all $m,m' \in M$

(5.1.1) $\phi(m,m) = t\alpha m$;

(5.1.2) $\phi(m,m') = 0$ if $\alpha m = \alpha m' = 0$;

(5.1.3) $\phi(m,m') = \phi(m',m)$.

It is shown in *loc. cit.* that such a biadditive extension exists assuming C is finitely generated. (In fact they do not assume the symmetry condition (5.1.3), but their proof of existence yields such a ϕ .)

Let $K = Ker\ \alpha$ and let L be the product group $D \times K$. Let M act on L on the left by

$$^{m}(d,k) = (d + \phi(m,k),k) ,$$

for $m \in M$, $d \in D$, $k \in K$. Define $\lambda,\lambda' : L \longrightarrow M$ by

$$\lambda(d,k) = -k \ , \ \lambda'(d,k) = k \ ,$$

for $(d,k) \in L$, and let M act trivially on itself. Then λ

and λ' are M-morphisms, and (5.1.2) shows that they are also crossed modules. Define $h : M \times M \to L$ by

$$h(m,m') = (\phi(m,m'),0)$$

for $m,m' \in M$. A straightforward check shows that we have defined a crossed square G say. The symmetry condition, or even the weaker condition that $\phi(m,m') = \phi(m',m)$ if m or m' lies in K, is used to verify that

$$h(\lambda(d,k),m) = (d,k) - {}^{m}(d,k) .$$

The homotopy groups of BG are computed as the homology groups of the chain complex

$$L \xrightarrow{(-\lambda,\lambda')} M \times M \xrightarrow{\psi} M$$

where $\psi(m,m') = m - m'$. Thus $\pi_2 BG \cong M/K \cong C$, $\pi_3 BG \cong D$. Further $h(m,m) = (\phi(m,m),0) = (t\omega m,0)$. This proves the final assertion of the theorem. □

Note that in the proof of this theorem, while the groups are Abelian, the actions are in general non-trivial. So the associated cat^2-group in general has non-Abelian big group.

Acknowledgments

I would like to thank J.-L. Loday for conversations on the material of this paper. The work was supported by: the British Council; the Université Louis Pasteur, Strasbourg; and the SERC.

References

M.G.Barratt, J.H.C.Whitehead, 'The first non-vanishing group of an $(n + 1)$-ad', *Proc. London Math. Soc.* (3) 6, 417-439, 1956.

H.J.Baues, *Algebraic homotopy*, Cambridge University Press, 1989.

H.J.Baues, *Combinatorial homotopy and 4-dimensional complexes*, De Gruyter, 1991.

H.J.Baues, R.Brown, 'On the relative homotopy groups of the product filtration and a formula of Hopf', *J. Pure Appl. Algebra* (to appear), *UCNW Math. Preprint* 90.25, 1990.

H.J.Baues, D.Conduché, 'The tensor algebra of a non-Abelian group', Max Planck Institut-für-Mathematatik Preprint, 1991.

A.L.Blakers, 'On the relations between homotopy and homology groups', *Ann. Math.*, 49, 428-461, 1948.

R.Brown, 'Groupoids and Van Kampen's theorem', *Proc. London Math. Soc.* (3) 17, 385-401, 1967.

R.Brown, 'Coproducts of crossed P-modules: applications to second homotopy groups and to the homology of groups', *Topology* 23, 337-345, 1984.

R.Brown, *Elements of Modern Topology,* McGraw Hill, Maidenhead, 1968; *Topology: a geometric account of general topology, homotopy types and the fundamental groupoid,* Ellis Horwood, Chichester, 1988.

R. Brown, 'Symme t ry, groupoids, and higher dimensional analogues', *Symmetry II,* Ed. I.Hargittai, *Computers Math. Applic.* 17, 49-57, 1989a.

R.Brown, 'Triadic Van Kampen theorems and Hurewicz theorems', *Proc. Int. Conf. Algebraic Topology, Evanston, 1988* ed. M.Mahowald, *Cont. Math.* 96, 113-134, 1989b.

R.Brown, 'Some problems in non-Abelian homological and homotopical algebra', *Homotopy theory and related topics: Proceedings Kinosaki, 1988,* Edited M.Mimura, Springer Lecture Notes in Math. 1418, 105-129, 1990a.

R.Brown, N.D.Gilbert, 'Algebraic models of 3-types and automorphism structures for crossed modules', *Proc. London Math. Soc.* (3) 59, 51-73, 1989.

R.Brown, N.D.Gilbert, H.-J.Hoehnke, J.Shrimpton, 'Non-abelian tensor products of groups; groupoids; quadratic forms; and higher order symmetry', in *The heritage of C.F.Gauss,* ed. G.M.Rassias, World Scientific (to appear), Bangor Maths Preprint 91.14 (1991).

R.Brown, P.J.Higgins, 'On the connection between the second rela tive homotopy groups of some related spaces', *Proc. London Math. Soc.* (3), 36 , 193-212, 1978.

R.Brown, P.J.Higgins, 'The algebra of cubes', *J. Pure Appl. Algebra,* 21, 233-260, 1981a.

R.Brown, P.J.Higgins, 'Colimit theorems for relative homotopy groups', *J. Pure Appl. Algebra,* 22, 11-41, 1981b.

R.Brown, P.J.Higgins, 'Tensor products and homotopies for ω-groupoids and crossed complexes', *J. Pure Appl. Algebra,* 47, 1-33, 1987.

R.Brown, P.J.Higgins, 'Crossed complexes and chain complexes with operators', *Math. Proc. Camb. Phil. Soc.* 107, 33-57, 1989.

R.Brown, P.J.Higgins, 'The classifying space of a crossed complex', *Math. Proc. Camb. Phil. Soc.* (to appear), 1991.

R.Brown, D.L.Johnson, E.F.Robertson, 'Some computations of non-abelian tensor product of groups', *J. Algebra,* 111, 177-202, 1987.

R.Brown, J.-L.Loday, 'Van Kampen theorems for diagrams of spaces', *Topology,* 26, 311-335, 1987a.

R.Brown, J.-L.Loday, 'Homotopical excision, and Hurewicz theorems, for n-cubes of spaces', *Proc. London Math. Soc.* (3), 54, 176-192, 1987b.

R.Brown, C.B.Spencer, 'Double groupoids and crossed modules', *Cah. Top. Geom. Diff.* , 17, 343-362, 1976.

E.Cech, 'Höherdimensionale homotopiegruppen', *Verhandlungen des Internationalen Mathematiker-Kongresses,* Zürich, 1932, Band 2, p.203, 1932.

J.-M.Cordier, T.Porter, 'Fibrant diagrams, rectification and a construction of Loday', *J. Pure Appl. Algebra* 67, 111-124, 1990.

J.Dieudonné, *A history of algebraic and differential topology 1900-1960*, Birkhäuser, Boston-Basel, 1989.

D.A.Edwards, H.M.Hastings, *Cech and Steenrod homotopy theories with applications to geometric topology*, Lecture Notes in Mathematics 542, Springer, Berlin, 1976.

C.Ehresmann, 'Catégories structurées', *Ann. Sci. Ec. Norm. Sup.* 80, 349-426, 1963; (*Oeuvres completes et commentees, Partie* III-1, pp24-98).

S.Eilenberg, S.Mac Lane, 'On the groups $H(\pi,n)$ III', *Annals of Math.*, 60, 513-557, 1954.

G.J.Ellis, 'The non-abelian tensor product of finite groups is finite', *J. Algebra*, 111, 203-205, 1987.

G.J.Ellis, R.Steiner, 'Higher-dimensional crossed modules and the homotopy groups of $(n+1)$-ads', *J. Pure Appl. Algebra*, 46, 117-136, 1987.

N.D.Gilbert, 'The fundamental cat^n-group of an n-cube of spaces', *Proc. Conf. Algebraic Topology, Barcelona, 1986* (ed. J.Aguadé and R.Kane) Springer Lecture Notes in Math. 1298, 124-139, 1988.

N.D.Gilbert, P.J.Higgins, 'The non-Abelian tensor product of groups and related constructions', *Glasgow Math. J.*, 31, 17-29, 1989.

A.Grothendieck, *Pursuing stacks,* 600pp, (distributed from Bangor), 1983.

D.Holt, 'An interpretation of the cohomology groups $H^n(G,M)$', *J. Algebra*, 16, 307-318, 1979.

S.-T.Hu, 'The Homotopy Addition Theorem', *Ann. Math.*58, 108-122, 1953.

J.Huebschmann, 'Crossed n-fold extensions and cohomology', *Comm. Math. Helv.* 55, 302-314, 1980.

J.Huebschmann, 'Automorphisms of group extensions and differentials in the Lyndon-Hochschild-Serre spectral sequence', *J. Algebra* 72, 296-334, 1981.

W.Hurewicz, 'Beiträge zur Topologie der deformationen I-IV', *Nederl. Akad. Wetensc. Proc.* Ser. A, 38, 112-119, 521-528, 1935; 39, 117-126, 215-224, 1936.

D.W.Jones, *Polyhedral T-complexes*, University of Wales PhD Thesis, 1984; published as *A general theory of polyhedral sets and their corresponding T-complexes*, Diss. Math. 266, 1988.

J.-L.Loday, 'Spaces with finitely many non-trivial homotopy groups', *J. Pure Appl. Algebra*, 24, 179-202, 1982.

A.S.-T.Lue, 'Cohomology of groups relative to a variety', *J. Algebra*, 69, 155-174, 1981.

S.Mac Lane, 'Historical note', *J. Algebra*, 60, 319-320, 1979.

S.Mac Lane, J.H.C.Whitehead, 'On the 3-type of a complex', *Proc. Nat. Acad. Sci. Washington* 36, 41-58, 1950.

T.Porter, 'A combinatorial definition of n-types', *UCNW Math. Preprint* 90.10, 49pp, 1990.

R.Steiner, 'Resolutions of spaces by n-cubes of fibrations', *J. London Math. Soc.*(2), 34, 169-176, 1986.

G.W.Whitehead, *Elements of homotopy theory*, Springer, Berlin, 1978.

J.H.C.Whitehead, 'Combinatorial homotopy II', *Bull. Amer. Math. Soc.*, 55, 453-496, 1949.

ON ORTHOGONAL PAIRS IN CATEGORIES AND LOCALISATION

Carles Casacuberta, Georg Peschke and Markus Pfenniger

In memory of Frank Adams

0 Introduction

Special forms of the following situation are often encountered in the literature: Given a class of morphisms $\mathcal{M}$ in a category $\mathcal{C}$, consider the full subcategory $\mathcal{D}$ of objects $X \in \mathcal{C}$ such that, for each diagram

$$\begin{array}{ccc} A & \xrightarrow{f} & B \\ {\scriptstyle g}\downarrow & & \\ X & & \end{array}$$

with $f \in \mathcal{M}$, there is a unique morphism $h: B \to X$ with $hf = g$. The *orthogonal subcategory problem* [13] asks whether $\mathcal{D}$ is reflective in $\mathcal{C}$, i.e., under which conditions the inclusion functor $\mathcal{D} \to \mathcal{C}$ admits a left adjoint $E: \mathcal{C} \to \mathcal{D}$; see [17]. Many authors have given conditions on the category $\mathcal{C}$ and the class of morphisms $\mathcal{M}$ ensuring the reflectivity of $\mathcal{D}$, sometimes even providing an explicit construction of the left adjoint $E: \mathcal{C} \to \mathcal{D}$; see for example Adams [1], Bousfield [3, 4], Deleanu-Frei-Hilton [9, 10], Heller [15], Yosimura [22], Dror-Farjoun [11], Kelly [12]. The functor E is often referred to as a *localisation functor* of $\mathcal{C}$ at the subcategory $\mathcal{D}$. Most of the known existence results of left adjoints work well when the category $\mathcal{C}$ is cocomplete [12] or complete [19]. Unfortunately, these methods cannot be directly applied to the homotopy category of CW-complexes, as it is neither complete nor cocomplete. This difficulty is often circumvented by resorting to semi-simplicial techniques.

In this paper we offer a construction of localisation functors depending only on the availability of certain weak colimits in the category $\mathcal{C}$. From a technical point of view, the existence of such weak colimits reduces our arguments essentially to the situation in cocomplete categories. From a practical point of view, however, our result is a simple recipe for the explicit construction of localisation functors. It unifies a number of constructions created for specific purposes; cf. [4, 18, 20]. In fact, its scope goes beyond these applications: For example, it can be used to show that there is a whole family of functors extending P-localisation of nilpotent homotopy types to the homotopy category of all CW-complexes. We deal with this issue in [7], where we discuss the geometric significance of these functors as well as their interdependence.

Section 1 of the present paper contains background, followed by the statement and proof of our main result: the affirmative solution of the orthogonal subcategory problem in a wide range of cases. In Section 2 we discuss extensions of a localisation functor in a category $\mathcal{C}$ to localisation functors in supercategories of $\mathcal{C}$. Our results allow us to give, in Section 3, a uniform existence proof for various localisation functors and also to explain their interrelation. The basic features of our project have been outlined in [8].

Acknowledgements. We are indebted to Emmanuel Dror-Farjoun, discussions with whom significantly helped the present development. We are also grateful to the CRM of Barcelona for the hospitality extended to the authors.

1 Orthogonal pairs and localisation functors

We begin by explaining the basic categorical notions we shall use. Our main sources are [1, 3, 4, 13].

A morphism $f: A \to B$ and an object X in a category $\mathcal{C}$ are said to be *orthogonal* if the function

$$f^*: \mathcal{C}(B,X) \to \mathcal{C}(A,X)$$

is bijective, where $\mathcal{C}(\ ,\ )$ denotes the set of morphisms between two given objects of $\mathcal{C}$. For a class of morphisms $\mathcal{M}$, we denote by $\mathcal{M}^\perp$ the class of objects orthogonal to each $f \in \mathcal{M}$. Similarly, for a class of objects $\mathcal{O}$, we denote by $\mathcal{O}^\perp$ the class of morphisms orthogonal to each $X \in \mathcal{O}$.

Definition 1.1 An *orthogonal pair* in $\mathcal{C}$ is a pair $(\mathcal{S},\mathcal{D})$ consisting of a class of morphisms $\mathcal{S}$ and a class of objects $\mathcal{D}$ such that $\mathcal{D}^\perp = \mathcal{S}$ and $\mathcal{S}^\perp = \mathcal{D}$.

If (E, η) is an idempotent monad [1] in $\mathcal{C}$, then the classes

$$\mathcal{S} = \{f: A \to B \mid Ef: EA \cong EB\}$$

$$\mathcal{D} = \{X \mid \eta_X: X \cong EX\}$$

form an orthogonal pair (note that these could easily be proper classes). The morphisms in $\mathcal{S}$ are then called *E-equivalences* and the objects in $\mathcal{D}$ are said to be *E-local*. Not every orthogonal pair $(\mathcal{S}, \mathcal{D})$ arises from an idempotent monad in this way; cf. [19]. If so, we call E the *localisation functor* associated with $(\mathcal{S}, \mathcal{D})$. Then the full subcategory of objects in $\mathcal{D}$ is reflective and E is left adjoint to the inclusion $\mathcal{D} \to \mathcal{C}$. The following proposition enables us to detect localisation functors.

Proposition 1.2 *Let $\mathcal{C}$ be a category and $(\mathcal{S}, \mathcal{D})$ an orthogonal pair in $\mathcal{C}$. If for each object X there exists a morphism $\eta_X: X \to EX$ in $\mathcal{S}$ with EX in $\mathcal{D}$, then*

(i) *η_X is terminal among the morphisms in $\mathcal{S}$ with domain X;*

(ii) *η_X is initial among the morphisms of $\mathcal{C}$ from X to an object of $\mathcal{D}$;*

(iii) *The assignment $X \mapsto EX$ defines a localisation functor on $\mathcal{C}$ associated with $(\mathcal{S}, \mathcal{D})$.*

For each class of morphisms $\mathcal{M}$, the pair $(\mathcal{M}^{\perp\perp}, \mathcal{M}^{\perp})$ is orthogonal. We say that this pair is *generated* by $\mathcal{M}$ and call $\mathcal{M}^{\perp\perp}$ the *saturation* of $\mathcal{M}$. If $\mathcal{M}^{\perp\perp} = \mathcal{M}$, then $\mathcal{M}$ is said to be *saturated*. This terminology applies to objects as well. Note that if $(\mathcal{S}, \mathcal{D})$ is an orthogonal pair then both $\mathcal{S}$ and $\mathcal{D}$ are saturated. The next properties of saturated classes are easily checked and well-known in a slightly more general context [3, 13].

Lemma 1.3 *If a class of morphisms $\mathcal{S}$ is saturated, then*

(i) *$\mathcal{S}$ contains all isomorphisms of $\mathcal{C}$.*

(ii) *If the composition gf of two morphisms is defined and any two of f, g, gf are in $\mathcal{S}$, then the third is also in $\mathcal{S}$.*

(iii) *Whenever the coproduct of a family of morphisms of $\mathcal{S}$ exists, it is in the class $\mathcal{S}$.*

(iv) *If the diagram*

$$\begin{array}{ccc} A & \xrightarrow{s} & B \\ \downarrow & & \downarrow \\ C & \xrightarrow{t} & D \end{array}$$

is a push-out in which $s \in \mathcal{S}$, then $t \in \mathcal{S}$.

(v) *If α is an ordinal and $F: \alpha \to \mathcal{C}$ is a directed system with direct limit T, such that for each $i < \alpha$ the morphism $s_i: F(0) \to F(i)$ is in $\mathcal{S}$, then $s_\alpha: F(0) \to T$ is in $\mathcal{S}$.*

We call a class of morphisms $\mathcal{S}$ *closed* in $\mathcal{C}$ if it satisfies (i), (ii) and (iii) in Lemma 1.3 above. We restrict attention to closed classes from now on.

We proceed with the statement of our main result. Recall that a *weak colimit* of a diagram is defined by requiring only existence, without insisting on uniqueness, in the defining universal property [17].

Theorem 1.4 *Let $\mathcal{C}$ be a category with coproducts and let $\mathcal{S}$ be a closed class of morphisms in $\mathcal{C}$. Suppose that:*

(C1) *There is a set $\mathcal{S}_0 \subseteq \mathcal{S}$ with $\mathcal{S}_0^\perp = \mathcal{S}^\perp$.*

(C2) *For every diagram $C \overset{f}{\leftarrow} A \overset{s}{\rightarrow} B$ with $s \in \mathcal{S}$ there exists a weak push-out*

$$\begin{array}{ccc} A & \overset{s}{\rightarrow} & B \\ {\scriptstyle f}\downarrow & & \downarrow \\ C & \overset{t}{\rightarrow} & Z \end{array}$$

with $t \in \mathcal{S}$.

(C3) *There is an ordinal α such that, for every $\beta \leq \alpha$, every directed system $F\colon \beta \to \mathcal{C}$ in which the morphisms $s_i\colon F(0) \to F(i)$ are in $\mathcal{S}$ for $i < \beta$ admits a weak direct limit T satisfying*

(a) *the morphism $s_\beta\colon F(0) \to T$ is in $\mathcal{S}$;*

(b) *for each $s\colon A \to B$ in $\mathcal{S}_0$, every morphism $f\colon A \to T$ factors through $f'\colon A \to F(i)$ for some $i < \alpha$;*

(c) *if two morphisms $g_1, g_2\colon B \to T$ satisfy $g_1 s = g_2 s$ with $s\colon A \to B$ in $\mathcal{S}_0$, then they factor through $g_1', g_2'\colon B \to F(i)$ for some $i < \alpha$, in such a way that $g_1' s = g_2' s$.*

Then the class $\mathcal{S}$ is saturated and the orthogonal pair $(\mathcal{S}, \mathcal{S}^\perp)$ admits a localisation functor E. Furthermore, for each object X, the localising morphism $\eta_X : X \to EX$ can be constructed by means of a weak direct limit indexed by α.

PROOF. For each morphism $s\colon A \to B$ in $\mathcal{S}_0$ fix a weak push-out

$$\begin{array}{ccc} A & \overset{s}{\rightarrow} & B \\ {\scriptstyle s}\downarrow & & \downarrow{\scriptstyle t_2} \\ B & \overset{t_1}{\rightarrow} & Z_s \end{array}$$

in which $t_1 \in \mathcal{S}$. Then also $t_2 \in \mathcal{S}$ because $\mathcal{S}$ is closed.

Remark 1.5 With applications in mind, it is worth observing that part (c) of hypothesis (C3) in Theorem 1.4 is satisfied if each map $f\colon Z_s \to T$ factors through $f'\colon Z_s \to F(i)$ for some $i < \alpha$.

Choose next a morphism $u_s\colon Z_s \to B$ rendering commutative the diagram

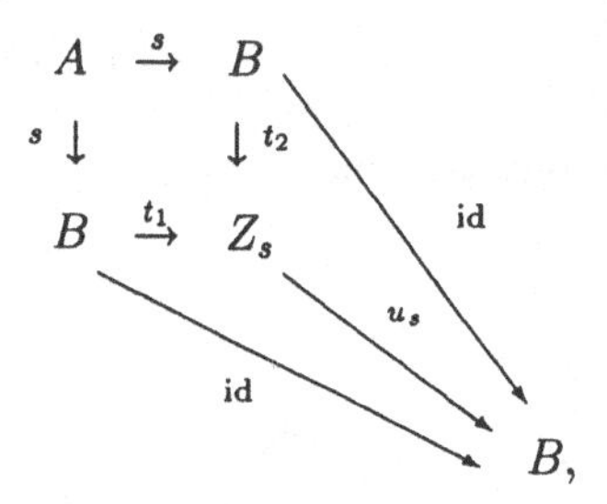

and note that $u_s \in \mathcal{S}$. Write $\mathcal{D}$ for $\mathcal{S}^{\perp}$. We shall construct, for each object $X \in \mathcal{C}$, a morphism $\eta_X : X \to EX$ with $EX \in \mathcal{D}$ and $\eta_X \in \mathcal{S}$. Set $X_0 = X$. Given $i < \alpha$, assume that X_i has been constructed, together with a morphism $X \to X_i$ belonging to $\mathcal{S}$. Define a morphism $\sigma_i\colon X_i \to X_{i+1}$ as follows: For each $s\colon A \to B$ in the set $\mathcal{S}_0$, consider all morphisms $\varphi\colon A \to X_i$ and $\psi : Z_s \to X_i$ for which no factorisation through $s : A \to B$, resp. $u_s : Z_s \to B$, exists (if there are no such morphisms, then $X_i \in \mathcal{D}$ and we may set $EX = X_i$). Choose a weak push-out

$$\begin{array}{ccc} \coprod_{s\in\mathcal{S}_0}((\coprod_{\varphi} A)\amalg(\coprod_{\psi} Z_s)) & \xrightarrow{\phi} & \coprod_{s\in\mathcal{S}_0}((\coprod_{\varphi} B)\amalg(\coprod_{\psi} B)) \\ {\scriptstyle f}\downarrow & & \downarrow \\ X_i & \xrightarrow{\sigma_i} & X_{i+1} \end{array}$$

with $\sigma_i \in \mathcal{S}$, in which f is the coproduct morphism and ϕ is the corresponding coproduct of copies of $s : A \to B$ and $u_s : Z_s \to B$ (which is therefore a morphism in $\mathcal{S}$). Iterate this procedure until reaching the ordinal α. If $\beta \le \alpha$ is a limit ordinal, define X_β by choosing a weak direct limit of the system $\{X_i,\, i < \beta\}$, according to (C3). Set $EX = X_\alpha$. The construction guarantees that the composite morphism $\eta_X : X \to EX$ is in $\mathcal{S}$. We claim that $EX \in \mathcal{D}$. Since $\mathcal{D} = \mathcal{S}_0^{\perp}$, it suffices to check that EX is orthogonal to each morphism in $\mathcal{S}_0$. Take a diagram

$$\begin{array}{ccc} A & \xrightarrow{s} & B \\ {\scriptstyle f}\downarrow & & \\ EX & & \end{array}$$

with $s \in \mathcal{S}_0$. Then f factors through $f'\colon A \to X_i$ for some $i < \alpha$ and hence, either f' factors through $s\colon A \to B$, or there is a commutative diagram

$$\begin{array}{ccc} A & \xrightarrow{s} & B \\ {\scriptstyle f'}\downarrow & & \downarrow{\scriptstyle g'} \\ X_i & \xrightarrow{\sigma_i} & X_{i+1} \end{array}$$

which provides a morphism $g\colon B \to EX$ such that $gs = f$. Now suppose that there are two maps $g_1, g_2\colon B \to EX$ with $g_1 s = g_2 s = f$. Then we can choose an object X_i with $i < \alpha$, and morphisms $g_1', g_2'\colon B \to X_i$ such that $g_1' s = g_2' s$. Using the weak push-out property of Z_s, we obtain a morphism $h\colon Z_s \to X_i$ rendering commutative the diagram

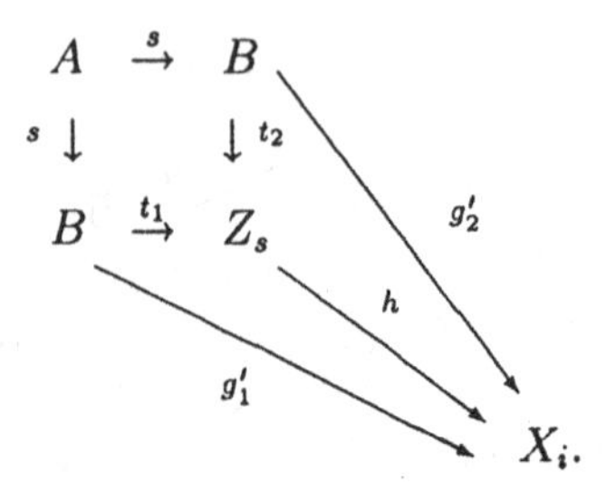

Then, either h factors through $u_s : Z_s \to B$ and $g_1' = g_2'$, or there is a commutative diagram

$$\begin{array}{ccc} Z_s & \xrightarrow{u_s} & B \\ h\downarrow & & \downarrow k \\ X_i & \xrightarrow{\sigma_i} & X_{i+1} \end{array}$$

which yields

$$\sigma_i g_1' = \sigma_i h t_1 = k u_s t_1 = k = k u_s t_2 = \sigma_i h t_2 = \sigma_i g_2'$$

and hence $g_1 = g_2$. This shows that $EX \in \mathcal{D}$.

To complete the proof it remains to show that $\mathcal{S}^{\perp\perp} = \mathcal{S}$. The inclusion $\mathcal{S} \subseteq \mathcal{S}^{\perp\perp}$ is trivial. For the converse, let $f : A \to B$ be orthogonal to all objects in $\mathcal{D}$. Since $\eta_A : A \to EA$ is in $\mathcal{S}$ and $EB \in \mathcal{D}$, there is a unique morphism Ef rendering commutative the diagram

$$\begin{array}{ccc} A & \xrightarrow{f} & B \\ \eta_A\downarrow & & \downarrow \eta_B \\ EA & \xrightarrow{Ef} & EB. \end{array}$$

But $\eta_B f$ is orthogonal to EA and this provides a morphism $g\colon EB \to EA$ which is two-sided inverse to Ef. Hence Ef is an isomorphism and $f \in \mathcal{S}$ because $\mathcal{S}$ is closed. □

Given an orthogonal pair $(\mathcal{S}, \mathcal{D})$, the class $\mathcal{S}$ is saturated and, a fortiori, closed. Therefore

Corollary 1.6 *Let $\mathcal{C}$ be a category with coproducts and $(\mathcal{S}, \mathcal{D})$ an orthogonal pair in $\mathcal{C}$. Suppose that some set $\mathcal{S}_0 \subseteq \mathcal{S}$ generates the pair $(\mathcal{S}, \mathcal{D})$ and that the class $\mathcal{S}$ satisfies conditions* (C2) *and* (C3) *in Theorem* 1.4. *Then the pair $(\mathcal{S}, \mathcal{D})$ admits a localisation functor E.*

Moreover, if the category $\mathcal{C}$ is cocomplete, then it follows from Lemma 1.3 that for each orthogonal pair $(\mathcal{S}, \mathcal{D})$ condition (C2) and part (a) of condition (C3) are automatically satisfied. This leads to Corollary 1.7 below. An object X has been called *presentable* [14] or *s-definite* [3] if, for some sufficiently large ordinal α, the functor $\mathcal{C}(X, \ \ )$ preserves direct limits of directed systems $F\colon \alpha \to \mathcal{C}$. For example, all groups are presentable [3]. For finitely presented groups it suffices to take α to be the first infinite ordinal.

Corollary 1.7 [3] *Let $\mathcal{C}$ be a cocomplete category. Let $(\mathcal{S}, \mathcal{D})$ be the orthogonal pair generated by an arbitrary set $\mathcal{S}_0$ of morphisms of $\mathcal{C}$. Suppose that the domains and codomains of morphisms in $\mathcal{S}_0$ are presentable. Then $(\mathcal{S}, \mathcal{D})$ admits a localisation functor.*

Since any colimit of presentable objects is again presentable, the following definition together with the results of [19] imply Corollary 1.9 below.

Definition 1.8 A set $\{E_\alpha\}$ of objects of a category $\mathcal{C}$ is a *cogenerator set* of $\mathcal{C}$ if any morphism $f\colon X \to Y$ of $\mathcal{C}$ inducing bijections $f_*\colon \mathcal{C}(E_\alpha, X) \cong \mathcal{C}(E_\alpha, Y)$ for each α, is an isomorphism.

Corollary 1.9 *Let $\mathcal{C}$ be a cocomplete category. Suppose that $\mathcal{C}$ has a cogenerator set whose elements are presentable. Then any orthogonal pair generated by an arbitrary set of morphisms of $\mathcal{C}$ admits a localisation functor.*

2 Extending localisation functors

Let E be a localisation functor on the subcategory $\mathcal{C}'$ of $\mathcal{C}$. We wish to discuss extensions of E over $\mathcal{C}$. Familiar examples include the extension of P-localisation of abelian groups to nilpotent groups and further to all groups. Two problems arise here: existence —for which we often refer to Theorem 1.4— and uniqueness. An appropriate setting for discussing the latter is obtained by partially ordering the collection of all orthogonal pairs in $\mathcal{C}$ as follows: For two given orthogonal pairs $(\mathcal{S}_1, \mathcal{D}_1)$, $(\mathcal{S}_2, \mathcal{D}_2)$ in $\mathcal{C}$ we write $(\mathcal{S}_1, \mathcal{D}_1) \geq (\mathcal{S}_2, \mathcal{D}_2)$ if $\mathcal{D}_1 \supseteq \mathcal{D}_2$ (or, equivalently, if $\mathcal{S}_1 \subseteq \mathcal{S}_2$).

Remark 2.1 If E_1, E_2 are localisation functors associated to $(\mathcal{S}_1, \mathcal{D}_1)$ and $(\mathcal{S}_2, \mathcal{D}_2)$ respectively, and if $(\mathcal{S}_1, \mathcal{D}_1) \geq (\mathcal{S}_2, \mathcal{D}_2)$, then there is a natural transformation of functors $E_1 \to E_2$. In fact, the restriction of E_2 to $\mathcal{D}_1$ is left adjoint to the inclusion $\mathcal{D}_2 \to \mathcal{D}_1$.

An orthogonal pair $(\mathcal{S},\mathcal{D})$ of $\mathcal{C}$ is said to *extend* the orthogonal pair $(\mathcal{S}',\mathcal{D}')$ of the subcategory $\mathcal{C}'$ if both $\mathcal{S}'\subseteq\mathcal{S}$ and $\mathcal{D}'\subseteq\mathcal{D}$. The collection of all extensions of $(\mathcal{S}',\mathcal{D}')$ is partially ordered. Moreover we have

Proposition 2.2 *Let $\mathcal{C}'$ be a subcategory of $\mathcal{C}$ and $(\mathcal{S}',\mathcal{D}')$ an orthogonal pair in $\mathcal{C}'$. If $(\mathcal{S},\mathcal{D})$ is an extension of $(\mathcal{S}',\mathcal{D}')$ to $\mathcal{C}$, then*

$$((\mathcal{S}')^{\perp\perp},(\mathcal{S}')^{\perp})\geq(\mathcal{S},\mathcal{D})\geq((\mathcal{D}')^{\perp},(\mathcal{D}')^{\perp\perp}),$$

where orthogonality is meant in $\mathcal{C}$.

In this situation, we call the orthogonal pair in $\mathcal{C}$ generated by the class $\mathcal{S}'$ the *maximal extension* of $(\mathcal{S}',\mathcal{D}')$, and the one generated by $\mathcal{D}'$ the *minimal extension*. A convenient tool for recognising such extremal extensions is given in the next proposition.

Proposition 2.3 *Let $\mathcal{C}'$ be a subcategory of $\mathcal{C}$, $(\mathcal{S}',\mathcal{D}')$ an orthogonal pair in $\mathcal{C}'$, and $(\mathcal{S},\mathcal{D})$ an extension of $(\mathcal{S}',\mathcal{D}')$ to $\mathcal{C}$. Then*

(a) *$(\mathcal{S},\mathcal{D})$ is the maximal extension of $(\mathcal{S}',\mathcal{D}')$ if and only if there is a subclass $\mathcal{S}_0\subseteq\mathcal{S}'$ such that $\mathcal{S}_0^{\perp}\subseteq\mathcal{D}$.*

(b) *$(\mathcal{S},\mathcal{D})$ is the minimal extension of $(\mathcal{S}',\mathcal{D}')$ if and only if there is a subclass $\mathcal{D}_0\subseteq\mathcal{D}'$ such that $\mathcal{D}_0^{\perp}\subseteq\mathcal{S}$.*

Of course $(\mathcal{S}',\mathcal{D}')$ admits a unique extension to $\mathcal{C}$ if and only if the minimal and the maximal extensions coincide.

Example 2.4 Let $\mathcal{C}$ be the category of finite groups and $\mathcal{C}'$ the subcategory of finite nilpotent groups. Fix a prime p and consider the orthogonal pair $(\mathcal{S}',\mathcal{D}')$ in $\mathcal{C}'$ associated to p-localisation [16]. The class $\mathcal{D}'$ consists of all p-groups, and the orthogonal pair $(\mathcal{S},\mathcal{D})=((\mathcal{D}')^{\perp},\mathcal{D}')$ in $\mathcal{C}$ is both the maximal and the minimal extension of $(\mathcal{S}',\mathcal{D}')$ to $\mathcal{C}$. The pair $(\mathcal{S},\mathcal{D})$ admits a localisation functor —namely, mapping each finite group G onto its maximal p-quotient—, which is therefore the unique extension to all finite groups of the p-localisation of finite nilpotent groups.

3 Applications of the basic existence result

Examples 3.1, 3.2 and 3.3 below discuss well-known functors, each of whose constructions may be viewed as particular cases of Theorem 1.4. Examples 3.4 to 3.7 are new.

Example 3.1 Let $\mathcal{H}_1$ be the pointed homotopy category of simply-connected CW-complexes, and P a set of primes. The P-localisation functor described by Sullivan [21] is associated to the orthogonal pair $(\mathcal{S}, \mathcal{D})$ generated by the set

$$\mathcal{S}_0 = \{\rho_n^k \colon S^k \to S^k \mid \deg \rho_n^k = n,\ k \geq 2,\ n \in P'\},$$

where P' denotes the set of primes not in P. Objects in $\mathcal{D}$ are simply-connected CW-complexes whose homotopy groups are $\mathbf{Z}_P$-modules. Morphisms in $\mathcal{S}$ are $H_*(\ ;\mathbf{Z}_P)$-equivalences. The hypotheses of Corollary 1.6 are fulfilled by taking α to be the first infinite ordinal and using homotopy colimits.

Example 3.2 Let $\mathcal{H}$ denote the pointed homotopy category of connected CW-complexes and h_* an additive homology theory. Take $\mathcal{S}$ to be the class of morphisms $f \colon X \to Y$ inducing an isomorphism $f_* \colon h_*(X) \cong h_*(Y)$. We know from [4] that $\mathcal{S}$ satisfies the hypotheses of Theorem 1.4: Choose α to be the smallest infinite ordinal whose cardinality is bigger than the cardinality of $h_*(\mathrm{pt})$; the collection of all CW-inclusions $\varphi \colon A \to B$ with $h_*(\varphi) = 0$ and $\mathrm{card}(B) < \mathrm{card}(\alpha)$ represents a set $\mathcal{S}_0$ with $\mathcal{S}_0^\perp = \mathcal{S}^\perp$.

In the case $h_* = H_*(\ ;\mathbf{Z}_P)$, the corresponding orthogonal pair $(\mathcal{S}, \mathcal{D})$ extends the pair $(\mathcal{S}', \mathcal{D}')$ associated with P-localisation of nilpotent spaces (see [4]). It is indeed the *minimal* extension of $(\mathcal{S}', \mathcal{D}')$, because the spaces $K(\mathbf{Z}_P, n)$, $n \geq 1$, belong to $\mathcal{D}'$ (cf. Proposition 2.3).

Example 3.3 Let $\mathcal{G}$ be the category of groups and P a set of primes. The P-localisation functor described by Ribenboim [20] is associated to the orthogonal pair $(\mathcal{S}, \mathcal{D})$ generated by the set

$$\mathcal{S}_0 = \{\rho_n \colon \mathbf{Z} \to \mathbf{Z} \mid \rho_n(1) = n,\ n \in P'\}.$$

Groups in $\mathcal{D}$ are those in which P'-roots exist and are unique. Such groups have been studied for several decades (see [2, 20] and the references there). The hypotheses of Theorem 1.4 are readily checked (use Corollary 1.7). We may choose α to be the first infinite ordinal. We denote by $l \colon G \to G_P$ the P-localisation homomorphism.

If $(\mathcal{S}', \mathcal{D}')$ is the orthogonal pair corresponding to P-localisation of nilpotent groups, then, since $\mathcal{S}_0 \subset \mathcal{S}'$, Proposition 2.3 implies that $(\mathcal{S}, \mathcal{D})$ is the *maximal* extension of $(\mathcal{S}', \mathcal{D}')$. In particular, for each group G there is a natural homomorphism from G_P to the Bousfield $H\mathbf{Z}_P$-localisation of G (cf. [5]).

Example 3.4 Example 3.3 can be generalised to the category $\mathcal{C}$ of π-groups for a fixed group π; that is, objects are groups with a π-action and morphisms are action-preserving group homomorphisms. Let $F(\xi)$ be the free π-group on one generator (it can be explicitly described as the free group on the symbols ξ^x, $x \in \pi$, with the obvious left π-action; cf. [18]). Define a π-homomorphism $\rho_{n,x}\colon F(\xi) \to F(\xi)$ for each $x \in \pi$, $n \in \mathbf{Z}$, by the rule

$$\rho_{n,x}(\xi) = \xi(x \cdot \xi)(x^2 \cdot \xi) \dots (x^{n-1} \cdot \xi)$$

and consider the set of morphisms

$$\mathcal{S}_0 = \{\rho_{n,x}\colon F(\xi) \to F(\xi) \mid x \in \pi,\ n \in P'\}.$$

By Corollary 1.7, the orthogonal pair $(\mathcal{S}, \mathcal{D})$ generated by $\mathcal{S}_0$ admits a localisation functor. It again suffices to take the first infinite ordinal as α in the construction. Example 3.3 is the special case $\pi = \{1\}$.

We extend the term *P-local* to the π-groups in $\mathcal{D}$ and the term *P-equivalences* to the morphisms in $\mathcal{S}$. They are particularly relevant to the next example.

Example 3.5 This example is extracted from [7]. Let $\mathcal{H}$ be the pointed homotopy category of connected CW-complexes and P a set of primes. We consider the class $\mathcal{D}$ of those spaces X in $\mathcal{H}$ for which the power map $\rho_n\colon \Omega X \to \Omega X$, $\rho_n(\omega) = \omega^n$ is a homotopy equivalence for all $n \in P'$. Then there exists a set of morphisms $\mathcal{S}_0$ such that $\mathcal{S}_0^\perp = \mathcal{D}$, namely

$$\mathcal{S}_0 = \{\rho_n^k\colon S^1 \wedge (S^k \cup \mathrm{pt}) \to S^1 \wedge (S^k \cup \mathrm{pt}) \mid k \geq 0,\ n \in P'\},$$

where $\rho_n^k = \rho_n \wedge \mathrm{id}$, $\rho_n : S^1 \to S^1$ denotes the standard map of degree n, and pt denotes a disjoint basepoint. Morphisms in $\mathcal{S} = \mathcal{D}^\perp$ turn out to be those $f : X \to Y$ for which $f_* : \pi_1(X) \to \pi_1(Y)$ is a P-equivalence of groups and $f_*\colon H_*(X; A) \to H_*(Y; A)$ is an isomorphism for each abelian $\pi_1(Y)_P$-group A which is P-local in the sense of Example 3.4. The conditions of Corollary 1.6 are satisfied. One can take α to be the first infinite ordinal. Spaces in $\mathcal{D}$ will be called *P-local* and maps in $\mathcal{S}$ will be called *P-equivalences.* We denote the P-localisation map by $l\colon X \to X_P$. The pair $(\mathcal{S}, \mathcal{D})$ extends the pair $(\mathcal{S}', \mathcal{D}')$ corresponding to P-localisation of nilpotent spaces.

Since the orthogonal pair corresponding to $H_*(\ ;\mathbf{Z}_P)$-localisation is minimal among those pairs extending P-localisation of nilpotent spaces (see Example 3.2), for each space X there is a natural map from X_P to the $H_*(\ ;\mathbf{Z}_P)$-localisation of X.

Example 3.6 Let $\mathcal{H}$ denote the pointed homotopy category of connected CW-complexes and P a set of primes. Consider the orthogonal pair $(\mathcal{S}, \mathcal{D})$ generated by the set

$$\mathcal{S}_0 = \{\rho_n^k \colon S^k \to S^k \mid \deg \rho_n^k = n,\ k \geq 1,\ n \in P'\}.$$

The class $\mathcal{D}$ consists of spaces whose homotopy groups are P-local, and one finds, with the same methods as in [7, 9], that $\mathcal{S}$ consists of morphisms $f : X \to Y$ such that $f_* : \pi_1(X) \to \pi_1(Y)$ is a P-equivalence of groups and $f^* : H^k(Y; A) \to H^k(X; A)$ is an isomorphism for $k \geq 2$ and every $\mathbf{Z}_P[\pi_1(Y)_P]$-module A. This class $\mathcal{S}$ is not closed under homotopy colimits, because the natural map from S^1 to $K(\mathbf{Z}_P, 1)$, which is the homotopy colimit of a certain direct system of maps $\rho_n^1, n \in P'$, fails to induce an isomorphism in H^2 with coefficients in the group ring $\mathbf{Z}_P[\mathbf{Z}_P]$, and hence does not belong to $\mathcal{S}$. Thus, Corollary 1.6 does not apply in this case. In fact, the orthogonal pair $(\mathcal{S}, \mathcal{D})$ does not admit a localisation functor [7].

On the other hand, if we delete from $\mathcal{S}_0$ the maps ρ_n^1, $n \in P'$, then the resulting class $\mathcal{D}$ consists of spaces whose higher homotopy groups are P-local, and $\mathcal{S}$ consists of morphisms $f \colon X \to Y$ inducing an isomorphism of fundamental groups and such that $f^* \colon H^k(Y; A) \to H^k(X; A)$ is an isomorphism for all k and every $\mathbf{Z}_P[\pi_1(Y)]$-module A. This orthogonal pair $(\mathcal{S}, \mathcal{D})$ is the maximal extension to $\mathcal{H}$ of the pair described in Example 3.1. Now Corollary 1.6 provides a localisation functor associated to $(\mathcal{S}, \mathcal{D})$. This functor induces an isomorphism of fundamental groups and P-localises the higher homotopy groups, i.e., corresponds to fibrewise localisation with respect to the universal covering fibration $\tilde{X} \to X \to K(\pi_1(X), 1)$.

Example 3.7 Fix a group G and let $\mathcal{H}(G)$ be the category whose objects are maps $X \to K(G, 1)$ in $\mathcal{H}$ and whose morphisms are homotopy commutative triangles. Given an abelian G-group A, let $\mathcal{S}(A)$ be the class of morphisms f such that $f_* \colon H_*(X; A) \to H_*(Y; A)$ is an isomorphism. Then $\mathcal{S}(A)$ satisfies the conditions of Theorem 1.4. Example 3.2 corresponds to the particular case $G = \{1\}$. In [7] we show that several idempotent functors on $\mathcal{H}$ extending P-localisation of nilpotent spaces can be obtained by splicing localisation functors with respect to twisted homology in a suitable way. In fact, Example 3.5 can be alternatively obtained as a special case of this procedure.

References

[1] J. F. ADAMS, *Localisation and Completion*, Lecture Notes University of Chicago (1975).

[2] G. BAUMSLAG, Some aspects of groups with unique roots, *Acta Math.* **104** (1960), 217–303.

[3] A. K. BOUSFIELD, Constructions of factorization systems in categories, *J. Pure Appl. Algebra* **9** (1977), 207–220.

[4] A. K. BOUSFIELD, The localization of spaces with respect to homology, *Topology* **14** (1975), 133–150.

[5] A. K. BOUSFIELD, Homological localization towers for groups and π-modules, *Mem. Amer. Math. Soc.* **10** (1977), no. 186.

[6] A. K. BOUSFIELD and D. M. KAN, *Homotopy Limits, Completions and Localizations*, Lecture Notes in Math. **304**, Springer-Verlag (1972).

[7] C. CASACUBERTA and G. PESCHKE, Localizing with respect to self maps of the circle, *Trans. Amer. Math. Soc.* (to appear).

[8] C. CASACUBERTA, G. PESCHKE and M. PFENNIGER, Sur la localisation dans les catégories avec une application à la théorie de l'homotopie, *C. R. Acad. Sci. Paris Sér. I Math.* **310** (1990), 207–210.

[9] A. DELEANU and P. HILTON, On Postnikov-true families of complexes and the Adams completion, *Fund. Math.* **106** (1980), 53–65.

[10] A. DELEANU, A. FREI and P. HILTON, Generalized Adams completion, *Cahiers Top. Geom. Diff.* **15** (1974), no. 1, 61–82.

[11] E. DROR-FARJOUN, Homotopical localization and periodic spaces (unpublished manuscript, 1988).

[12] G. M. KELLY, A unified treatment of transfinite constructions for free algebras, free monoids, colimits, associated sheaves, and so on, *Bull. Austral. Math. Soc.* **22** (1980), 1–83.

[13] P. J. FREYD and G. M. KELLY, Categories of continuous functors I, *J. Pure Appl. Algebra* **2** (1972), 169–191.

[14] P. GABRIEL and F. ULMER, *Lokal präsentierbare Kategorien*, Lecture Notes in Math. **221**, Springer-Verlag (1971).

[15] A. HELLER, Homotopy Theories, *Mem. Amer. Math. Soc.* **71** (1988), no. 383.

[16] P. HILTON, G. MISLIN and J. ROITBERG, *Localization of Nilpotent Groups and Spaces*, North-Holland Math. Studies **15** (1975).

[17] S. MACLANE, *Categories for the Working Mathematician*, Graduate Texts in Math. **5**, Springer-Verlag (1975).

[18] G. PESCHKE, Localizing groups with action, *Publ. Mat.* **33** (1989), no. 2, 227–234.

[19] M. PFENNIGER, Remarks related to the Adams spectral sequence, U.C.N.W. Maths Preprint **91.19**, Bangor (1991).

[20] P. RIBENBOIM, Torsion et localisation de groupes arbitraires, Lecture Notes in Math. **740**, Springer-Verlag (1978), 444–456.

[21] D. SULLIVAN, Genetics of homotopy theory and the Adams conjecture, *Ann. of Math.* **100** (1970), 885–887.

[22] Z. YOSIMURA, Localization of Eilenberg-MacLane G-spaces with respect to homology theory, *Osaka J. Math.* **20** (1983), 521–537.

Carles CASACUBERTA
Centre de Recerca Matemàtica
Institut d'Estudis Catalans
Apartat 50
E–08193 Bellaterra, Barcelona
Spain

Georg PESCHKE
Department of Mathematics
University of Alberta
Edmonton T6G 2G1
Canada

Markus PFENNIGER
School of Mathematics
University of Wales
Dean Street
Bangor, LL57 1UT
United Kingdom

A note on extensions of nilpotent groups

By
CARLES CASACUBERTA and PETER HILTON

Dedicated to the memory of Frank Adams, close friend and inspiring teacher

0 Introduction

In studying the Mislin genus for finitely generated nilpotent groups and its relation to the genus of groups-with-operators (see [1,2]) we have been led to the following question. Suppose that

$$N \rightarrowtail G \overset{\kappa}{\twoheadrightarrow} Q \qquad (0.1)$$

is a short exact sequence of nilpotent groups which splits at every prime p, that is, such that

$$N_p \rightarrowtail G_p \overset{\kappa_p}{\twoheadrightarrow} Q_p$$

admits a right splitting $\tau(p) : Q_p \to G_p$ for each p. Does it follow that the original sequence (0.1) itself splits? In Theorem 2.1 of [1] we collected together the known results, which we restate here.

Theorem 0.1 *If $N \rightarrowtail G \overset{\kappa}{\twoheadrightarrow} Q$ splits at every prime, then it splits provided*

(a) *Q is finitely generated and N is commutative, or*

(b) *Q is finitely generated and N is finite.*

Moreover, in case (b), *given the splitting maps $\tau(p) : Q_p \to G_p$, we may choose the splitting map $\tau: Q \to G$ so that $\tau_p = \tau(p)$.*

A further advance—though it was not presented as such—was contained in Theorem 1.6 of [1], which effectively gave an example showing that the answer to our question is not always affirmative.

The main purpose of this paper is to generalize both Theorem 0.1 and the counterexample contained in Theorem 1.6 of [1]. Thus we prove (cf. Theorems 2.1 and 2.3, which reproduce part (b) and part (a) of Theorem 0.2 respectively)

Theorem 0.2 *If $N \rightarrowtail G \overset{\kappa}{\twoheadrightarrow} Q$ splits at every prime, then it splits provided*

(a) *Q is finitely generated and N_p is commutative for almost all primes p, being torsion for the exceptional values of p, or*

(b) *N is torsion and almost torsionfree.*

Moreover, in case (b), *given the splitting maps $\tau(p) : Q_p \to G_p$, we may choose the splitting map $\tau : Q \to G$ so that $\tau_p = \tau(p)$.*

Recall that N is *almost torsionfree* if $TN_p = \{1\}$ for almost all primes p. Of course a finite (nilpotent) group is almost torsionfree.

This theorem actually presented itself as a byproduct of a study of the Pull-back Theorem [4, Theorem I.3.9] which asserts that *(i)* for any nilpotent group G and any partition (P_1, P_2) of the family of all primes, the square

$$\begin{array}{ccc} G & \to & G_{P_1} \\ \downarrow & & \downarrow \\ G_{P_2} & \to & G_0 \end{array}$$

is a pull-back; we also know that *(ii)* every element of G_0 is expressible as x_1x_2, where x_1 is the image of an element of G_{P_1} and x_2 is the image of an element of G_{P_2}. We need to generalize the pull-back property (i) to any finite partition of the family of primes. In doing this, we also present a different generalization of Theorem I.3.9 of [4], in which we no longer require that (P_1, P_2) be a partition. We also generalize the supplementary conclusion (ii) in Section 1 below—it is interesting that the hypotheses for the generalizations of the properties (i) and (ii) in fact diverge. The arguments used extend in a natural way to the case that the groups involved are *locally nilpotent*, i.e. such that every finitely generated subgroup is nilpotent; we state the results in Section 1 in this generality. Obviously all the generalizations so far referred to have implications for the homotopy theory of nilpotent spaces, which we plan to pursue in a subsequent paper.

Section 2 closes with a broad generalization of the counterexample implicit in Theorem 1.6 of [1]. We study the group $\mathrm{Ext}(Q, N)$ when Q and N are commutative, and establish conditions under which $\mathrm{Ext}(Q, N)$ is uncountable although $\mathrm{Ext}(Q_p, N_p) = 0$ for all p. A more detailed study of this situation is currently being undertaken by Robert Militello.

1 On the pull-back and quasi-push-out properties

We start with Theorem I.3.9 of [4] which asserts that if G is a nilpotent group and (P_1, P_2) is a partition of the set of all primes, then the square

$$\begin{array}{ccc} G & \stackrel{e_{P_1}}{\rightarrow} & G_{P_1} \\ e_{P_2} \downarrow & & \downarrow r_{P_1} \\ G_{P_2} & \stackrel{r_{P_2}}{\rightarrow} & G_0 \end{array} \tag{1.1}$$

is a pull-back. Moreover, the Remark following Theorem I.3.9 asserts that every element of G_0 is expressible as $(r_{P_1}x_1)(r_{P_2}x_2)$, $x_i \in G_{P_i}$. If G is commutative this last is equivalent to the square being a push-out (in the category of commutative groups). We will describe this property, in general, by saying that (1.1) is a *quasi-push-out*. We will generalize these two properties in two directions.

We say that $(P_1, P_2, \ldots, P_k)$ is a (finite) *presentation* of the family P of primes if $P = \bigcup_i P_i$. It is a *thin* presentation if $\bigcap_i P_i = \emptyset$. Thus a partition coincides with a thin presentation in the case $k = 2$. However, observe the divergence of the conditions stated in Theorems 1.1 and 1.2 below.

Theorem 1.1 *Let G be a locally nilpotent group and let $(P_1, P_2, \ldots, P_k)$ be a finite partition of the family P of primes. Then G_P is the pull-back of $G_{P_1}, G_{P_2}, \ldots, G_{P_k}$ over G_0.*

Theorem 1.2 *Let G be a locally nilpotent group and let $(P_1, P_2, \ldots, P_k)$ be a thin presentation of the family P of primes. Then G_0 is the quasi-push-out of $G_{P_1}, G_{P_2}, \ldots, G_{P_k}$, in the sense that every element of G_0 can be expressed as $\prod_{i=1}^{k} r_{P_i}x_i$, $x_i \in G_{P_i}$.*

Before giving the proofs, we explain the second direction of our generalization. In the next theorem, we no longer insist that (P_1, P_2) be a partition.

Theorem 1.3 *Let P_1, P_2 be any families of primes and let $P = P_1 \cup P_2$, $P_0 = P_1 \cap P_2$. Let G be a locally nilpotent group. Then the square*

$$\begin{array}{ccc} G_P & \stackrel{e^P_{P_1}}{\rightarrow} & G_{P_1} \\ e^P_{P_2} \downarrow & & \downarrow e^{P_1}_{P_0} \\ G_{P_2} & \stackrel{e^{P_2}_{P_0}}{\rightarrow} & G_{P_0}, \end{array}$$

where all arrows are localization maps, is a pull-back and a quasi-push-out.

The strategy of proof is based on the following lemmas, as in [4]. We use the notation $G \in \mathcal{P}$ to indicate that G has the pull-back and quasi-push-out properties asserted by Theorem 1.3.

Lemma 1.4 *Let $G' \rightarrowtail G \overset{\kappa}{\twoheadrightarrow} G''$ be a central extension of nilpotent groups. Then $G \in \mathcal{P}$ if $G', G'' \in \mathcal{P}$.*

PROOF. We first check the pull-back property for G. Note that it suffices to prove that if $x_i \in G_{P_i}$ with $e^{P_1}_{P_0} x_1 = e^{P_2}_{P_0} x_2 = x_0$ then there exists $x \in G_P$ with $e^P_{P_i} x = x_i$, for the uniqueness of x follows from the fact that the kernel of $e^P_{P_i}\colon G_P \to G_{P_i}$ consists of elements which are $(P \setminus P_i)$-torsion. Thus, with the given elements x_i, we have $e^{P_1}_{P_0} \kappa_{P_1} x_1 = e^{P_2}_{P_0} \kappa_{P_2} x_2 = \kappa_{P_0} x_0$. Since $G'' \in \mathcal{P}$, there exists an element $x'' \in G''_P$ with $e^P_{P_i} x'' = \kappa_{P_i} x_i$, $i = 1, 2$. Let $x'' = \kappa_P \bar{x}$. Then $\kappa_{P_i} e^P_{P_i} \bar{x} = e^P_{P_i} x'' = \kappa_{P_i} x_i$, so that $x_i = (e^P_{P_i} \bar{x}) x'_i$ with $x'_i \in G'_{P_i}$. Moreover $x_0 = e^{P_i}_{P_0} x_i = (e^P_{P_0} \bar{x})(e^{P_i}_{P_0} x'_i)$, so that $e^{P_1}_{P_0} x'_1 = e^{P_2}_{P_0} x'_2$. Since $G' \in \mathcal{P}$, there exists $x' \in G'_P$ with $e^P_{P_i} x' = x'_i$, $i = 1, 2$. Thus $e^P_{P_i}(\bar{x} x') = (e^P_{P_i} \bar{x}) x'_i = x_i$, and the first part of the lemma is proved (note that we did not use here the assumption that the extension is central). Now we turn to the quasi-push-out property. Let $x_0 \in G_{P_0}$. Then $\kappa_{P_0} x_0 = (e^{P_1}_{P_0} x''_1)(e^{P_2}_{P_0} x''_2)$, $x''_i \in G''_{P_i}$, since $G'' \in \mathcal{P}$. Let $x''_i = \kappa_{P_i} x_i$, $x_i \in G_{P_i}$, so that $\kappa_{P_0} x_0 = \kappa_{P_0}(e^{P_1}_{P_0} x_1)(e^{P_2}_{P_0} x_2)$. Thus $x_0 = (e^{P_1}_{P_0} x_1)(e^{P_2}_{P_0} x_2) x'_0$ with $x'_0 \in G'_0$. Since $G' \in \mathcal{P}$, $x'_0 = (e^{P_1}_{P_0} x'_1)(e^{P_2}_{P_0} x'_2)$, $x'_i \in G'_{P_i}$. Since G' is central in G, we conclude finally that

$$x_0 = (e^{P_1}_{P_0}(x_1 x'_1))(e^{P_2}_{P_0}(x_2 x'_2)). \qquad \square$$

Lemma 1.5 *Let $\{G^\alpha, \varphi^{\alpha\beta}\}$ be a directed system of nilpotent groups such that $G^\alpha \in \mathcal{P}$ for all α. Then $G = \varinjlim \{G^\alpha, \varphi^{\alpha\beta}\}$ belongs to $\mathcal{P}$.*

PROOF. Recall first that localization commutes with direct limits over directed sets. Now let $x_i \in G_{P_i}$, $i = 1, 2$, with $e^{P_1}_{P_0} x_1 = e^{P_2}_{P_0} x_2 = x_0$. Then there exists α and $x_i^\alpha \in G^\alpha_{P_i}$ such that $\varphi^\alpha_{P_i} x_i^\alpha = x_i$, $i = 1, 2$, where φ^α is the canonical homomorphism $\varphi^\alpha\colon G^\alpha \to G$. Moreover, since $\varphi^\alpha_{P_0} e^{P_1}_{P_0} x_1^\alpha = \varphi^\alpha_{P_0} e^{P_2}_{P_0} x_2^\alpha$, we may find β and $x_i^\beta \in G^\beta_{P_i}$ such that $\varphi^\beta_{P_i} x_i^\beta = x_i$ and $e^{P_1}_{P_0} x_1^\beta = e^{P_2}_{P_0} x_2^\beta$. Since $G^\beta \in \mathcal{P}$, there exists $x^\beta \in G^\beta_P$ with $e^P_{P_i} x^\beta = x_i^\beta$, whence $e^P_{P_i} \varphi^\beta x^\beta = x_i$, $i = 1, 2$, establishing the pull-back property for G. The quasi-push-out property is obvious. $\square$

Lemma 1.6 *The groups* $\mathbf{Z}$*,* $\mathbf{Z}/p^n$ *belong to* $\mathcal{P}$*.*

PROOF. If $G = \mathbf{Z}$, we must show that if the reduced fractions (with positive denominators) $m_1/n_1 \in \mathbf{Z}_{P_1}$, $m_2/n_2 \in \mathbf{Z}_{P_2}$ are equal as rational numbers, then $m_1 = m_2 = m$, $n_1 = n_2 = n$, and n is a P'-number. This, however, is clear since $P_1' \cap P_2' = P'$. We must also consider the reduced fraction m/n where n is a P_0'-number. Write n as uvw, where u is a $(P_1 \setminus P_0)$-number, v is a $(P_2 \setminus P_0)$-number, and w is a P'-number. Since u, v are mutually prime, $1 = au + bv$ for some $a, b \in \mathbf{Z}$. Thus $m/n = (ma/vw) + (mb/uw)$, and vw is a P_1'-number while uw is a P_2'-number. This shows that $\mathbf{Z}$ belongs to $\mathcal{P}$. If $G = \mathbf{Z}/p^n$, both conclusions are obvious if $p \notin P$ or if $p \in P_0$. If $p \in P_1 \setminus P_0$, then $G_{P_2} = G_{P_0} = \{0\}$ and $e^P_{P_1} = \mathrm{id}$, so, again, both conclusions are clear. □

PROOF OF THEOREM 1.3. Starting from Lemma 1.6 and repeatedly using Lemma 1.4, we conclude that $G \in \mathcal{P}$ if G is finitely generated commutative. Lemma 1.5 then allows us to drop the condition of finite generation. Assume next G nilpotent and proceed by induction on the nilpotency class of G. For if $\operatorname{nil} G = c$, so that $\Gamma^c G = \{1\}$, we set $\Gamma = \Gamma^{c-1} G$ and we have the central extension $\Gamma \rightarrowtail G \twoheadrightarrow G/\Gamma$, with $\operatorname{nil} G/\Gamma = c - 1$. Thus Lemma 1.4 allows us to infer that $G \in \mathcal{P}$. Finally, the proof in the case of G locally nilpotent is achieved again using Lemma 1.5. □

The above argument applies *mutatis mutandi* to provide proofs of Theorems 1.1 and 1.2. However, it is more economic now to infer Theorems 1.1 and 1.2 from Theorem 1.3 as follows.

Lemma 1.7 *If the diagrams*

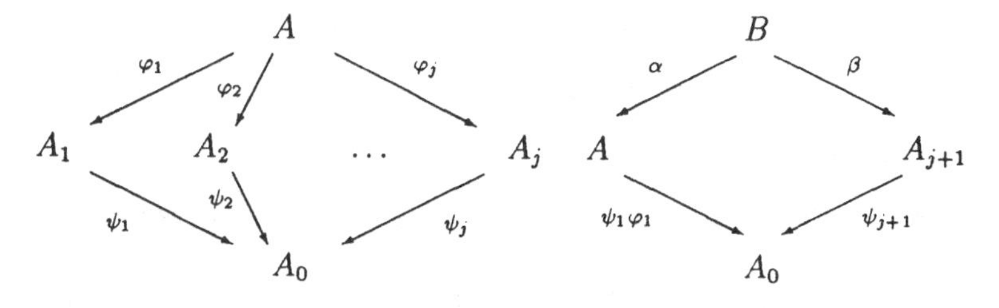

are pull-backs in an arbitrary category, so is the diagram

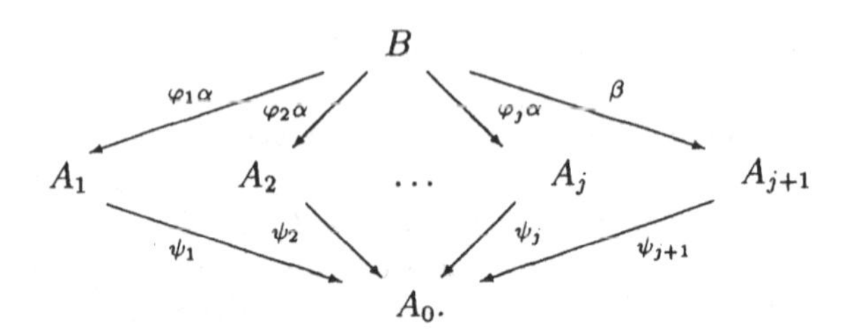

Proof. This only requires a routine checking of the characteristic universal property. □

Proof of Theorem 1.1. We argue by induction on k. By Theorem 1.3, $G_{P_1 \cup P_2}$ is the pull-back of G_{P_1} and G_{P_2} over G_0, since $P_1 \cap P_2 = \emptyset$. Assume inductively that $G_{P_1 \cup \ldots \cup P_i}$ is the pull-back of $G_{P_1}, \ldots, G_{P_i}$ over G_0. Then, again by Theorem 1.3, $G_{P_1 \cup \ldots \cup P_{i+1}}$ is the pull-back of $G_{P_1 \cup \ldots \cup P_i}$ and $G_{P_{i+1}}$ over G_0. Therefore, by Lemma 1.7, $G_{P_1 \cup \ldots \cup P_{i+1}}$ is the pull-back of $G_{P_1}, \ldots, G_{P_{i+1}}$ over G_0. □

Theorem 1.8 *Let G be a locally nilpotent group and let $(P_1, P_2, \ldots, P_k)$ be any family of sets of primes with intersection P_0. Then G_{P_0} is the quasi-push-out of $G_{P_1}, G_{P_2}, \ldots, G_{P_k}$.*

Proof. We argue again by induction on k. Set $Q = P_2 \cap \ldots \cap P_k$. Then, given $x \in G_{P_0}$, by Theorem 1.3 we may write

$$x = (e_{P_0}^{P_1} x_1)(e_{P_0}^{Q} y), \quad x_1 \in G_{P_1}, \quad y \in G_Q.$$

But, by the inductive hypothesis, we have

$$y = \prod_{i=2}^{k} e_Q^{P_i} x_i, \quad x_i \in G_{P_i},$$

and hence

$$e_{P_0}^{Q} y = \prod_{i=2}^{k} e_{P_0}^{P_i} x_i,$$

which proves the theorem. □

Of course, Theorem 1.2 is a special case of Theorem 1.8. It is easy to see that Theorem 1.8 in fact remains true if we assume the family $(P_1, P_2, \ldots)$ to be infinite, provided we understand that a product $\prod_i e_{P_0}^{P_i} x_i$ has only finitely many nontrivial factors.

Notice, however, that there is no useful analog of Theorem 1.8 in the case of the pull-back property, for we would have to assume a presentation $(P_1, \ldots, P_k)$ of P such that $P_i \cap P_j$, $i \neq j$, is independent of (i, j). This only seems sensible in the case of a partition.

2 The splitting problem

We consider a short exact sequence of nilpotent groups

$$N \rightarrowtail G \overset{\kappa}{\twoheadrightarrow} Q, \tag{2.1}$$

which we suppose splits at every prime p. Indeed, let $\tau(p)\colon Q_p \to G_p$ be a splitting at the prime p, so that $\kappa_p\tau(p) = \mathrm{id}_{Q_p}$.

Theorem 2.1 *If N is torsion but almost torsionfree, then (2.1) splits. Moreover, there is a splitting $\tau\colon Q \to G$ such that $\tau_p = \tau(p)$.*

Proof. Let $P = \{p_1, p_2, \dots, p_k\}$ be the (finite) set of primes at which N has torsion. We consider the partition $(p_1, p_2, \dots, p_k, P')$ of the family of all primes. For each $i = 1, 2, \dots, k$ we write $\tau_i = \tau(p_i)\colon Q_{p_i} \to G_{p_i}$. Now $N_{P'} = \{1\}$ so that $\kappa_{P'}\colon G_{P'} \cong Q_{P'}$. Set $\tau' = (\kappa_{P'})^{-1}\colon Q_{P'} \to G_{P'}$. Note that each of $\tau_1, \tau_2, \dots, \tau_k, \tau'$ rationalizes to a section of $\kappa_0\colon G_0 \to Q_0$; but κ_0 is an isomorphism so that the only section of κ_0 is $(\kappa_0)^{-1}$. Thus the splitting maps $\tau_1, \tau_2, \dots, \tau_k, \tau'$ agree over Q_0. We may therefore invoke Theorem 1.1 to infer the existence of $\tau\colon Q \to G$ such that

$$\tau_{p_i} = \tau_i,\ i = 1, 2, \dots, k; \qquad \tau_{P'} = \tau'.$$

Then $e_{p_i}\kappa\tau = \kappa_{p_i}\tau_i e_{p_i} = e_{p_i}$ and $e_{P'}\kappa\tau = \kappa_{P'}\tau' e_{P'} = e_{P'}$. Thus by the uniqueness assertion implicit in a pull-back, $\kappa\tau = \mathrm{id}_Q$ and τ is a splitting of (2.1). By construction, $\tau_{p_i} = \tau(p_i)$, $i = 1, 2, \dots, k$; and, if $p \in P'$, then τ_p is the unique section of the isomorphism κ_p so that $\tau_p = \tau(p)$. This completes the proof of the theorem. □

This result generalizes part (b) of Theorem 0.1. Furthermore, we may use it, together with the following observation, to yield a generalization of part (a).

Lemma 2.2 *Let $N \rightarrowtail G \overset{\kappa}{\twoheadrightarrow} Q$ be an extension of nilpotent groups and $\Gamma \subseteq N$ a subgroup which is normal in G. Assume given a splitting $s\colon Q \to G/\Gamma$ of the extension*

$$N/\Gamma \rightarrowtail G/\Gamma \overset{\varepsilon}{\twoheadrightarrow} Q$$

and a homomorphism $t\colon G/\Gamma \to G$ such that $\varepsilon\pi t = \varepsilon$ in the diagram

$$\begin{array}{ccccc} \Gamma & \rightarrowtail & G & \overset{\pi}{\twoheadrightarrow} & G/\Gamma \\ {\scriptstyle\iota}\downarrow & & {\scriptstyle =}\downarrow & & {\scriptstyle\varepsilon}\downarrow \\ N & \rightarrowtail & G & \overset{\kappa}{\twoheadrightarrow} & Q. \end{array}$$

Then the composite ts is a section of κ.

PROOF. $\kappa ts = \varepsilon\pi ts = \varepsilon s = \mathrm{id}_Q$. □

Theorem 2.3 *Assume that Q is finitely generated, N_p is commutative for all primes p except for a finite number $p_1, p_2, \dots, p_k$, and N_{p_i} is torsion for $i = 1, 2, \dots, k$. Then (2.1) splits.*

PROOF. Write $P = \{p_1, p_2, \dots, p_k\}$, and set $\Gamma = N_{p_1} N_{p_2} \cdots N_{p_k}$. Then Γ is a direct factor in TN and certainly normal in G. If π denotes the projection of G onto G/Γ, then for each prime p the homomorphism $\pi_p \tau(p)$ splits the extension

$$N/\Gamma \rightarrowtail G/\Gamma \overset{\varepsilon}{\twoheadrightarrow} Q, \tag{2.2}$$

at the prime p, since $\varepsilon_p \pi_p \tau(p) = \kappa_p \tau(p) = \mathrm{id}_{Q_p}$. But N/Γ is commutative because $(N/\Gamma)_p$ is commutative for all p. Hence, (2.2) splits by part (a) of Theorem 0.1. We next check that

$$\Gamma \rightarrowtail G \overset{\pi}{\twoheadrightarrow} G/\Gamma \tag{2.3}$$

also splits, so that the theorem will follow from Lemma 2.2. Since Γ is torsion and almost torsionfree, by Theorem 2.1 it suffices to check that (2.3) splits at every prime p. If $p \in P$ then the p-localization of (2.3) coincides with the p-localization of (2.1) and hence splits. If $p \in P'$, then $\Gamma_p = \{1\}$ and hence π_p also admits an obvious section. This completes the argument. □

It remains to show that *some* restriction on the sequence (2.1) is necessary in order to deduce from the splitting of its localizations that it itself splits.

We confine ourselves to the situation in which both N and Q are commutative and will further assume that Q is a group of *pseudo-integers* [3], i.e. such that $Q_p \cong C_p$ for all primes p, where C is cyclic infinite.

Lemma 2.4 *Let Q be a group of pseudo-integers. Then $H_n(Q) = 0$ for $n \geq 2$.*

PROOF. If $n \geq 2$ we have

$$H_n(Q)_p \cong H_n(Q_p) \cong H_n(C_p) \cong H_n(C)_p = 0$$

for all p. It follows that $H_n(Q) = 0$. □

Proposition 2.5 *Let N be commutative and Q a group of pseudo-integers. Then, for every central extension $N \rightarrowtail G \twoheadrightarrow Q$, G is commutative.*

PROOF. Since $\mathrm{Hom}(H_2(Q), N) = 0$, we have $H^2(Q; N) = \mathrm{Ext}(Q, N)$. □

We will thus concentrate on abelian extensions

$$N \rightarrowtail G \twoheadrightarrow Q$$

with Q a group of pseudo-integers.

Proposition 2.6 *Every such abelian extension splits at every prime.*

PROOF. $N_p \rightarrowtail G_p \twoheadrightarrow Q_p$ is an extension of $\mathbf{Z}_p$-modules with Q_p free. □

We will now study $\mathrm{Ext}(Q, N)$, writing our abelian groups additively. Then, if we enumerate the primes, Q is characterized by a set of non-negative integers $(n_1, n_2, \ldots)$, in the sense that $Q = \langle 1/p_1^{n_1}, 1/p_2^{n_2}, \ldots \rangle$. Thus there is an exact sequence

$$\mathbf{Z} \rightarrowtail Q \twoheadrightarrow \bigoplus_i \mathbf{Z}/p_i^{n_i},$$

giving rise to an exact sequence

$$\mathrm{Hom}(\mathbf{Z}, N) \to \mathrm{Ext}(\bigoplus_i \mathbf{Z}/p_i^{n_i}, N) \to \mathrm{Ext}(Q, N) \to 0. \qquad (2.4)$$

Now $\mathrm{Ext}(\bigoplus_i \mathbf{Z}/p_i^{n_i}, N) = \prod_i \mathrm{Ext}(\mathbf{Z}/p_i^{n_i}, N) = \prod_i N/p_i^{n_i}N$, so that (2.4) becomes

$$N \xrightarrow{\theta} \prod_i N/p_i^{n_i}N \twoheadrightarrow \mathrm{Ext}(Q, N).$$

Moreover it is easy to see that the ith component of θ is just the standard projection of N onto $N/p_i^{n_i}N$. We conclude that

$$\mathrm{Ext}(Q, N) \cong \mathrm{coker}\,\theta;$$

and it only remains to describe conditions under which $\mathrm{coker}\,\theta \neq \{0\}$. Obviously a sufficient condition is that N be countable and $\prod_i N/p_i^{n_i}N$ uncountable. Now $\prod_i N/p_i^{n_i}N$ is uncountable if $N/p_i^{n_i}N \neq \{0\}$ for infinitely many i. This will occur if there are infinitely many i such that $n_i > 0$ and N is not p_i-divisible. Thus we infer

Theorem 2.7 *Let N be a countable abelian group and let*

$$Q = \langle 1/p_i^{n_i},\ i = 1, 2, \ldots \rangle$$

be a group of pseudo-integers. Then $\mathrm{Ext}(Q, N)$ is uncountable, provided there exist infinitely many i such that $n_i > 0$ and N is not p_i-divisible. □

Notice that we may take as suitable groups N

(i) any countable abelian group having a $\mathbf{Z}$-summand;

(ii) any countable abelian torsion group such that N_p is non-zero and reduced for infinitely many primes p.

In case (i) any group of pseudo-integers not isomorphic to $\mathbf{Z}$ would be a suitable Q; in case (ii) Q would need to be chosen more carefully. Of course, case (i) immediately allows us to generalize to the following result.

Corollary 2.8 *Let N be an abelian group having a $\mathbf{Z}$-summand and let Q be a group of pseudo-integers, $Q \not\cong \mathbf{Z}$. Then* $\mathrm{Ext}(Q, N)$ *is uncountable.*

In fact, we know that the conclusion of Corollary 2.8 holds so long as Q is $\mathbf{Z}^k$-like but $Q \not\cong \mathbf{Z}^k$, with the same assumption on N.

References

[1] C. CASACUBERTA and P. HILTON, On special nilpotent groups, *Bull. Soc. Math. Belg. Sér. B* **49** (1989), no. 3, 261–273.

[2] P. HILTON, On the genus of groups with operators, *Topology Appl.* **25** (1987), 113–120.

[3] P. HILTON, On groups of pseudo-integers, *Acta Math. Sinica (N. S.)* **4** (1988), no. 2, 189–192.

[4] P. HILTON, G. MISLIN and J. ROITBERG, *Localization of Nilpotent Groups and Spaces*, North-Holland Math. Studies **15** (1975).

Centre de Recerca Matemàtica, Institut d'Estudis Catalans, Apartat 50, E-08193 Bellaterra, Barcelona, Spain

Department of Mathematical Sciences, State University of New York at Binghamton, Binghamton, New York 13902–6000, USA

On the Swan Subgroup of Metacyclic Groups

Victor P. Snaith*

Abstract

A geometrical technique is given to construct relations among Swan modules of metacyclic groups. The method is based upon maps between representation spheres which were originally introduced by Frank Adams.

1 Swan modules

Let G be a finite group of order n. Let k be any integer which is coprime to n, and denote by $S(k)$ the $Z[G]$-module which is given by the ideal

1.1 $$S(k) = ideal\{k, \sigma\} \triangleleft Z[G]$$

where $\sigma = \sum_{g \in G} g$. The $Z[G]$-module of 1.1 will be called a *Swan module.* Clearly $S(k)$ is free when $k = \pm 1$. In general $S(k)$ is a locally free (and hence projective) $Z[G]$-module which is often non-trivial in $\mathcal{CL}(Z[G])$, the class group given by the (reduced) Grothendieck group of finitely generated, projective $Z[G]$-modules. In terms of algebraic K-theory

$$\mathcal{CL}(Z[G]) = Ker\{K_0(Z[G]) \longrightarrow K_0(Z)\}.$$

Sending the integer k to the class of $S(k)$ defines a homomorphism

$$S : (Z/n)^*/\{\pm 1\} \longrightarrow \mathcal{CL}(Z[G])$$

whose image is called the *Swan subgroup* and is denoted by $T(G)$.

There is an analogy between the image of the J-homomorphism within the stable homotopy groups of spheres and the image of S within the class-group.

When G is a p-group the calculation of $T(G)$ is given in [2] (see also [3] II, §54). Most of the cases are due to M.J. Taylor, using determinantal congruences. This manner of detection, which uses Adams operations ψ^l (in the

*Research partially supported by an NSERC grant

form of a quotient map ψ^l/l) to give a lower bound for $T(G)$, is reminiscent of J.F. Adams's approach to the detection of the image of the J-homomorphism in the stable homotopy groups of spheres [1]. The more difficult part of the determination of the image of the J-homomorphism was the construction of an upper bound. This led to the celebrated Adams conjecture [1] and its ultimate proof by D. Sullivan and D.G. Quillen ([4], [5]) . The Adams conjecture concerns the maps between spheres of vector bundles. This motivated the method which we shall now employ to prove our main result (Theorem 2.2). We shall study maps between unit spheres of representation spaces. This approach will involve a digression into the topology of group actions, for which I must ask the reader's indulgence, before we return to the Swan subgroup. The maps which we use were originally employed by Frank Adams in his work on the J-homomorphism. We shall use their equivariant geometry to obtain some new relations between Swan modules of metacyclic groups.

I am very grateful to the referee for his suggestions on how to condense my original proof.

2 Relations Between Swan Modules

Throughout this section we shall assume that G is a finite group in an extension (not necessarily split) of the form

2.1 $$\{1\} \longrightarrow J \longrightarrow G \longrightarrow Z/p \longrightarrow \{1\}$$

where J is cyclic and p is prime.

We shall use a geometrical technique, which to my knowledge is new but which is very much in the geometrical spirit which motivated R.G. Swan's original introduction of the Swan modules $S(k)$ ([6], p.278).

Theorem 2.2 *Let G be as in 2.1 and let* $\mathrm{hcf}(k, \#(G)) = 1$. *Then*

$$S(k^p) - \sum_g Ind^G_{N_G<g>}(S(k)) = 0 \in \mathcal{CL}(Z[G])$$

where $g \in G - J$ runs over generators for the distinct conjugacy classes of subgroups $< g >$ of order p, and $N_G < g >$ denotes the normaliser of $< g >$ in G.

This result will be proved in 2.9. Before proceeding to the proof let us pause to examine some applications.

Example 2.3 Let p be an odd prime and let Z/p act on Z/p^n $(n \geq 2)$ by $\tau(z) = z^{(1+p)^{p^{n-2}}}$ where $z \in Z/p^n$ and τ generates Z/p. From ([3] II, p.365)

$$T(Z/p \propto Z/p^n) \cong Z/p^n.$$

If we set $G = Z/p \propto Z/p^n$ and $J = Z/p^n$ we may apply Theorem 2.2 to any integer k which is not divisible by p. Hence

$$0 = S(k^p) - \sum_g Ind^G_{N_G<g>}(S(k))$$

where the sum runs over some of the $g \in G - J$ of order p. However, for each of these, $N_G < g >= Z/p \propto Z/p^{n-1}$, so that if we set $k = t^{p^{n-1}}$ then each $Ind^G_{N_G<g>}(S(k))$ vanishes, by induction on the order of G, and we find that

$$S(t^{p^n}) = 0 \in T(Z/p \propto Z/p^n)$$

as expected.

Example 2.4 Let $G = Q(8n)$, the generalised quaternion group of order $8n$,

$$Q(8n) = \{X, Y \mid Y^2 = X^{2n}, Y^4 = 1, YXY^{-1} = X^{-1}\}.$$

Setting $J =< X >$, $p = 2$ and observing that there are no elements of order two in $G - J$, we find that for all n,

2.5 $$2T(Q(8n)) = 0.$$

In [2] one finds some very difficult and technical calculations of Swan subgroups which are related to 2.5 . En route to the main theorem of [2] one finds a calculation of $T(Q(8p))$ when p is a prime (see [2], Theorem 3.5). The preceding application of Theorem 2.2 gives a new manner in which to derive an upper bound for $T(Q(8n))$ which is less computational than the method of [2].

Notice that the Artin exponent of $Q(8p)$ is 4 ([2], p.462) so that 2.5 improves upon the estimate given by the fact that the Artin exponent annihilates the Swan subgroup ([3] II, p.346, Theorem 53.12).

Let us recall part of ([2], Theorem 3.5).

Theorem 2.6 *Let p be an odd prime such that either $p \equiv \pm 3$(modulo 8) or $p \equiv 1$(modulo 8) and $ord_p(2)$ is even, where $ord_p(t)$ is the order of t in F_p^*. Then*

$$T(Q(8p)) \cong Z/2 \oplus Z/2.$$

Corollary 2.7 *Let $Q(8n))$ denote the generalised quaternion group of order $8n$. Suppose that $n = 2^\alpha p_1^{m_1} \ldots p_r^{m_r}$ with $m_i \geq 1$, $\alpha \geq 0$ and let $\{p_1, p_2, \ldots, p_r\}$ be a set of distinct primes such that either $p_i \equiv \pm 3$(modulo 8)* *or* $p_i \equiv 1(\text{modulo } 8)$ *and* $ord_{p_i}(2)$ *is even, where* $ord_p(t)$ *is the order of* t *in* F_p^*. *Then*

$$T(Q(8n)) \cong (Z/2)^{r+1}.$$

Proof By 2.5, $T(Q(8n))$ is a quotient of

$$\begin{aligned}&(((Z/2^{\alpha+3})^* \times \prod_{i=1}^r (Z/p_i^{m_i})^*/\{\pm 1\}) \otimes Z/2\\ &\cong Z/2 \times \prod_{i=1}^r F_{p_i}^*/(F_{p_i}^*)^2\\ &\cong (Z/2)^{r+1}.\end{aligned}$$

However, inspection of ([2], Theorem 3.5) shows that the natural maps to the Swan groups of the subquotients $\{Q(8p_i) \ : 1 \leq i \leq r\}$ detect $(Z/2)^{r+1}$ in $T(Q(8n))$, which completes the proof.

The following result is well-known.

Lemma 2.8 *Let* $\epsilon : Z[G] \longrightarrow Z/r$ *denote the reduction modulo* r *of the augmentation map. If* $\text{hcf}(r, \#(G)) = 1$ *then*

$$[Ker(\epsilon : Z[G] \longrightarrow Z/r)] = -S(r) \in \mathcal{CL}(Z[G]).$$

2.9 *Proof of Theorem 2.2*

Let V be any one-dimensional complex representation of J such that J acts freely away from the origin. Let V' denote the k-th tensor power of V so that the map $z \longmapsto z^k$ is a J-equivariant map from V to V'. Let W and W' be the induced G-representations and let $f : W \longrightarrow W'$ be the G-map given by $(z_1, \ldots, z_p) \longmapsto (z_1^k, \ldots, z_p^k)$. After renorming, f restricts to a map of unit spheres $f : SW \longrightarrow SW'$. Let $C_* = C_*(f)$ denote the cellular chain complex of the mapping cone of f and let $D_* = C_*(f^S)$ denote the cellular chains of the singular set of f.

Hence C_*/D_* is a complex of finitely generated, free $Z[G]$-modules so that (if H_* denotes *reduced* homology)

$$[H_*(C_*)] = [H_*(D_*)] \in \mathcal{CL}(Z[G])$$

providing that both classes make sense as class-group elements.

The singular sets in SW and SW' are unions of circles, one for each splitting $Z/p \subset G$, and f maps them via maps of degree k. Hence, as G-representations,

$$H_*(D_*) = \sum_g Ind_{N_G<g>}^G(Z/k),$$

where the sum is taken over the indexing set of 2.2 and Z/k has the trivial $N_G < g >$-action. Also $H_*(C_*) \cong Z/k^p$ with the trivial G-action. The formula of 2.2 follows immediately from 2.8.

2.10 Research Problems

(i) Generalise the result of 2.2 to the case of a group G which is given by an extension of the form

$$\{1\} \longrightarrow Z/n \longrightarrow G \longrightarrow (Z/p)^m \longrightarrow \{1\}.$$

One would expect the relations to take the form of an alternating sum involving $m+1$ terms, rather like an *Euler characteristic*. Some modification of the geometric proof of 2.2 might well yield such a generalisation.

(ii) Generalise the result of 2.2 to the case of a group G which is given by an extension of the form

$$\{1\} \longrightarrow H \longrightarrow G \longrightarrow Z/p \longrightarrow \{1\}.$$

More precisely, attempt to prove that the relation

$$S(k^p) - \sum_g Ind^G_{N_G<g>}(S(k)) = 0 \in \mathcal{CL}(Z[G])$$

holds for any integer k for which $\mathrm{hcf}(k, \#(G)) = 1$ and $0 = S(k) \in \mathcal{CL}(Z[H])$.

The proof of Theorem 2.2 can be extended to the case when H acts freely on two G-spheres which admit an intertwining G-map of degree k.

It should be admitted that the conjectured relation is suggested merely by the evidence of the metacyclic case (Theorem 2.2), the p-group case and its aesthetically pleasant form.

References

[1] J.F. Adams: On the groups J(X), I, Topology 2 (1963), 181-195 .

[2] S. Bentzen and I. Madsen: On the Swan subgroup of certain periodic groups, Math. Ann. 264 (1983), 447-474 .

[3] C.W. Curtis and I. Reiner: *Methods of Representation Theory II*, Wiley 1987.

[4] D.G. Quillen: The Adams conjecture, Topology 10 (1971), 67-80.

[5] D. Sullivan: Genetics of homotopy theory and the Adams conjecture, Ann. Math. 100 (1974), 1-79.

[6] R.G. Swan: Periodic resolutions for finite groups, Ann. Math. (1960), 267-291.

Department of Mathematics,
McMaster University,
Hamilton, Ontario L8S 4K1,
Canada.
(e-mail: SNAITH @ McMaster.bitnet)

Fields of spaces

M. C. Crabb, J. R. Hubbuck and Kai Xu

1. Introduction

Throughout this paper p will be a fixed prime number. All spaces will be based, and we write H_* for reduced homology with $\mathbf{F}_p$-coefficients.

Let X be a finite complex with p-torsion integral homology, which we call a finite p-torsion complex. The set of stable self-maps $\{X;\ X\}$ is a finite $\hat{\mathbf{Z}}_p$-algebra.

Proposition 1.1.

(i) *The ring* $\{X;\ X\}/\mathrm{rad}\,\{X;\ X\}$ *is a semi-simple* $\mathbf{F}_p$*-algebra and so is a finite product of matrix algebras over finite fields of characteristic* p.

(ii) *The group* $\mathcal{E}^{\bullet}(X)/O_p(\mathcal{E}^{\bullet}(X))$ *is isomorphic to a finite product of general linear groups over finite fields of characteristic* p.

The notations are explained below. The *Jacobson radical* $\mathrm{rad}(R)$ of a ring R (with identity) is usually defined to be the two-sided ideal $\{r \in R \mid 1-rs$ is a unit for all $s \in R\}$. For a *finite* ring R, an alternative formulation of the definition in terms of nilpotent elements is more illuminating for our purposes:

$$(1.2) \qquad r \in \mathrm{rad}(R) \iff rs \text{ is nilpotent for all } s \in R.$$

See (5.5)(iv).

We write $R^{\bullet}$ for the group of units in a ring R. In Proposition 1.1(ii), $\mathcal{E}^{\bullet}(X)$ denotes the group $\{X;\ X\}^{\bullet}$ of stable self-equivalences of X. For a finite group G, the *Fitting p-subgroup* $O_p(G)$ is classically defined as the intersection of the Sylow p-subgroups of G. It is the unique maximal normal p-subgroup of G, and admits another description, somewhat analogous to (1.2):

$$(1.3) \qquad g \in O_p(G) \iff \begin{cases} gh \text{ has order a power of } p \\ \text{for each } h \in G \text{ with order a power of } p. \end{cases}$$

Proposition 1.1 is just an algebraic consequence of the fact that $\{X;\, X\}$ is a finite $\hat{\mathbb{Z}}_p$-algebra, so not a deep result. It is the associated question of realizability which concerns us. We establish in Section 2:

Theorem 1.4. *Any finite product of matrix algebras over finite fields of characteristic p (that is, any semi-simple finite-dimensional* $\mathbb{F}_p$*-algebra) can be realized as* $\{X;\, X\}/\mathrm{rad}\,\{X;\, X\}$ *for some finite p-torsion complex* X.

This implies that any finite product of general linear groups over finite fields of characteristic p arises as $\mathcal{E}^\bullet(X)/O_p(\mathcal{E}^\bullet(X))$ for some finite p-torsion complex X.

We investigate unstable analogues of these results. Consider a connected space Y of the homotopy type of a CW-complex with each homotopy group $\pi_i(Y)$ a finite p-group and $\pi_i(Y) = 0$ for all but finitely many i. Then the homotopy classes of self-maps of Y, $[Y;\, Y]$, form a finite set with a zero and an identity under composition.

By analogy with (1.2), let

$$(1.5) \qquad \mathcal{N} = \{f \in [Y;\, Y] \mid fg \text{ is nilpotent for all } g \in [Y;\, Y]\}.$$

We now assume that Y is an H-space. It is shown in [2] that the relation on $[Y;\, Y]$ defined by "$f \sim g$ if and only if $f = g + n$ for some $n \in \mathcal{N}$" is an equivalence relation and that the quotient $[Y;\, Y]/\mathcal{N}$ has a natural ring-structure and is a semi-simple finite-dimensional $\mathbb{F}_p$-algebra.

In Section 3 we establish:

Theorem 1.6. *Any semi-simple finite-dimensional* $\mathbb{F}_p$*-algebra can be realized as* $[Y;\, Y]/\mathcal{N}$ *for some connected H-space* Y *with each* $\pi_i Y$ *a finite p-group, zero for all but finitely many i.*

In Section 5 we study the group $\mathcal{E}(Y)$ of homotopy classes of self-equivalences of Y as above (but not necessarily an H-space). To explain the result and the strategy of the proofs used in this paper more generally, it is convenient to return to Proposition 1.1.

For X a finite p-torsion complex, we consider

$$(1.7) \qquad \alpha : \{X;\, X\} \to \prod_{i \geq 0} \operatorname{End} H_i(X)$$

with components given by the induced maps on homology. It detects a number of ring-theoretic properties.

Lemma 1.8. *Let $f \in \{X; X\}$. Then*
(i) *f is a unit $\Longleftrightarrow$ $\alpha(f)$ is a unit;*
(ii) *f is nilpotent $\Longleftrightarrow$ $\alpha(f)$ is nilpotent;*
(iii) *a unit f has (multiplicative) order a power of p $\Longleftrightarrow$ $\alpha(f)$ has order a power of p.*

Proof. The first statement is a stable Whitehead theorem. Parts (ii) and (iii) are purely algebraic consequences of (i); see (5.6).

From Lemma 1.8, (i) or (ii), we see that α induces an isomorphism

$$(1.9) \qquad \{X; X\}/\mathrm{rad}\,\{X; X\} \xrightarrow{\cong} \mathrm{im}\,\alpha/\mathrm{rad}(\mathrm{im}\,\alpha),$$

and so Proposition 1.1(i) follows from Wedderburn's theorem.

To prove Proposition 1.1(ii) one uses the fact that

$$(1.10) \qquad O_p(GL_n(\mathsf{F}_q)) = \{1\}, \qquad q = p^k.$$

This is true as upper, and lower, triangular matrices with diagonal entries 1 give Sylow p-subgroups with trivial intersection. Then elementary algebra, (5.6)(ii), shows that the projection $\{X; X\} \to \{X; X\}/\mathrm{rad}\{X; X\}$ induces an isomorphism

$$\mathcal{E}^{\bullet}(X)/O_p \xrightarrow{\cong} (\{X; X\}/\mathrm{rad})^{\bullet}/O_p.$$

Now let Y be as above and consider

$$(1.11) \qquad \beta : [Y; Y] \to \prod_{i \geq 1} \mathrm{End}\, PH_i(Y)$$

given by the induced maps on the primitives in mod p homology. (Recall that $PH_i(Y) = \ker\{\Delta_* : H_i(Y) \to H_i(Y \wedge Y)\}$, where Δ is the diagonal inclusion of Y in $Y \wedge Y$.) Analogous to Lemma 1.8 we have:

Proposition 1.12. *Let $f \in [Y; Y]$. Then*
(i) *f is a unit $\Longleftrightarrow$ $\beta(f)$ is a unit;*
(ii) *f is nilpotent $\Longleftrightarrow$ $\beta(f)$ is nilpotent;*
(iii) *a unit f has (multiplicative) order a power of p $\Longleftrightarrow$ $\beta(f)$ has order a power of p.*

This seems to be less well known than Lemma 1.8 and we include a proof of a more general result in Section 4 which is applicable to p-complete spaces of finite type.

We deduce from this that $\mathcal{E}(Y)/O_p(\mathcal{E}(Y))$ is determined homologically:

Theorem 1.13. *Let Y be a connected space with each $\pi_i Y$ a finite p-group, zero for all but finitely many i. Then β, (1.11), induces an isomorphism of groups*

$$\mathcal{E}(Y)/O_p(\mathcal{E}(Y)) \xrightarrow{\cong} \beta(\mathcal{E}(Y))/O_p(\beta(\mathcal{E}(Y))).$$

When Y is an H-space (or, more generally, if $\operatorname{im}\beta$ is a subring), the quotient $\beta(\mathcal{E}(Y))/O_p(\beta(\mathcal{E}(Y)))$ can be identified with the group of units in $\operatorname{im}\beta/\operatorname{rad}(\operatorname{im}\beta)$ and so is a finite product of general linear groups over finite fields of characteristic p.

It follows that any finite product of general linear groups over finite fields of characteristic p can be realized as $\mathcal{E}(Y)/O_p(\mathcal{E}(Y))$ for some H-space Y as above.

The starting point for almost everything in this note was a correspondence between Frank Adams and the first two authors on general questions of atomicity. In particular in Section 2, we adapt the method appearing in [**1**], Example 2.8, replacing p-completed spheres by mod p Moore spaces, and give a uniform construction for all primes p (rather than the case $p = 2$ considered in [**1**]).

We are indebted to Professor A. Kono for numerous conversations on the subject of this paper, especially on Section 4.

2. Realizing finite fields stably

In this section we establish Theorem 1.4. We shall prove, first of all, that for given $q = p^k$, $k \geq 1$, there is a space X with $\{X; X\}/\mathrm{rad} \cong \mathbf{F}_q$. It will be convenient, in the course of the proof, to use instead of α, (1.7), the representation

$$\alpha' : \{X; X\} \to \prod_{i \geq 0} \operatorname{End}(H_i(X; \mathbf{Z}) \otimes \mathbf{F}_p) \tag{2.1}$$

on integral homology. We can replace α by α' in (1.8) and (1.9).

Let C_p be the mod p Moore space $S^1 \cup_p e^2$. We recall that

$$\{C_p; C_p\} = \begin{cases} (\mathbf{Z}/4)1 & \text{if } p = 2 \\ (\mathbf{Z}/p)1 & \text{if } p \text{ is odd.} \end{cases} \tag{2.2}$$

Set $m = 2p - 3$.

Lemma 2.3. *We have $\{\Sigma^m C_p;\, C_p\} = (\mathbf{Z}/p)x \oplus (\mathbf{Z}/p)y$, where the generators x and y are as described below.*

Proof. We consider the stable homotopy exact sequence:

$$\cdots \xrightarrow{p} \{\Sigma^m C_p;\, S^1\} \to \{\Sigma^m C_p;\, C_p\} \to \{\Sigma^m C_p;\, S^2\} \xrightarrow{p} \cdots$$

of the cofibre sequence

$$S^1 \xrightarrow{p} S^1 \to C_p \to S^2 \to \cdots .$$

The lefthand group $\{\Sigma^m C_p;\, S^1\}$ is cyclic of order 4 if $p = 2$, of order p if p is odd. So a generator maps to an element x, say, of order p in $\{\Sigma^m C_p;\, C_p\}$. Let y be the class $\eta \wedge 1$, if $p = 2$, $\alpha_1 \wedge 1$, if p is odd, where η or α_1 is the standard element of order p in the stable $(2p-3)$-stem $\{S^m;\, S^0\}$, and 1 is the identity in $\{C_p;\, C_p\}$. By looking at the cofibre sequences on the left, one sees that $\{\Sigma^m C_p;\, S^2\}$ is cyclic of order p and generated by the image of y. This establishes the lemma.

For $n > 1$ we write M^n for the mod p Moore space $\Sigma^{n-2} C_p$ with top cell in dimension n. We fix $n > 2p+2$, which will be large enough to ensure that later computations are performed in the stable range.

If P and Q are finite complexes then the stable maps from the k-fold wedge $\bigvee_k P$ to the l-fold wedge $\bigvee_l Q$ are naturally identified with $l \times k$-matrices with entries in $\{P;\, Q\}$:

$$\{\textstyle\bigvee_k P;\, \bigvee_l Q\} = M_{l,k}(\{P;\, Q\}) = M_{l,k}(\mathbf{Z}) \otimes \{P;\, Q\}. \tag{2.4}$$

(Note the order, chosen so that composition of maps corresponds to matrix multiplication.)

Let $V = \bigvee_k M^{m+n}$ and $W = \bigvee_k M^n$. Now fix a matrix $A \in M_k(\mathbf{F}_p)$. Let $\phi : V \to W$ be a map whose class in $\{V;\, W\}$ is given under the correspondence (2.4) above by

$$I_k \otimes x + \tilde{A} \otimes y \in M_k(\mathbf{Z}) \otimes \{M^{m+n};\, M^n\}, \tag{2.5}$$

where $\tilde{A}$ is any matrix reducing (mod p) to A. We define X to be the cofibre of ϕ:

$$V \xrightarrow{\phi} W \xrightarrow{\iota} X \xrightarrow{\delta} \Sigma V \to \Sigma W \to \cdots . \tag{2.6}$$

From the homology exact sequence we obtain: $H_i(X;\, \mathbf{Z}) = (\mathbf{Z}/p)^k$ if $i = n-1$ or $n+m$, and 0 otherwise.

Lemma 2.7. *The image of*

$$\alpha' : \{X;\, X\} \to \operatorname{End}(H_{n-1}(X;\, \mathbf{Z}) \otimes \mathbf{F}_p) \times \operatorname{End}(H_{n+m}(X;\, \mathbf{Z}) \otimes \mathbf{F}_p)$$

is $\{(L, M) \in M_k(\mathbf{F}_p) \times M_k(\mathbf{F}_p) \mid L = M,\ AL = MA\}$.

Proof. Let $L,\, M \in M_k(\mathbf{F}_p)$. Suppose that $\tilde{L},\, \tilde{M} \in M_k(\mathbf{Z})$ reduce (mod p) to L and M, respectively, and let $\lambda : V \to V$, $\mu : W \to W$ be maps whose stable classes are $\tilde{L} \otimes 1 \in M_k(\mathbf{Z}) \otimes \{M^{m+n};\, M^{m+n}\}$, $\tilde{M} \otimes 1 \in M_k(\mathbf{Z}) \otimes \{M^n;\, M^n\}$. The compositions $\mu.\phi$ and $\phi.\lambda$ will be homotopic if and only if

$$\tilde{M} \otimes x + \tilde{M}\tilde{A} \otimes y = \tilde{L} \otimes x + \tilde{A}\tilde{L} \otimes y \in M_k(\mathbf{Z}) \otimes \{M^{m+n};\, M^n\},$$

that is, if $L = M$ and $AL = MA$.

A choice of homotopy $\mu.\phi \simeq \phi.\lambda$ determines a map $f : X \to X$ of cofibres which fits into a homotopy commutative ladder:

$$\begin{array}{ccccccccccc}
V & \xrightarrow{\phi} & W & \longrightarrow & X & \longrightarrow & \Sigma V & \xrightarrow{\Sigma\phi} & \Sigma W & \longrightarrow & \cdots \\
{\scriptstyle\lambda}\downarrow & & {\scriptstyle\mu}\downarrow & & {\scriptstyle f}\downarrow & & {\scriptstyle\Sigma\lambda}\downarrow & & {\scriptstyle\Sigma\mu}\downarrow & & \\
V & \xrightarrow[\phi]{} & W & \longrightarrow & X & \longrightarrow & \Sigma V & \xrightarrow[\Sigma\phi]{} & \Sigma W & \longrightarrow & \cdots
\end{array}$$

Then $\alpha'(f) = (L, M)$, and we have shown that $(L, L) \in \operatorname{im} \alpha'$ whenever $AL = LA$.

To establish the converse, consider a map $f : X \to X$. Then the composition $\delta.f.\iota : W \to X \to X \to \Sigma V$ (in the notation of (2.6)) is null-homotopic, because the dimension of W is less than or equal to the connectivity of ΣV, $n \leq m + n - 1$. (Note that this is just achieved when $p = 2$.) So $\delta.f$ factors, up to homotopy, through ΣV, and a choice of homotopy gives a homotopy commutative diagram:

$$\begin{array}{ccccccccccc}
W & \xrightarrow{\iota} & X & \xrightarrow{\delta} & \Sigma V & \xrightarrow{\Sigma\phi} & \Sigma W & \longrightarrow & \Sigma X & \longrightarrow & \cdots \\
 & & {\scriptstyle f}\downarrow & & {\scriptstyle\lambda'}\downarrow & & {\scriptstyle\mu'}\downarrow & & {\scriptstyle\Sigma f}\downarrow & & \\
W & \xrightarrow[\iota]{} & X & \xrightarrow[\delta]{} & \Sigma V & \xrightarrow[\Sigma\phi]{} & \Sigma W & \longrightarrow & \Sigma X & \longrightarrow & \cdots
\end{array}$$

Describing the stable classes of the maps λ' and μ' by matrices $\tilde{L},\, \tilde{M}$, as above, we find that $\alpha'(f) = (L, M)$ with $L = M$, $AL = MA$.

Remark 2.8. If p is odd, it is not too difficult to compute the ring $\{X;\, X\}$ itself by using the representation

$$\alpha'' : \{X;\, X\} \to \operatorname{End} K_{n-1}(X)$$

to complex K-theory. The generator y, (2.3), maps to zero in K-theory. So as far as K-theory is concerned X decomposes as a wedge of k pieces. Moreover, the generator x in K-theory corresponds under Bott periodicity to a non-trivial map $C_p \to \Sigma C_p$. It follows that K-theoretically X is a k-fold wedge of mod p^2 Moore spaces and $K_{n-1}(X) \cong (\mathbf{Z}/p^2)^k$. The group $\{X; X\}$ is easily computed up to extension by looking at the stable homotopy exact sequences on the left and right for the defining cofibration (2.6). The map α'' is checked to be injective with image isomorphic to the ring of all $k \times k$ matrices over $\mathbf{Z}/p^2$ whose reduction (mod p) commutes with A.

For a suitable choice of A the centralizer $\{L \in M_k(\mathbf{F}_p) \mid AL = LA\}$ will be isomorphic to $\mathbf{F}_q$. (To be definite, let A be the matrix, with respect to some basis over $\mathbf{F}_p$ of $\mathbf{F}_q$, of the linear transformation $\mathbf{F}_q \to \mathbf{F}_q$ given by multiplication by some generator of the group of units $\mathbf{F}_q^\bullet$.) Then we have constructed a space X, as required, with $\{X; X\}/\mathrm{rad} \cong \mathbf{F}_q$. Indeed, varying n we have constructed infinitely many such spaces.

Finally, we prove Theorem 1.4. Let F_j, $1 \le j \le s$, be a family of finite fields of characteristic p, and $r_j \ge 1$ corresponding positive integers. By the construction above, we can find non-equivalent complexes X_j with $\{X_j; X_j\}/\mathrm{rad} \cong F_j$. Then by taking r_j copies of X_j for $1 \le j \le s$ and forming the wedge, we obtain a finite complex X with

$$\{X; X\}/\mathrm{rad}\{X; X\} \cong \prod_{1 \le j \le s} M_{r_j}(F_j),$$

as asserted by the theorem.

3. Realizing finite fields unstably

In this section we consider a (connected) H-space Y with each homotopy group a finite p-group and all but finitely many zero. We write '+' for the operation on $[Y; Y]$ given by some H-multiplication, and refer back to (1.11) and (1.5) for the definitions of β and $\mathcal{N}$.

In [**2**] an equivalence relation $\sim$ was defined on $[Y; Y]$ by

$$f \sim g \iff f = g + n \quad \text{for some } n \in \mathcal{N}.$$

To see that this in an equivalence relation one shows that

$$f \sim g \iff \beta(f) - \beta(g) \in \mathrm{rad}(\mathrm{im}\,\beta).$$

In one direction this is clear, because β is compatible with addition and composition. On the other hand, given f and g we can write $f = g +$

n for some $n \in [Y; Y]$. If $\beta(f) - \beta(g) \in \mathrm{rad}(\mathrm{im}\,\beta)$, then by (1.2) and (1.12) we have $n \in \mathcal{N}$. The set of equivalence classes, written as $[Y; Y]/\mathcal{N}$, then clearly acquires a ring-structure (not depending upon the choice of H-multiplication) from the addition and composition on $[Y; Y]$.

For the purposes of the construction below it is useful to replace β by:

$$\beta' : [Y; Y] \to \prod_{i \geq 1} \mathrm{End}(\pi_i(Y) \otimes \mathbf{F}_p), \tag{3.1}$$

given by the action on homotopy groups. (For a non-abelian finite p-group G, $G \otimes \mathbf{F}_p$ is to be read as G/G^* where G^* is the subgroup generated by all commutators and pth-powers.) See [**1**], p 475.

Lemma 3.2. *We have:*

$$f \sim g \iff \beta'(f) - \beta'(g) \in \mathrm{rad}(\mathrm{im}\,\beta').$$

Proof. One shows, following the argument in [**2**], that $f \in [Y; Y]$ is nilpotent if and only if $\beta'(f)$ is nilpotent. Compare (5.5)(iii).

Corollary 3.3. *The ring* $[Y; Y]/\mathcal{N}$ *is isomorphic to* $\mathrm{im}\,\beta'/\mathrm{rad}(\mathrm{im}\,\beta')$.

We wish to show that, given finite fields F_j of characteristic p and positive integers r_j, $1 \leq j \leq s$, there is an H-space Y with $[Y; Y]/\mathcal{N}$ isomorphic to $\prod_{1 \leq j \leq s} M_{r_j}(F_j)$. As in the preceding section it is sufficient to consider the case of a single field, and this time consider direct products of the resulting spaces for the general case.

Our construction will be formally dual to that in the stable case, replacing Moore spaces by Eilenberg-MacLane spaces, cofibres by fibres. We fix $k \geq 1$ and set $m = 2p - 1$ this time. Choose $n > 2p + 1$, which will ensure that looping

$$\Omega : [K(\mathbf{Z}/p, n); K(\mathbf{Z}/p, m+n)] \to [K(\mathbf{Z}/p, n-1); K(\mathbf{Z}/p, m+n-1)]$$

is an isomorphism.

Lemma 3.4. *We have* $[K(\mathbf{Z}/p, n); K(\mathbf{Z}/p, m+n)] = (\mathbf{Z}/p)x \oplus (\mathbf{Z}/p)y$, *where* $x = \mathcal{P}^1\beta_p$, *(*$Sq^2 Sq^1$ *if* $p = 2$*),* $y = \beta_p \mathcal{P}^1$, *(*$Sq^1 Sq^2$ *if* $p = 2$*).*

Let $V = \prod^k K(\mathbf{Z}/p, m+n)$ and $W = \prod^k K(\mathbf{Z}/p, n)$ be the k-fold products. Restriction to the wedge $\bigvee_k K(\mathbf{Z}/p, n) \subseteq W$ gives, by connectivity, an isomorphism: $[W; V] \to [\bigvee_k K(\mathbf{Z}/p, n); V]$, that is,

$$[W; V] \xrightarrow{\cong} M_k(\mathbf{F}_p) \otimes [K(\mathbf{Z}/p, n); K(\mathbf{Z}/p, m+n)]. \tag{3.5}$$

Fix a matrix $A \in M_k(\mathbf{F}_p)$, and let $\phi : W \to V$ be a map corresponding to $I_k \otimes x + A \otimes y$ under (3.5). We define Y to be the homotopy fibre of ϕ:

$$\cdots \to \Omega W \to \Omega V \to Y \to W \xrightarrow{\phi} V. \tag{3.6}$$

As ϕ is a loop map, Y is an H-space. The homotopy exact sequence computes its homotopy groups: $\pi_i(Y) = (\mathbf{Z}/p)^k$ if $i = n$ or $m+n+1$, and 0 otherwise. An argument formally identical to that given for (2.7) yields the image of the map β'.

Lemma 3.7. *The image of*

$$\beta' : [Y;\, Y] \to \mathrm{End}(\pi_n(Y) \otimes \mathbf{F}_p) \times \mathrm{End}(\pi_{m+n-1}\,(Y) \otimes \mathbf{F}_p)$$

is $\{(M, L) \in M_k(\mathbf{F}_p) \times M_k(\mathbf{F}_p) \mid L = M,\ AM = LA\}$.

Remark 3.8. The space Y is, in fact, an infinite loop space. From the homotopy exact sequence on the right, and Serre exact sequence on the left, it is straightforward to see that every self-map of Y is an infinite loop map. The set $[Y;\, Y]$ is thus seen to be a ring, and $\mathcal{N}$ is the Jacobson radical.

Finally, one can choose A such that $\mathrm{im}\,\beta'$, and hence $[Y;\, Y]/\mathcal{N}$, is a field of order p^k. This completes the proof of Theorem 1.6.

4. Self-maps of p-complete spaces

We need to review some facts about p-complete spaces of finite type, following the approach of Sullivan, [**5**].

Let $\mathfrak{F}$ denote the homotopy category of connected spaces F (of the homotopy type of CW-complexes) with each homotopy group $\pi_i F$ a finite p-group and all but finitely many homotopy groups trivial. These spaces admit the following, more practical description.

Lemma 4.1. *A space F is in $\mathfrak{F}$ if and only if there is a finite sequence of spaces $* = F_0, F_1, \ldots, F_m \simeq F$ with each F_i, $1 \le i \le m$, the homotopy fibre of a map $F_{i-1} \to K(C_i, n_i + 1)$ to an Eilenberg-Maclane space $K(C_i, n_i + 1)$ where C_i is an elementary abelian p-group, $n_i \ge 1$.*

In other words, the objects of $\mathfrak{F}$ can be constructed inductively from a point by means of principal fibrations of the form:

$$\cdots \to K(C, n) \to F \to F' \to K(C, n+1), \tag{4.2}$$

where C is an elementary abelian p-group, $n \geq 1$. This is a standard result (although a complete proof is hard to find in the literature); see [5], [6].

Let Y be a connected space with mod p homology $H_*(Y)$ of finite type. Write $\mathfrak{F}(Y)$ for the category whose objects are homotopy classes $\xi : Y \to F_\xi$ of maps to spaces in $\mathfrak{F}$ and morphisms $\xi \to \eta$ homotopy classes $\phi : F_\xi \to F_\eta$ with $\phi.\xi = \eta$. Then Y is *p-complete of finite type* if the natural map

$$(4.3) \qquad [X;\, Y] \to \varprojlim_{\mathfrak{F}(Y)} [X;\, F_\xi]$$

is a bijection for each connected finite complex X.

In that case, the map (4.3) will be a bijection for any complex X with mod p homology of finite type, and the set $[X;\, Y]$ acquires a natural profinite topology as the inverse limit of the finite (discrete) sets $[X;\, F_\xi]$.

If X, Y and Z are p-complete of finite type, then composition

$$[Y;\, Z] \times [X;\, Y] \to [X;\, Z]$$

is continuous. It follows (by taking $X = Y$ and $Z = K(\mathbf{Z}/p, n)$) that

$$(4.4) \qquad \beta : [Y;\, Y] \to \prod_{i \geq 1} \operatorname{End} PH_i(Y),$$

given by the induced maps on primitives, is continuous. (We always give finite sets the discrete topology and form topological products.)

Proposition 1.12 generalizes to p-complete spaces of finite type as follows.

Proposition 4.5. *Let a space Y be p-complete of finite type, $f \in [Y;\, Y]$. Then*

(i) *f is a unit $\iff$ $\beta(f)$ is a unit;*
(ii) *f is topologically nilpotent $\iff$ $\beta(f)$ is topologically nilpotent;*
(iii) $f^{p^\infty} = 1 \iff \beta(f)^{p^\infty} = 1$.

Part (i) is a standard Whitehead theorem, [5]. In (ii), f *topologically nilpotent* means that $f^n \to 0$ as $n \to \infty$; and in (iii) $f^{p^\infty} = 1$ means that $f^{p^r} \to 1$ as $r \to \infty$.

The proof will depend on an elementary algebraic lemma. Let S and T be finite sets with base-points $*$, G a finite group acting on S. Write $\lambda : G \to S$ for the map $\lambda(g) = g.*$. Suppose that $\mu : S \to T$ is a base-point preserving map such that the sequence

$$(4.6) \qquad G \xrightarrow{\lambda} S \xrightarrow{\mu} T$$

is exact, in the sense that $\mu(x) = \mu(y)$ if and only if x and y lie in the same G-orbit of S.

Suppose that $\gamma : G \to G$ is a group homomorphism and that $\sigma : S \to S$ and $\tau : T \to T$ are base-point preserving maps, such that $\sigma(g.x) = \gamma(g).\sigma(x)$ for $g \in G$, $x \in S$, and $\mu\sigma = \tau\mu$. We let 0 denote both the constant map to the base-point on S or T and the trivial endomorphism of G; and 1 will be the identity map.

Lemma 4.7.

(i) *If γ and τ are bijective, then so is σ.*
(ii) *If $\gamma^m = 0$ and $\tau^n = 0$, then $\sigma^{m+n} = 0$.*
(iii) *Suppose that G is a p-group of order p^m. If $\gamma^{p^n} = 1$ and $\tau^{p^n} = 1$, then $\sigma^{p^{m+n}} = 1$.*

Proof. The verification is straightforward diagram-chasing. Part (i) depends, of course, on the finiteness of the sets involved. We can now proceed to the proof of Proposition 4.4. The three parts are similar and we consider only part (ii) in detail.

Suppose that $\beta(f)$ is topologically nilpotent. We wish to prove that f is topologically nilpotent. (The converse follows from the continuity of β.) First we apply (4.7)(ii) with

$$PH_i(Y) \to H_i(Y) \xrightarrow{\Delta} H_i(Y \wedge Y)$$

as (4.6) and γ, σ, τ induced by f to deduce, by induction on i, that $f_* : H_i(Y) \to H_i(Y)$ is nilpotent.

So $f^* : [Y;\, K(C,n)] \to [Y;\, K(C,n)]$ is nilpotent for any elementary abelian p-group C, $n \geq 1$. Now apply (4.7)(ii) again to the homotopy exact sequence

$$[Y;\, K(C,n)] \to [Y;\, F] \to [Y;\, F']$$

of the fibration (4.2) to deduce that, if f is nilpotent on $[Y;\, F']$, then it is nilpotent on $[Y;\, F]$. From (4.1), we see that $f^* : [Y;\, F] \to [Y;\, F]$ is nilpotent for each space F in the category $\mathfrak{F}$.

Hence, for each object $\xi : Y \to F_\xi$ in $\mathfrak{F}(Y)$ there is an integer $n \geq 1$ such that $(f^*)^n(\xi) = \xi.f^n = 0 \in [Y;\, F_\xi]$. From (4.3) with $X = Y$ and the definition of the profinite topology on $[Y;\, Y]$, f is topologically nilpotent.

The argument in the other two cases is entirely similar, using clauses (i) and (iii) of (4.7). If the space Y is in $\mathfrak{F}$, then $[Y;\, Y]$ and $\operatorname{im}\beta$ are finite sets with the discrete topology and (4.5) simplifies to the statement (1.12).

5. The group of self-equivalences

In this section we study the group of self-equivalences $\mathcal{E}(Y)$ of a p-complete space Y of finite type. This group is topologized as a subspace of $[Y; Y]$, and we observe first:

Lemma 5.1. *The group of self-equivalences $\mathcal{E}(Y)$ is a profinite topological group.*

Proof. One can follow the proof in [**1**] that $[Y; Y]$ is a profinite monoid. We prefer to exploit the group structure as follows. Since composition is continuous, $\{(g,h) \in [Y; Y] \times [Y; Y] \mid gh = 1 = hg\}$ is a closed subspace of the compact Hausdorff space $[Y; Y] \times [Y; Y]$. By elementary compactness arguments, we find that $\mathcal{E}(Y)$ is a compact Hausdorff topological group. It is totally disconnected, being a subspace of $[Y; Y]$, and hence is a profinite group. (For standard results on profinite groups we refer to [**3**], [**4**].) We need to recall the Sylow theory for profinite groups. Let G be a profinite group. For $g \in G$ we write $g^{p^\infty} = 1$ if $g^{p^r} \to 1$ as $r \to \infty$. A closed subgroup $H \leq G$ is called a p-subgroup if $h^{p^\infty} = 1$ for all $h \in H$. A Sylow p-subgroup is a maximal p-subgroup. From the theory for finite groups one deduces that:

(5.2) (i) *every p-subgroup of G lies in a Sylow p-subgroup, and*
(ii) *any two Sylow p-subgroups are conjugate.*

We now extend the definition of the Fitting p-subgroup to profinite groups, defining $O_p(G)$ to be the intersection of the Sylow p-subgroups of G. It follows at once from the properties (i) and (ii) above that $O_p(G)$ is the unique maximal normal p-subgroup of G.

Lemma 5.3. *The Fitting p-subgroup, $O_p(G)$, of a profinite group G is equal to*

$$N = \{g \in G \mid h^{p^\infty} = 1 \Rightarrow (gh)^{p^\infty} = 1\}.$$

Proof. First, suppose $g \in O_p(G)$ and $h^{p^\infty} = 1$. By (5.2)(i), h lies in some Sylow p-subgroup P. By definition, $g \in P$. Hence gh is an element of the p-subgroup P and $(gh)^{p^\infty} = 1$. For the converse, one checks easily that N is a subgroup of G; it is then clearly a normal p-subgroup. Equality follows.

Proposition 5.4. *Let Y be p-complete of finite type. Then β, (4.4), induces an isomorphism*

$$\mathcal{E}(Y)/O_p(\mathcal{E}(Y)) \xrightarrow{\cong} \beta(\mathcal{E}(Y))/O_p(\beta(\mathcal{E}(Y)))$$

of profinite groups.

Proof. This is immediate from (5.1) and (4.5)(iii).

This proves the first half of Theorem 1.13. To complete the proof we need some, mostly well known, results on profinite rings.

Lemma 5.5. *Let R be a profinite ring.*

(i) *Let $r \in R$. Then there is a unique idempotent in the closure of $\{r^n \mid n \geq 1\}$.*

(ii) *The group $R^\bullet$ of units in R is closed in R and is a profinite topological group.*

(iii) *An element $r \in R$ is topologically nilpotent $\iff$ $1 - r^n$ is a unit for all $n \geq 1$.*

(iv) *The Jacobson radical* $\operatorname{rad} R$ *is closed in R, and $r \in \operatorname{rad} R \iff rs$ is topologically nilpotent for all $s \in R$.*

Proof. (i) One reduces easily to the case in which R is a finite ring. If $r^m = r^{m+n}$ with $m, n \geq 1$, then r^{mn} is idempotent. On the other hand, if r^k and r^l are idempotent, then $r^k = r^{kl} = r^l$.
(ii) Follow the proof of (5.1).
(iii) If r is topologically nilpotent, then $1 - r^n$ has inverse $1 + r^n + r^{2n} + \ldots$. Conversely, let e be the unique idempotent, given by (i), in the closure of $\{r^n \mid n \geq 1\}$. If each $1 - r^n$ is a unit, then by (ii) $1 - e$ is a unit and idempotent. So $e = 0$, and hence r is topologically nilpotent.
(iv) Closure follows from (ii). We see from (iii) that, for $r \in R$, $1 - rs$ is a unit for all $s \in R$ if and only if rs is topologically nilpotent for all s.

Lemma 5.6. *Suppose that $\rho : R \to S$ is a continuous homomorphism of profinite rings such that: $r \in R$ is a unit if and only if $\rho(r) \in S$ is a unit (or, in other words, such that* $\ker \rho \subseteq \operatorname{rad} R$*).*

(i) *Then $r \in R$ is topologically nilpotent $\iff$ $\rho(r) \in S$ is topologically nilpotent.*

(ii) *Suppose further that $p^n r \to 0$ as $n \to \infty$ for each $r \in R$ (so that R becomes a topological $\hat{\mathbf{Z}}_p$-algebra). Then, for a unit $r \in R$ we have: $r^{p^\infty} = 1 \iff \rho(r)^{p^\infty} = 1$.*

Proof. Part (i) is immediate from (5.5)(iii). For (ii) one reduces to the case in which R and S are finite. If $\rho(r)^{p^k} = 1$, then $r^{p^k} = 1 + x$ with $x \in \operatorname{rad} R$. Since x is nilpotent and p-torsion, we have $(1 + x)^{p^l} = 1$ for large l. Hence $r^{p^{k+l}} = 1$. Let Y now be an H-space. From [**2**], the ring $[Y; Y]/\mathcal{N}$ is an (at most) countable product of matrix algebras over finite fields of characteristic p.

Proposition 5.7. *Let Y be a p-complete H-space of finite type. Then the profinite group $\mathcal{E}(Y)/O_p(\mathcal{E}(Y))$ is naturally identified with the group of units in the profinite ring $[Y; Y]/\mathcal{N}$ and so is isomorphic to a countable product of general linear groups over finite fields of characteristic p.*

Proof. Write R for the profinite $\mathbf{F}_p$-algebra $\operatorname{im}\beta$. We have to show that $R^\bullet/O_p(R^\bullet)$ is isomorphic to $S^\bullet/O_p(S^\bullet)$, where $S = R/\operatorname{rad}R$. But this follows at once from (5.6)(ii).

Finally, $O_p(S^\bullet)$ is trivial, by (1.10).

References

[1] J. F. Adams and N. J. Kuhn, *Atomic Spaces and Spectra,* Proc. Edinburgh Math. Soc. **32** (1989), 473-481.

[2] J. R. Hubbuck, *Self Maps of H-spaces,* in Advances in Homotopy Theory, London Math. Soc. Lecture Note Series **139** (1989), 105-110.

[3] J.-P. Serre, Cohomologie Galoisienne, Lecture Notes in Math. 5, Springer, Berlin, (1965).

[4] S. S. Shatz, Profinite Groups, Arithmetic and Geometry, Princeton University Press, Princeton, (1972).

[5] D. Sullivan, *Genetics of Homotopy Theory and the Adams Conjecture,* Ann. of Math. **100** (1974), 1-79.

[6] Kai Xu, Endomorphisms of complete spaces, Ph. D thesis, University of Aberdeen, (1991).

Department of Mathematical Sciences
University of Aberdeen
Aberdeen AB9 2TY
Scotland

Maps between p-completed classifying spaces. III.

Zdzisław Wojtkowiak

Dedicated to the memory of
Professor J. Frank Adams.

0. Introduction.

Let G and G' be connected, compact Lie groups with maximal tori T and T'. In the celebrated paper, [AM], Adams and Mahmud demonstrated the importance of the maximal torus in the study of maps between classifying spaces. They showed that for any map $f : BG \to BG'$ there is a homomorphism $\phi : T \to T'$ such that the following diagram of cohomology algebras commutes.

$$\begin{array}{ccc} H^*(BG', \mathbb{Q}) & \xrightarrow{f^*} & H^*(BG, \mathbb{Q}) \\ \downarrow i'^* & & \downarrow i^* \\ H^*(BT', \mathbb{Q}) & \xrightarrow{(B\phi)^*} & H^*(BT, \mathbb{Q}) \end{array}$$

(The maps i and i' are induced by the inclusions of maximal tori.) Later Zabrodsky showed that any map $g : BT \to BG$ is induced by a homomorphism from T to G, [Z]. From this it follows that the cohomology diagram is induced by a homotopy commutative diagram of spaces. Similarly it was shown in [AW], that for any map between the p-completions $f : (BG)_p \to (BG')_p$, there is a map $\Phi : (BT)_p \to (BT')_p$ such that the following diagram commutes up to homotopy

$$
(*) \qquad \begin{array}{ccc}
(BG)_p & \xrightarrow{f} & (BG')_p \\
\uparrow {\scriptstyle i_p} & & \uparrow {\scriptstyle i'_p} \\
(BT)_p & \xrightarrow{\Phi} & (BT')_p \,.
\end{array}
$$

One might wonder if the p-local version of the result just stated is also true. The following result due to C. Mc Gibbon, shows that it is not the case. We shall quote an extract from his letter (2 January 1991) to the author.

... Given a map, say $f:(BG)_{(p)} \to (BG')_{(p)}$, between the p-localizations of classifying spaces of two compact Lie grops, *one cannot in general*, lift this map to a map of p-local maximal tori. In other words, the map $\overline{f}$, that would make the following diagram

$$
(**) \qquad \begin{array}{ccc}
(BT)_{(p)} & \xrightarrow{\overline{f}} & (BT')_{(p)} \\
\downarrow {\scriptstyle B_i} & & \downarrow {\scriptstyle B_{i'}} \\
(BG)_{(p)} & \xrightarrow{f} & (BG')_{(p)}
\end{array}
$$

commute, *does not necessarily exist.*

Here is a simple example that proves this claim. Take $G = G' = SU(2)$ and take $p = 7$. ($BG = HP^\infty$). Let $f : (HP^\infty)_{(7)} \to (HP^\infty)_{(7)}$ have deg $= 2$ in dimension 4. In other words, the induced homomorphism $f_* : H_4((HP^\infty)_{(7)}; \mathbb{Q}) \to H_4((HP^\infty)_{(7)}; \mathbb{Q})$ is multiplication by 2. This map, f, exists because 2 is the square of a 7-adic integer (see [R]). The number 2 is the square of a 7-adic integer because $2 \equiv 3^2 \bmod 7$ and basic number theory. However, 2 is not the square of a 7-local integer ...

Now we shall describe the contents of the paper. In section 1 to any map $f : (BG)_{(p)} \to (BG')_{(p)}$ we associate a map $\Phi : (BT)_p \to (BT')_p$ between p-completions of the classifying spaces of maximal tori such that the diagram $(**)$ commutes after p-completions. We study the properties of such maps Φ. We show that they are defined over a henselisation R_p of $Z_{(p)}$.

We use them to describe the set of homotopy classes $[BG_{(p)};\ BG'_{(p)}]$ if p does not divide the order of the Weyl group of G.

Let $\chi_{Z_p} : [(BG)_p, (BG')_p] \to [(BT_p), (BT')_p]$ be a multivalued map which to any $f : (BG)_p \to (BG')_p$ associates $\Phi : (BT)_p \to (BT')_p$ such that the diagram $(*)$ commutes. In order to get some information about $[(BG)_p,\ (BG')_p]$ one can try to understand $im\ \chi_{Z_p}$. Notice that in some cases the map χ_{Z_p} is even injective. In section 2 we shall discuss some examples where we are able to give some non-trivial restrictions on $im\ \chi_{Z_p}$.

In section 3 we study the classification of p-local spaces whose p-completion is homotopy equivalent to $(BG)_p$. To raise the interest of the reader we shall state the main result of this section.

Theorem. If X is a p-local, nilpotent space of finite $Z_{(p)}$-type such that its p-completion X_p is homotopy equivalent to $(BS^3)_p$ then X is homotopy equivalent to $(BS^3)_{(p)}$. If the rank of a simple Lie group G is bigger than 1 then there are uncountably many p-local, nilpotent homotopy types X of finite $Z_{(p)}$-type such that X_p is homotopy equivalent to $(BG)_p$.

1. Admissible maps.

Let W and W' be Weyl groups of G and G' respectively.

For a given map between the p-completions of classifying spaces, $f : (BG)_p \to (BG')_p$, let $\Phi : (BT)_p \to (BT')_p$ be such that the diagram $(*)$ from section 0 commutes up to homotopy.

Let

$$\begin{array}{cccc} K := \pi_2(\Phi) : & \pi_2\big((BT)_p\big) & \to & \pi_2\big((BT')_p\big) \\ & \wr\wr & & \wr\wr \\ & \pi_1(T)\otimes Z_p & & \pi_1(T')\otimes Z_p \end{array}$$

be the map induced by Φ on fundamental groups of maximal tori tensored by Z_p.

Definition 1. (see [AW]) *Let R be a free Z-module. We say that a homomorphism $\varphi : \pi_1(T)\otimes R \to \pi_1(T')\otimes R$ is admissible if for any $w \in W$ there is $w' \in W'$ such that $\varphi \circ w = w' \circ \varphi$.*

Observe that the map

$$K : \pi_1(T)\otimes Z_p \to \pi_1(T')\otimes Z_p$$

is admissible.

It follows from [AW] Proposition 1.3 that the correspondence $w \to w'$ from Definition 1 can be chosen a homomorphism.

Definition 2. *We say that a homomorphism $\varphi : \pi_1(T) \otimes R \to \pi_1(T') \otimes R$ is α-admissible if there is a homomorphism $\alpha : W \to W'$ such that for any $w \in W$ we have $\varphi \circ w = \alpha(w) \circ \varphi$.*

Let $Adm(T,T')(R)$ be the set of admissible homomorphisms from $\pi_1(T) \otimes R$ to $\pi_1(T') \otimes R$. We define a multivalued map

$$\chi_{Z_p} : [(BG)_p, (BG')_p] \to Adm(T,T')(Z_p)$$

setting $\chi_{Z_p}(f) = \left\{ \begin{array}{c} K \mid K = \pi_2(\Phi) \text{ and the diagram } (*) \\ \text{commutes up to homotopy} \end{array} \right\}$.

Similarly we define a map

$$\chi_Z : [BG, BG'] \to Adm(T,T')(Z).$$

Now we shall define a map $\chi_{Z_{(p)}}$. Let R_p be a henselization of $Z_{(p)}$ (see [M] §4). R_p can be view as a subring of Z_p consisting of all algebraic elements in Z_p.

Proposition 1. *Let $f : (BG)_{(p)} \to (BG')_{(p)}$ be a map. Then there is a homomorphism of algebras $F : H^*(BT', R_p) \to H^*(BT, R_p)$ such that the diagram*

$$\begin{array}{ccc}
H^*((BG')_{(p)}, Z_{(p)}) & \xrightarrow{f^*} & H^*((BG)_{(p)}, Z_{(p)}) \\
\downarrow i'^* & & \downarrow i^* \\
H^*((BT')_{(p)}, Z_{(p)}) & & H^*((BT)_{(p)}, Z_{(p)}) \\
\downarrow \iota' & & \downarrow \iota \\
H^*(BT', R_p) & \xrightarrow{F} & H^*(BT, R_p)
\end{array}$$

commutes where ι and ι' are induced by the inclusion $Z_{(p)} \hookrightarrow R_p$.

Proof. There is a map $\rho : (BT)_p \to (BT')_p$ such that the diagram

$$\begin{array}{ccc} H^*((BG')_{(p)}, Z_p) & \xrightarrow{f^*} & H^*((BG)_{(p)}, Z_p) \\ \downarrow i'^* & & \downarrow i^* \\ H^*(BT', Z_p) & \xrightarrow{\rho^*} & H^*(BT, Z_p) \end{array}$$

commutes. The extensions $i^* : H^*(BG, Z_p) \to H^*(BT, Z_p)$ and $i'^* : H^*(BG', Z_p) \to H^*(BT', Z_p)$ are of a transcendental degree zero. Hence ρ^* is defined over R_p (i.e. there exists $F : H^*(BT', R_p) \to H^*(BT, R_p)$ such that $\rho^* = F \underset{R_p}{\otimes} id_{Z_p}$) because f^* is defined over R_p. This proves the proposition.

Let $F^2 : H^2(BT', R_p) \to H^2(BT, R_p)$ be the restriction of F to the second cohomology group. Observe that $\pi_1(T) \otimes R_p$ is canonically isomorphic to $Hom_{R_p}(H^2(BT, R_p), R_p)$.

Definition 3. *We set*

$\chi_{Z_{(p)}}(f) := Hom(F^2, R_p)$:

$Hom(H^2(BT, R_p), R_p) \approx \pi_1(T) \otimes R_p \to Hom(H^2(BT', R_p), R_p) \approx \pi_1(T') \otimes R_p$

So we have defined a multivalued map

$$\chi_{Z_{(p)}} : [(BG)_{(p)}; (BG')_{(p)}] \to Adm(T, T')(R_p).$$

Definition 4. *We say that two admissible maps ϕ and ψ from $\pi_1(T) \otimes R$ to $\pi_1(T') \otimes R$ are equivalent if there exists $w' \in W'$ such that $\phi = w' \circ \psi$. We shall denote by $Ahom(T, T')(R)$ the set of equivalence classes of admissible maps from $\pi_1(T) \otimes R$ to $\pi_1(T') \otimes R$.*

The composition of χ_R with the quotient map $Adm(T, T')(R) \to Ahom(T, T')(R)$ we shall denote also by χ_R.

Proposition 2. *Assume that p does not divide the order of the Weyl group of G. Then the map*

$$\chi_{Z_{(p)}} : [(BG)_{(p)}; (BG')_{(p)}] \to Ahom(T, T')(R_p)$$

is injective.

Proof. We have a commutative diagram

$$\begin{array}{ccc}[(BG)_{(p)},(BG')_{(p)}] & \xrightarrow{\chi_{Z_{(p)}}} & Ahom(T,T')(R_p) \\ {\scriptstyle c_p}\downarrow & & {\scriptstyle \iota}\downarrow \\ [(BG)_p,(BG')_p] & \xrightarrow{\chi_{Z_p}} & Ahom(T,T')(Z_p)\end{array}$$

where c_p is induced by the p-completion and ι is induced by the inclusion $R_p \hookrightarrow Z_p$. The maps c_p and χ_{Z_p} (see [W] Theorem 1.3) are injective. Hence the map $\chi_{Z_{(p)}}$ is also injective.

The map $\chi_{Z_{(p)}}$ is not surjective neither its image is contained in $Ahom(T,T')(Z_{(p)})$. This last fact follows from the example of McGibbon from section 0.

Let $\varphi : \pi_1(T) \otimes R_p \to \pi_1(T') \otimes R_p$ be an admissible map. Then φ induces a map

$$\varphi^* : H^*(BT', R_p) \to H^*(BT, R_p).$$

The restriction of φ^* to subalgebras fixed by W and W' we shall denote by

$$\varphi^W : \left(H^*(BT', R_p)\right)^{W'} \to \left(H^*(BT, R_p)\right)^W.$$

We say that φ^W is defined over $Z_{(p)}$ if there is $\psi : H^*(BT', Z_{(p)})^{W'} \to H^*(BT, Z_{(p)})^W$ such that $\varphi^W = \psi \otimes id_{R_p}$.

Let us set

$$\sqrt{Ahom(T,T')(R_p)} :=$$
$$\{\varphi \in Ahom(T,T')(R_p) \mid \exists \psi \textit{ such that } \varphi^W = \psi \otimes id_{R_p}\}.$$

Proposition 3. *Assume that p does not divide the order of the Weyl group of G. Then the map*

$$\chi_{Z_{(p)}} : [(BG)_{(p)}; (BG')_{(p)}] \to \sqrt{Ahom(T,T')(R_p)}$$

is bijective.

Proof. It is clear that the image of $\chi_{Z_{(p)}}$ is contained in $\sqrt{Ahom(T,T')(R_p)}$. If $\varphi \in \sqrt{Ahom(T,T')(R_p)}$ then we can define a map

$h : (BG)_0 \to (BG')_0$ between rationalizations of BG and BG' which on cohomology coincides with $\psi \otimes id_{\mathbf{Q}}$. It follows from [W] Theorem 1.3 that there is $f : (BG)_p \to (BG')_p$ such that $\chi_{Z_p}(f) = \varphi \underset{R_p}{\otimes} id_{Z_p}$. The maps h and f induce the same map on the cohomology with $\mathbf{Q}_p$-coefficients. Hence by the Sullivan arithmetic square there is $g : (BG)_{(p)} \to (BG')_{(p)}$ such that the rationalization of g is homotopic to h and the p-completion of g is homotopic to f. This implies that $\chi_{Z_{(p)}}(g) = \varphi$.

2. Realization of admissible maps.

Let $Real(T,T')(Z_p)$ (resp. $Real(T,T')(Z_{(p)}), Real(T,T')(Z)$) be the image of χ_{Z_p} (resp. $\chi_{Z_{(p)}}, \chi_Z$). We say that an admissible map is realizable if it belongs to some $Real(T,T')(R)$.

Observe that $Adm(T,T')(Z_p)$ has a natural p-adic topology ($Adm(T,T')(Z_p)$ is a linear subspace of $Hom_{Z_p}(\pi_1(T) \otimes Z_p, \pi_1(T') \otimes Z_p)$).

It is natural to ask what are topological properties of the subspace $Real(T,T')(Z_p)$ of $Adm(T,T')(Z_p)$.

Proposition 4. *$Real(T,T')(Z_p)$ is a closed subset of $Adm(T,T')(Z_p)$.*

Proof. For any p-complete space X of finite Z_p-type, the set of homotopy classes $[Y,X]$ has a pro-finite topology (see [S]).
Consider the diagram

$$[(BG)_p,(BG')_p] \xrightarrow{i^*} [(BT)_p,(BG')_p] \xleftarrow{i'_*} [(BT)_p,(BT')_p].$$

It is a trivial observation that

$$Real(T,T')(Z_p) = (i'_*)^{-1}\Big(i^*\big([(BG)_p,(BG')_p]\big)\Big).$$

The maps i_* and i'^* are continuous. The space $[(BG)_p,(BG')_p]$ is compact, hence $A = i^*\big([(BG)_p,(BG')_p]\big)$ is compact. Therefore $(i'_*)^{-1}(A)$ is closed in $[(BT)_p,(BT')_p]$. The topology on $[(BT)_p,(BT')_p]$ coincides with the topology on $Hom(\pi_1(T)\otimes Z_p, \pi_1(T')\otimes Z_p)$, hence $Real(T,T')(Z_p)$ is closed in $Adm(T,T')(Z_p)$.

Now we shall try to look at the set $Real(T,T')(Z_p)$ more closely.

While in our previous works we were trying to generalize the result of Hubbuck (see [H]) this section arose from our attempts to generalize the result of Ishiguro (see [I]). We shall discuss several examples. They were contained in the last letters we sent to Professor J. Frank Adams.

Notation. For any torus T, the solutions in T of $t^{p^n} = 1$ make up a subgroup $T(n)$; let $T(\infty) = \bigcup_n T(n)$.

Let $G = U(n)$ and $G' = U(n)$. let T be a maximal torus of $U(n)$. The Weyl group of $U(n)$ is Σ_n. Up to conjugation there are three homomorphisms from Σ_n to Σ_n :

a) the trivial homomorphism $e : \Sigma_n \to \Sigma_n, e(\sigma) = Id$;

b) the sign of a permutation homomorphism $\tau : \Sigma_n \to Z/2 \hookrightarrow \Sigma_n$;

c) the identity homomorphism $Id : \Sigma_n \to \Sigma_n$.

If $n = 6$ there is a non-inner automorphism of Σ_6 however this automorphism does not preserve the standard representation of Σ_6 on $\pi_1(T)$.
The e-admissible and τ-admissible $f : (Z_p)^n \to (Z_p)^n$ have matrices $(a_{ij})_{\substack{i=1,\dots,n \\ j=1,\dots,n}}$ such that $a_{ij} = a$ for each pair (i,j).
The id-admissible maps $f : (Z_p)^n \to (Z_p)^n$ have matrices $(a_{ij})_{i,j}$ such that $a_{ii} = a$ for $i = 1,\dots,n$ and $a_{ij} = b$ for $i \neq j$. In this case we assume that $a \neq b$ because otherwise f is e-admissible.

Proposition 5. *Suppose that p divides $n!$.*

i) If an id-admissible map f belongs to $Real(T,T')(Z_p)$ then $a - b \not\equiv 0 (mod\ p)$.

ii) If $1/n \in Z_p$ then an id-admissible map belongs to $Real(T,T')(Z_p)$ if and only if $a - b \not\equiv 0 (mod\ p)$.

iii) Any e-admissible map belongs to $Real(T,T')(Z_p)$.

Proof. First we show i).

Let $F : (Z_p)^n \to (Z_p)^n$ be an id-admissible map such that $\chi_{Z_p}(F) = f$. Let $Z/p \subset \Sigma_n$. It follows from [DZ] Theorem 1.1 (see also [Z] Theorem A for a different proof) that the restriction of F to $B(T(\infty)\tilde{\times} Z/p)$ (Z/p acts by permuting the first p coordinates) is induced by a homomorphism $\varphi : T(\infty)\tilde{\times} Z/p \to T(\infty)\tilde{\times} Z/p$. Let g be a generator of Z/p and let $x = ((0,\dots,0), g)$. Then $\varphi(x) \neq 0$ where 0 is a neutral element of $T(\infty)\tilde{\times} Z/p$. Assume that $p > 2$. The element x, which we view now as a homomorphism from Z/p to $U(n)$ is conjugated to some element $y = (y_1,\dots,y_p,0,\dots,0) \in T(\infty)$. The equalities $c_i(x) = 0$ for $i \neq 0, p-1$ and $c_{p-1}(x) \neq 0$ in $H^{2(p-1)}(Z/p; Z/p)$ implies that

$$y_1 + \cdots + y_p = 0$$
$$y_1 y_2 + \cdots + y_{p-1} y_p = 0$$
$$\vdots$$
$$y_i y_2 \cdot \ldots \cdot y_p = 0$$

and

$$y_1 \cdot \dots \cdot y_{p-1} + y_1 \cdot \dots \cdot y_{p-2} \cdot y_p + \dots \neq 0.$$

Hence one gets $y = (1, 2, 3, \dots, p-1, 0, \dots, 0)$ up to permutation. The image of y by F is $\big(a + (p-1)b, 2a + (p-2)b, \dots\big)$. The fact that $\varphi(x) \neq 0$ implies that $a - b \not\equiv 0 (mod\ p)$. If $p = 2$ then x is conjugated to $y = (1, 0, \dots, 0) \in T(\infty)$. We have $\varphi(y) = (a, b, \dots, b)$ and $\varphi(y) \neq 0$. Hence if b is even then a is odd. Observe that $c_n(y) = 0$. This implies that if b is odd then a is even. Therefore $a - b \not\equiv 0(mod\ 2)$.

Now we prove ii).

Let $\Theta : SU(n) \times S^1 \to U(n)$ be given by $(A, t) \to t \cdot A$ where t is a diagonal matrix. The map Θ is a homomorphism. Hence it induces the map $BSU(n) \times BS^1 \to BU(n)$. This map after inverting n becomes a homotopy equivalence. Hence if $1/n \in Z_p$, $\big(BU(n)\big)_p \approx \big(BSU(n)\big)_p \times (BS^1)_p$. We set $F = \big(\psi^{a-b}, \psi^{a+(n-1)b}\big)$: $\big(BU(n)\big)_p \to \big(BU(n)\big)_p$. One checks that $\chi_{Z_p}(F)$ has a matrix $(a_{ij})_{i,j}$ such that $a_{ii} = a$ for $i = 1, \dots, n$ and $a_{ij} = b$ for $i \neq j$.

To show iii) we observe that the map $F = (B\ det)_p \circ (a)_p \circ \Delta : \big(BU(n)\big)_p \to \big(BU(n)\big)_p$, where $det : U(n) \to S^1$, $(a)_p : (BS^1)_p \to (BS^1)_p$ is induced by a raising to the power a homomorphism from Z/p^∞ to Z/p^∞ and $\Delta : (BS^1)_p \to \big(BU(n)\big)_p$ is induced the diagonal inclusion of S^1 into $U(n)$, is such that $\chi_{Z_p}(F) = (a_{ij})_{i,j}$ with $a_{ij} = a$ for any pair (ij).

Now we consider the case when $G = U(n)$ and $G' = SU(n)$. In this case an admissible map different from zero is an id_{Σ_n}-admissible (Σ_n is the Weyl group of G anf G') and it is given by $f : (Z_p)^n \to V_n \subset (Z_p)^n$ such that

$$f(x_1, \dots, x_n) = k\big((n-1)x_1 - x_2 - \dots - x_n, (n-1)x_2 - x_1 - x_3 - \dots - x_n, \dots\big)$$

where $V_n = \big\{(x_1, \dots, x_n) \in (Z_p)^n \mid \sum_{i=1}^{n} x_i = 0\big\}$.

Proposition 6. *Let $p \leq n$. An id-admissible map $f : (Z_p)^n \to V_n$ is realizable if and only if n and k are invertible in Z_p.*

Proof. Let us assume that f is realizable by $F: \big(BU(n)\big)_p \to \big(BSU(n)\big)_p$. The composition

$$\big(BSU(n)\big)_p \xrightarrow{i} \big(BU(n)\big)_p \xrightarrow{F} \big(BSU(n)\big)_p$$

is an Adams ψ^{kn}-map. Hence if p divides $n!$, $k \cdot n$ is invertible in Z_p. This implies that k and n are invertible in Z_p because $n \in Z$ and $k \in Z_p$.

If n is invertible in Z_p then $(BU(n))_p \approx (BS^1)_p \times (BSU(n))_p$. For k invertible in Z_p, $\psi^k : (BSU(n))_p \to (BSU(n))_p$ exists. Let

$$F : (BU(n))_p \approx (BS^1)_p \times (BSU(n))_p \to (BSU(n))_p \xrightarrow{\psi^k} (BSU(n))_p.$$

The map F realizes f.

Proposition 7. *Let $G = S^3$ and $G' = SO(3)$. If an admissible map $f : \pi_1(T) \otimes Z_2 \to \pi_1(T') \otimes Z_2$ is realizable then f is a multiplication by $2 \cdot k$ where k is a unit or zero in Z_2. The only realizable, admissible map $f : \pi_1(T') \otimes Z_2 \to \pi_1(T) \otimes Z_2$ is the zero map.*

Proof. This follows immediately from the fact that $\psi^k : (BSU(2))_2 \to (BSU(2))_2$ exists if and only if k is invertible in Z_2.

Proposition 8. *Let $G = U(n)$ and $G' = U(2n)$. Assume that p divides $n!$. Let $\alpha, \beta \in Z_p$ be different from zero. The admissible map*

$$\pi_1(T) \otimes Z_p \xrightarrow{\Delta} (\pi_1(T) \otimes Z_p) \oplus (\pi_1(T) \otimes Z_p) \approx \pi_1(T') \otimes Z_p \xrightarrow{(\cdot\alpha, \cdot\beta)}$$
$$(\pi_1(T) \otimes Z_p) \oplus (\pi_1(T) \otimes Z_p) \approx \pi_1(T') \otimes Z_p$$

(where Δ is the diagonal map, $\cdot\alpha$ (resp. $\cdot\beta$) is a multiplication by α (resp. β)) is realizable if and only if α and β are invertible in Z_p.

Proof is similar to the proof of Proposition 3 so we omit it.

We give one necessary condition for an admissible map to be realizable.

For any connected compact Lie group G let Γ be an extended Weyl group of G and let $\Gamma_0 = ker(\Gamma \to W)$. The group Γ_0 can be considered as a subgroup of $\pi_1(T)$ and then we have $\pi_1(T)/\Gamma_0 \approx \pi_1(G)$.

Proposition 9. *Let G and G' be connected, compact Lie groups. If an admissible map $f : \pi_1(T) \otimes Z_p \to \pi_1(T') \otimes Z_p$ is realizable then $f(\Gamma_0 \otimes Z_p) \subset \Gamma'_0 \otimes Z_p$.*

Proof. We have a homotopy commutative diagram

$$\begin{array}{ccc} (BG)_p & \xrightarrow{F} & (BG')_p \\ \uparrow i & & \uparrow i' \\ (BT)_p & \xrightarrow{\widetilde{F}} & (BT')_p. \end{array}$$

Observe that $\Gamma_0 \otimes Z_p \approx ker\Big(\pi_1\big((BT)_p\big) \to \pi_1\big((BG)_p\big)\Big)$. This implies that $f(\Gamma_0 \otimes Z_p) \subset \Gamma_0' \otimes Z_p$ where $\pi_1(\widetilde{F}) = f$.

3. Forms of $BG_{(p)}$.

We shall use the results about admissible maps to the solution of the following problem for the space BG_p.

Problem 10. *Let X be a nilpotent space of finite Z_p-type. Classify all nilpotent spaces Y of finite $Z_{(p)}$-type such that Y_p is homotopy equivalent to X.*

The related problems were considered by C. Wilkerson in [Wi] and Belfi, Wilkerson in [BWi]. He used minimal simplicial groups, while we are using Sullivan minimal models. It seems to us that the descent problem from $\mathbf{Q}_p$ to $\mathbf{Q}$ can be better expressed in the language of the Sullivan minimal models. Theorem 15 is taken from our second thesis presented in 1989 in the Universitat Autonoma de Barcelona.

Lemma 11. *Let X be a nilpotent space of finite Z_p-type and let X_0 be a rationalization of X. There exists a nilpotent space Y of finite type over $Z_{(p)}$ such that Y_p is homotopy equivalent to X if and only if there is a nilpotent rational space T of finite type over $\mathbf{Q}$ and a map $f : T \to X$, such that the induced map from the formal completion of T into X, $f_f : T_f \to X$ is a homotopy equivalence inducing an isomorphism of $\mathbf{Q}_p$-vector spaces on homotopy groups and on Lie algebras of fundamental groups.*

This Lemma follows immediately from the arithmetic cartesian square of Sullivan.

For each nilpotent space X of finite Z_p-type we denote by $G_p(X)$ the set of homotopy types of nilpotent spaces Y of finite $Z_{(p)}$-type such that Y_p is homotopy equivalent to X.

Proposition 12. *If X is a connected H-space of finite type over Z_p then the sets $G_p(X)$ and $G_p(BX)$ are not empty.*

This seems to be well known (see Theorem 3.1 in [BWi]) and it follows immediately from Lemma 11. It is sufficient to notice that X_0 is a product of Eilenberg-MacLane spaces $K(\mathbf{Q}_p, n)$ and the same holds for BX.

To give a complete answer to our original problem we must classify all rational spaces whose formal completion is homotopy equivalent to a given nilpotent space X of finite $\mathbf{Q}_p$-type. Below we make this requirement more precise.

We define homotopy category $H_{\mathbf{Q}_p}$ of nilpotent spaces of finite $\mathbf{Q}_p$-type. The objects of $H_{\mathbf{Q}_p}$ are spaces obtained by a rationalization of nilpotent spaces of finite type over Z_p. Homotopy groups and a Lie algebra of a fundamental group of any X in $H_{\mathbf{Q}_p}$ are finite dimensional vector spaces over $\mathbf{Q}_p$. Morphisms in $H_{\mathbf{Q}_p}$ are homotopy classes of maps which on homotopy groups and Lie algebras of fundamental group induce $\mathbf{Q}_p$-homomorphisms.

Let $DGA_{\mathbf{Q}_p}$ be a homotopy category of differential graded commutative algebras of finite type over $\mathbf{Q}_p$.

The categories $H_{\mathbf{Q}_p}$ and $DGA_{\mathbf{Q}_p}$ are equivalent. In particular for any X in $H_{\mathbf{Q}_p}$ there exists its minimal model $\mathcal{M}(X)$ in $DGA_{\mathbf{Q}_p}$ given up to an isomorphism.

The differentials of $\mathcal{M}(X)$ are given by expressions with coefficients a_i's in $\mathbf{Q}_p$. If $\sigma \in Gal(\mathbf{Q}_p : \mathbf{Q})$ then substituting in $\mathcal{M}(X)$ a_i's by $\sigma(a_i)$'s we get a differential graded commutative algebra which we denote by $\mathcal{M}(X)^\sigma$. If $f : \mathcal{M}(X) \to \mathcal{M}(Y)$ is a morphism of differential graded commutative algebras then in a similar way we define $(f)^\sigma : \mathcal{M}(X)^\sigma \to \mathcal{M}(Y)^\sigma$ conjugating coefficients of f.

Let $X \in H_{\mathbf{Q}_p}$. We denote by $G_0(X)$ the set of all homotopy types of nilpotent spaces Y of finite type over $\mathbf{Q}$ such that Y_f, the formal p-completion of Y is homotopy equivalent in $H_{\mathbf{Q}_p}$ to X.

Proposition 13. *Let $X \in H_{\mathbf{Q}_p}$. The set $G_0(X)$ is not empty if and only if for any $\sigma \in Gal(\mathbf{Q}_p : \mathbf{Q})$ there is a homotopy equivalence $f(\sigma) : \mathcal{M}(X) \to \mathcal{M}(X)^\sigma$ such that the function $Gal(\mathbf{Q}_p : \mathbf{Q}) \ni \sigma \to f(\sigma)$ satisfies the cocycle condition*

$$f(\sigma \cdot \tau) = f(\tau)^\sigma \cdot f(\sigma).$$

If $G_0(X)$ is not empty then we have a bijection

$$G_0(X) \approx H^1\Big(Gal(\mathbf{Q}_p : \mathbf{Q}); \mathcal{E}(\mathcal{M}(X))\Big),$$

where $\mathcal{E}(\mathcal{M}(X))$ is the group of self-equivalences of $\mathcal{M}(X)$ in $DGA_{\mathbf{Q}_p}$ and the group $Gal(\mathbf{Q}_p : \mathbf{Q}_p)$ acts on $\mathcal{E}(\mathcal{M}(X))$ by the following rule

$$(\sigma, f : \mathcal{M}(X) \to \mathcal{M}(X)) \to (f^\sigma : \mathcal{M}(X) \to \mathcal{M}(X)).$$

This proposition follows from the standard descent theory.

Proposition 14. (See also [BWi] p. 572.) *Let X be a nilpotent space of finite type over Z_p. If $G_p(X)$ is not empty then there is a bijection*

$$G_p(X) \approx \coprod_{Y \in G_0(X_0)} E(Y) \backslash \mathcal{E}(X_0)/E(X),$$

where $E(Y)$ and $E(X)$ are images of $\mathcal{E}(Y)$ and $\mathcal{E}(X)$ in $\mathcal{E}(X_0)$. ($\mathcal{E}(Z)$ is the group of homotopy classes of self homotopy equivalences of Z.)

Proof. Let $h : X_0 \to X_0$ be a homotopy equivalence compatible with $\mathbf{Q}_p$-structure on homotopy groups. We shall construct the space T_h in $G_p(X)$ as a homotopy pullback of the diagram

$$\begin{array}{ccccc} T_h & & \longrightarrow & & X \\ \downarrow & & & & \downarrow \\ Y & \longrightarrow & X_0 & \xrightarrow{h} & X_0. \end{array}$$

It is clear that T_h is of finite type over $Z_{(p)}$, its p-completion is X and its rationalization is Y.

Let T be such that there are homotopy equivalences $T_0 \overset{\approx}{\to} Y$ and $T_p \overset{\approx}{\to} X$. The map $T_0 \to Y \to X_0$ induces $\alpha : X_0 \to (T_p)_0$ and the map $T_p \to X \to X_0$ induces $\beta : X_0 \to (T_p)_0$. The composition $\beta^{-1} \circ \alpha : X_0 \to X_0$ is the map h. It is clear that T_h is homotopy equivalent to $T_{a \cdot h \cdot b}$ for $a \in E(Y)$ and $b \in E(X)$ and if T_h and $T_{h'}$ are homotopy equivalent then $h = a \cdot h' \cdot b$ for some $a \in E(Y)$ and $b \in E(X)$.

We shall calculate the set $G_p(X)$ if X is the p-completion of the classifying space of a compact, connected, simple Lie group. We shall apply our results about admissible maps.

Theorem 15. *Let G be a simple Lie group. If rank $G = 1$ i.e. $G = S^3$ or $G = SO(3)$ then the set $G_p(BG_p)$ contains one element.*

If rank G is bigger than 1 then $G_p(BG_p)$ is uncountable.

Proof. Let us assume that $rank\ G = 1$. Then there is a bijection

$$G_p(X) = \mathbf{Q}^* \backslash \mathbf{Q}_p^* / \{x^2 \mid x \in Z_p^*\}.$$

Any element x in $\mathbf{Q}_p^*$ can be written in the form $x = r\cdot(1+a_1\cdot p+a_2\cdot p^2+\cdots) = r\cdot e$ where $r \in \mathbf{Q}^*$ and e is a unit congruent to 1 $mod\ p$. If $p \neq 2$ then any such unit in Z_p is a square of some element in Z_p. If $p = 2$ then $\{x^2 \mid x \in Z_2^*\} = 1 + 8 \cdot Z_2$. We have

$$x = r\cdot(1+a_1\cdot 2+a_2\cdot 2^2+\cdots) = r(1+a_1\cdot 2+a_2\cdot 2^2)\cdot(1+a_1\cdot 2+a_2\cdot 2^2)^{-1}\cdot e = r'\cdot e'$$

where $r' \in \mathbf{Q}^*$ and e' is a square in Z_2^*. This implies that $G_p(X)$ contains only one element.

Let us assume now that $rank\ G$ is bigger that 1. Let $A(X) = \{x \in Z_p^*\}$ be the image of Adams ψ^α-maps in $E(X)$. Then it follows from [AW] Example 1.4 that $|E(X)/A(X)| \leq 6$. The image of the group $\mathcal{E}(X_0)$ in $Aut(\bigoplus_{i=2}^{\infty} \pi_i(X_0))$ contains the subgroup $(\mathbf{Q}_p^*)^n$ ($n = rank\ G > 1$) and the image of $A(X)$ in $Aut(\bigoplus_{i=2}^{\infty} \pi_i(X_0))$ is equal to $\{x^{m_1}, x^{m_2}, \ldots, x^{m_n} \mid x \in Z_p^*\}$ for some $m_1, m_2, \ldots, m_n$. It is clear that the quotient $\mathcal{E}(X_0)/E(X)$ is uncountable. Hence $G_p(X)$ is also uncountable.

Acknowledgement

We would like to thank very much the referee for his comments concerning the organization of the paper. We are also very grateful to him for pointing out an inaccuracy in the proof of the theorem 15. We would like to thank very much C. Mc Gibbon for letters on this subject.

References

[AM] J.F. Adams and Z. Mahmud, Maps between classifying spaces, Invent. Math. 35 (1976), 1-41.

[AW] J.F. Adams and Z. Wojtkowiak, Maps between p-completed classifying spaces, Proc. of the Royal Soc. of Edinburgh, 112A, 231-235, 1989.

[BWi] V. Belfi and C.W. Wilkerson, Some examples in the theory of p-completions, Indiana Math. Journal, 25 (1976), pp. 565-576.

[DZ] W.G. Dwyer and A. Zabrodsky, Maps between classifying spaces, in "Algebraic Topology Barcelona 1986", L.N. in Math. 1298, Springer-Verlag, 1987, pp. 106-119.

[H] J.R. Hubbuck, Mapping degree for classifying spaces I, Quart. J. Math. Oxford (2) 25 (1974), 113-133.

[I] K. Ishiguro, Unstable Adams operations on classifying spaces, Math. Proc. Camb. Phil. Soc., 102 (1987), pp. 71-75.

[M] J.S. Milne, Etale Cohomology, Princeton University Press, Princeton, New Jersey 1980.

[R] D. Rector, Loop structures on the homotopy type of S^3, in Loop structures on the Homotopy Type of S^3, L.N. in Math. 249, Springer-Verlag, 1971, pp. 99-105.

[S] D. Sullivan, Genetics of homotopy theory and Adams conjecture, Annals of Math. 100 (1974), 1-79.

[Wi] C.W. Wilkerson, Applications of Minimal Simplicial Groups, Topology 15 (1976), 111-130.

[W] Z. Wojtkowiak, Maps between p-completed classifying spaces II., Proc. of the Royal Soc. of Edinburgh, 118A (1991), 133-137.

[Z] A. Zabrodsky, Maps between classifying spaces, p-groups and tori, to appear.

Z. Wojtkowiak
I.H.E.S.
35, route de Chartres
91440 Bures-sur-Yvette (France)

RETRACTS OF CLASSIFYING SPACES

BY
KENSHI ISHIGURO

Given a compact connected Lie group G, it is well known that a retract of its p–completion $G_p^\wedge$ need not be p–equivalent to a compact group. In fact, such retracts are often p–equivalent to spheres or sphere bundles over spheres, [5,8,12]. For classifying spaces BG , however, the situation is very different. These spaces are much more rigid and controlled, as our main result indicates.

Theorem. *If G is a compact connected Lie group, any retract of the p–completed classifying space $BG_p^\wedge$ is homotopy equivalent to $BK_p^\wedge$ for some compact Lie group K .*

Corollary. *If a compact connected Lie group G is semi-simple, any retract of BG is 2–equivalent to BK for some compact connected semi-simple Lie group K .*

These results generalizes previous work of the author [6]. It is worth noting that these results are unstable. Indeed stable versions of our results are false. The classifying space BS^1 provides a good counter-example. After one suspension, its completion at $p > 3$ splits into $p-1$ indecomposable pieces, and none of which equal $(\Sigma BK)_p^\wedge$ for any compact Lie group K , [9, Proposition 2.2].

For a given retract, the compact Lie group K in our results is described precisely. To do this, in §1, we will define subgroups $Z(H,p)$ that are related to the center $Z(H)$ of a Lie group H which is a finite product of a torus and simply-connected simple Lie groups. In fact, if $p=2$, then $Z(H,2) = Z(H)$. Notice that for a compact connected Lie group G , there is a finite covering homomorphism $\widetilde{G} \longrightarrow G$ where $\widetilde{G} = \prod_i H_i$ is a finite product of a torus and simply-connected simple Lie groups. If G is semi-simple, then $\widetilde{G}$ is the

Typeset by $\mathcal{A}\mathcal{M}\mathcal{S}$-TEX

universal covering. For a retract X of $BG_p^\wedge$, we will show that there is a subgroup $\widetilde{K} = \prod_j H_{\alpha_j}$ of $\widetilde{G}$ such that the Lie group K is isomorphic to a quotient group of $N_{\widetilde{K}}Z(\widetilde{K},p)$ by a finite group D . Here $N_{\widetilde{K}}Z(\widetilde{K},p)$ denotes the normalizer of $Z(\widetilde{K},p)$ in $\widetilde{K}$. The classifying space $BN_{\widetilde{K}}Z(\widetilde{K},p)$ is p-equivalent to $B\widetilde{K}_p^\wedge$, and the retract $X \simeq BK_p^\wedge$ is determined by a fibration $BD \longrightarrow B\widetilde{K}_p^\wedge \longrightarrow X$ in the following diagram:

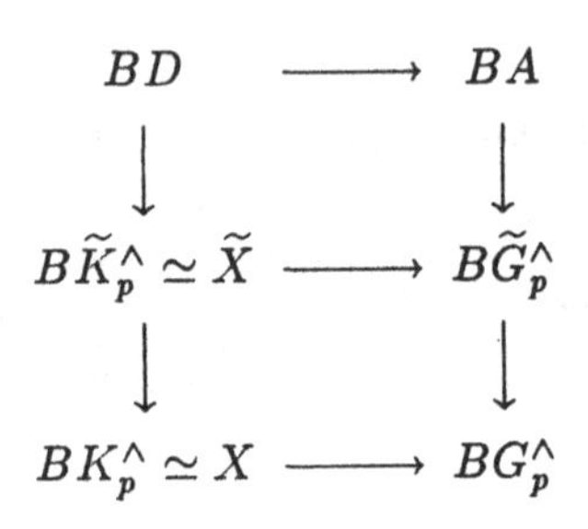

Since $\widetilde{K}$ is a finite product of a torus and simply-connected simple Lie groups , either $\pi_2 X = 0$ or $\pi_4 X = 0$. This implies that any 4−connected non-trivial retract of $G_p^\wedge$ doesn't come from a retract of $BG_p^\wedge$.

For example, if $G = SO(4)$, the above argument on the universal covering $Spin(4) \longrightarrow SO(4)$ implies $\widetilde{K} = S^3$ if a retract X is non-trivial. One can show that any non-trivial retract of $BSO(4)$ at $p = 2$ is homotopy equivalent to $BSO(3)$. If p is odd, it's easy to see that any non-trivial retract of $BSO(4)$ at p is homotopy equivalent to BS^3.

In §1, work of [7] on maps between classifing spaces is generalized, where the Lie group G is simple. In §2, G is arbitrary and the Theorem is proved. The proof uses the notion of a "topological kernel", defined in §1, of a map from BG .

I would like to thank Institut D'Estudis Catalans, Barcelona and E.T.H., Zurich for their hospitality during this research. I would also like to thank C. McGibbon for his suggestion.

§1 Classifying spaces of compact simple Lie groups and p–tori

Most of the work in this section can be regarded as the mod 2 version of [7], where the Lie group G is simple. Techniques used in this section are analogous to those in the paper, and some arguments are sketchy. The main work of [7] is to show that if a map $f : BG \longrightarrow X$ is trivial in mod p cohomology, the map f is null homotopic. Here the space X must satisfy some finiteness condition, and the odd prime p divides the order of the

Weyl group of the simple Lie group G . If $p = 2$ and $G = SU(2)$, the corresponding statement isn't true. The following lemma shows, however, that this is the only exception.

Lemma 1. *Let G be a compact connected simple Lie group with 2–$rank(G) > 1$. Suppose X is a 2–completed space such that $\pi_1 X$ is finite, that $H^*(X;\mathbb{F}_2)$ is of finite type, and that the component of the space of based maps containing the constant map, denoted by $map_*(B\mathbb{Z}/2, X)_0$, is weakly contractible. If $f : BG \longrightarrow X$ is trivial in mod 2 cohomology, then the map f is null homotopic.*

Proof. This result is the mod 2 version of [7, Theorem 1] where the prime p is assumed to be odd. The proof is analogous, and a covering-space argument shows that it's enough to consider when G is simply-connected. The role of $SU(p)$ in [7] will be played by $SO(3)$ and $Sp(2)$ at $p = 2$.

If $G = SO(3)$, the proof is in [7, Theorem 3]. In the proof, a sequence of 2–subgroups of $SO(3)$ consisting of dihedral groups D_{2^n} is used. Namely, if $f^* = 0$, one can inductively show $f|BD_{2^n} \simeq 0$ for any n and hence $f \simeq 0$. If $G = Sp(2)$, its subgroup $O(2)$ contains a similar sequence of dihedral groups, and an analogous proof is applicable. Thus the desired result also holds for $G = Sp(2)$.

The classification of simple Lie groups tells us that any simply-connected simple Lie group G with 2–$rank(G) > 1$ contains either $SO(3)$ or $Sp(2)$. This means that we can find a subtorus T^j with $j \geq 1$ of G such that $f|BT^j \simeq 0$. Consequently $f \simeq 0$ by [7, Lemma 4]. //

Suppose X is a p–completed space such that $\pi_1 X$ is finite, that $H^*(X;\mathbb{F}_p)$ is of finite type, and that the component of the space of based maps containing the constant map, denoted by $map_*(B\mathbb{Z}/p, X)_0$, is weakly contractible. The conditions on X are satisfied by the p–completion of BG and classifying spaces of connected mod p finite loop spaces, [4]. According to a result of Dwyer-Wilkerson [2], $map_*(B\mathbb{Z}/p, X)_0 \simeq_w 0$ if X is a 1-connected p–complete space such that $H^*(X;\mathbb{F}_p)$ is of finite type and that the module of indecomposables $Q(H^*(X;\mathbb{F}_p))$ is locally finite as a module over the Steenrod algebra. This condition is satisfied, of course, if $H^*(X;\mathbb{F}_p)$ is a finitely generated algebra. In particular, any retract of $BG_p^\wedge$ satisfies the condition.

Let T (or $T(G)$) be a maximal torus of the Lie group G. Suppose N_pT denotes the inverse image in the normalizer of a maximal torus T of a p–Sylow subgroup W_p of the Weyl group of G. We define a subgroup $N_{p^\infty}T$ of N_pT as follows:

$$\begin{array}{ccccccccc} 1 & \longrightarrow & T & \longrightarrow & N_pT & \longrightarrow & W_p & \longrightarrow & 1 \\ & & \uparrow & & \uparrow & & \| & & \\ 1 & \longrightarrow & \bigoplus \mathbb{Z}/p^\infty & \longrightarrow & N_{p^\infty}T & \longrightarrow & W_p & \longrightarrow & 1 \end{array}$$

Here $\bigoplus Z/p^\infty \subset T$ is the subgroup of elements whose order is a power of p . Finally we define the "kernel " of a map $f : BG \longrightarrow X$, where X is assumed to satisfy the above finiteness condition, as follows:

$$Ker\ f = \{ x \in N_{p^\infty}T \mid f|B < x > \simeq 0 \}$$

Here $< x >$ denotes the subgroup of $N_{p^\infty}T$ generated by x . Our definition generalizes the definition in [3,7]. We note that $Ker\ f$ is a group. If X is the p–completed classifying space of a compact Lie group, it's easy to see this, since $f|BN_{p^\infty}T$ is induced by a group homomorphism. The general case was proved by D.Notbohm,[11]. We warn that if a space X is not assumed to be a p–completed space which satisfies the finiteness condition, $Ker\ f$ would not be a group.

Next we define subgroups $Z(H,p)$ of H , where $H = \prod_i H_i$ is a finite product of a torus and simply-connected simple Lie groups. Let $Z(H)$ denote the center of H and $W(H)$ its Weyl group. If H is a simple Lie group other than the exceptional Lie group G_2 , we define

$$Z(H,p) = \begin{cases} Z(H) & \text{if } |W(H)| \equiv 0 \bmod p \\ \text{a maximal torus of } H & \text{if } |W(H)| \not\equiv 0 \bmod p. \end{cases}$$

If $H = G_2$, then

$$Z(G_2,p) = \begin{cases} Z(G_2) & \text{if } p = 2 \\ \text{the center of the subgroup } SU(3) \text{ of } G_2 & \text{if } p = 3 \\ \text{a maximal torus of } G_2 & \text{if } p \geq 5 \end{cases}$$

If $H = T^n$, then $Z(T^n,p) = T^n$. Finally if $H = \prod_i H_i$, then $Z(H,p) = \prod_i Z(H_i,p)$.

Lemma 2. *Suppose X is a p–completed space such that $\pi_1 X$ is finite, that $H^*(X;\mathbb{F}_p)$ is of finite type, and that the component of the space of based maps containing the constant map, $map_*(B\mathbb{Z}/p, X)_0$, is weakly contractible. Let G be a compact simply-connected simple Lie group. If a map $f : BG \longrightarrow X$ is essential, then $Ker\ f$ is contained in $Z(G,p)$.*

Remark. The immediate consequence is that $Ker\ f$ is a normal subgroup of $Z(G,p)$.

Proof. First assume p is odd. If p doesn't divide the order of the Weyl group $W(G)$, we see $Z(G,p) = T(G)$ by definiton. Hence $Ker\ f \subset \bigoplus \mathbb{Z}/p^\infty \subset Z(G,p)$. If $|W(G)| = 0 \bmod p$, the desired result follows from [7, Theorem 2']. Namely, if $Ker\ f \not\subset Z(G)$, one can find an element $a \in G$ of order p such that the weak closure of a in N_pT with respect to G is equal to the subgroup $N_{p^\infty}T$. Hence $f \simeq 0$.

It remains to discuss the case $p = 2$. Since $Z(G,2) = Z(G)$, it suffices to show that if $Ker\ f \not\subset Z(G)$, then $f \simeq 0$. To do this we will find a subtorus T^j with $j \geq 1$ of G such that $f|BT^j \simeq 0$. By [7, Lemma 4] , this implies $f \simeq 0$. According to the classification of compact 1-connected simple Lie groups, we need to consider the following 8 cases.

<u>*Case 1.*</u> $G = SU(n)$ with $n \geq 2$

If $n \geq 3$, the argument is analogous to [7, Theorem 2', Case 1.1]. Our assumption $Ker\ f \not\subset Z(G)$ implies that there is an element $a \in Ker\ f$ of order 2 such that $a \notin Z(G)$; the element a can be taken as the following diagonal matrix

$$a = \begin{pmatrix} -1 & & & & \\ & -1 & & & \\ & & 1 & & \\ & & & \ddots & \\ & & & & 1 \end{pmatrix} \in SU(n).$$

A subset of conjugacy classes of a generates the elementary abelian 2-subgroup of a maximal torus of $SU(n)$. Hence $H^*(f;\mathbb{F}_p) = 0$ and $f \simeq 0$ by Lemma 1. When $n = 2$ and $Ker\ f \not\subset Z(SU(2))$, the map $f : BSU(2) \longrightarrow X$ factors through $BSO(3)$, say $\overline{f} : BSO(3) \longrightarrow X$. There is $x \in Ker\ f$ of order at least 4. Let $\overline{x}$ denote the image of x under the universal covering map $SU(2) \longrightarrow SO(3)$. The order of $\overline{x}$ is at least 2 and $\overline{f}|B < \overline{x} > \simeq 0$. Hence $\overline{f} \simeq 0$ by [7, Theorem 4] and therefore $f \simeq 0$.

Case 2. $G = Sp(n)$ with $n \geq 2$
A proof similar to Case 1 with $n \geq 3$ is applicable; the element a can be taken as the following diagonal matrix

$$a = \begin{pmatrix} -1 & & & \\ & 1 & & \\ & & \ddots & \\ & & & 1 \end{pmatrix} \in Sp(n).$$

Case 3. $G = Spin(n)$ with $n \geq 5$
Suppose $x \in Ker\ f$ and $x \notin Z(Spin(n))$. The image $\overline{x} \in SO(n)$ under the covering homomorphism $Spin(n) \longrightarrow SO(n)$ can not be contained in the center $Z(SO(n))$. Thus the weak closure of $\overline{x}$ in $N_{2^\infty}T(SO(n))$, the subgroup of $N_{2^\infty}T(SO(n))$ generated by the set of elements $\{ y\overline{x}y^{-1} \in N_{2^\infty}T(SO(n)) \mid y \in SO(n) \}$, is equal to $N_{2^\infty}T(SO(n))$, by [7, Theorem 4]. Consequently $N_{2^\infty}T(SO(5))$ is included in the image of $Ker\ f$ under the covering homomorphism $Spin(n) \longrightarrow SO(n)$ for $n \geq 5$. We can find an element $y \in Spin(5) \cap Ker\ f$ such that $y \notin Z(Spin(5))$. Thus $f|BSpin(5) \simeq 0$ by Case 2, since $Spin(5) \cong Sp(2)$. Consequently $f \simeq 0$.

Case 4. $G = G_2$
Assume $Ker\ f \not\subset Z(G)$. Since $SU(3) \subset G_2$ and $rank(SU(3)) = 2$, an argument analogous to Case 1 with $n \geq 3$ is applicable to show $f|BSU(3) \simeq 0$ and hence $f \simeq 0$.

Case 5. $G = F_4$
Since $SU(3) \times SU(3)/\mathbb{Z}_3 \subset F_4$ and $rank(SU(3) \times SU(3)/\mathbb{Z}_3) = 4$, if $Ker\ f \not\subset Z(G)$ one can show $f|B(SU(3) \times SU(3)/\mathbb{Z}_3) \simeq 0$ and hence $f \simeq 0$

Case 6. $G = E_6$
Since $SU(3) \times SU(3) \times SU(3)/\mathbb{Z}_3 \subset E_6$ and $rank(SU(3) \times SU(3) \times SU(3)/\mathbb{Z}_3) = 6$, if Ker $f \not\subset Z(G)$ one can show $f|B(SU(3) \times SU(3) \times SU(3)/\mathbb{Z}_3) \simeq 0$ and hence $f \simeq 0$.

Case 7. $G = E_7$
The argument is analogous to [7, Proof of Theorem 2', Case 1.4]. We use the subgroup $SU(3) \times SU(6)/\mathbb{Z}_3 \subset E_7$. Let c be the generator of the center $Z(E_7) \cong \mathbb{Z}_2$. By [13, Theorem 2.27] we can find an elementary abelian 2-subgroup
$V = Z_2 < a > \times Z_2 < c > \subset SU(3) \times SU(6)/\mathbb{Z}_3$. Since $H^2(BV; \mathbb{Z}_3) = 0$, we may assume $a, c \in SU(3) \times SU(6)$. Say $a = [a_1, a_2]$ where $a_1 \in SU(3)$ and

$a_2 \in SU(6)$. Notice $a \notin Z(SU(3) \times SU(6))$. If $a_1 = 1$, then $f|BSU(3) \simeq 0$. And $a_2 = 1$ implies $f|BSU(6) \simeq 0$. Consequently $f \simeq 0$.

Case 8. $G = E_8$
Assume $Ker\ f \not\subset Z(G)$. Since $E_6 \times SU(3)/\mathbb{Z}_3 \subset E_8$ and $rank(E_6 \times SU(3)/\mathbb{Z}_3) = 8$, one shows $f|B(E_6 \times SU(3)/\mathbb{Z}_3) \simeq 0$ and hence $f \simeq 0$. //

§2 Retracts of BG

Lemma 3. *Let $G = \prod_i G_i$ be a finite product of a torus and simply-connected simple Lie groups and let X be as in Lemma 2 . If $Ker\ f$ of a map $f : BG \longrightarrow X$ is a finite group, then $Ker\ f$ is contained in $Z(G,p)$.*

Proof. Suppose there is $x \in Ker\ f$ such that $x = (x_i) \notin Z(G,p) = \prod_i Z(G_i,p)$. This means that we can find a simple group G_i such that $x_i \notin Z(G_i,p)$ with $|W(G_i)| = 0 \bmod p$. Let $T(G_i)$ denote a maximal torus of G_i . Just as in the proof of Lemma 2 we see that the weak closure of x_i in $N_pT(G_i)$ with respect to G_i is equal to the subgroup $N_{p^\infty}T(G_i)$. This implies that the weak closure of x_i in N_pT with respect to the group G includes an infinite group isomorphic to $\bigoplus^n Z/p^\infty$, where $n = rank(G_i)$. Hence $\bigoplus^n Z/p^\infty \subset Ker\ f$. This contradicts our assumption that $Ker\ f$ is a finite group. This completes the proof. //

Lemma 4. *Let G be a finite product of a torus and simply-connected simple Lie groups and let X be as in Lemma 2 . If $Ker\ f$ of a map f : $BG \longrightarrow X$ is a finite group, there is a compact Lie group $\overline{G}$ such that the map f factors through $B\overline{G}$ and that $Ker\overline{f}$ of the induced map $\overline{f} : B\overline{G} \longrightarrow X$ is trivial .*

Proof. Since $Ker\ f$ is finite, Lemma 3 shows $Ker\ f \subset Z(G,p)$. Recall that $BNT(G_i) \simeq_p BG_i$ if $|W(G_i)| = 0 \bmod p$, that $N(Z(G)) = G$, and that $BNZ(G_2, 3)$ is 3-equivalent to BG_2 . Consequently $BNZ(G,p)$ is p-equivalent to BG . Let $\overline{G}$ be the quotient group $NZ(G,p)/Ker\ f$. Then the map $f : BG \longrightarrow X$ factors through $B\overline{G}$. It is easy to see that $Ker\overline{f}$ is trivial for the map $\overline{f} : B\overline{G} \longrightarrow X$. //

Proof of Theorem. Since G is compact and connected, we can find a finite covering $\widetilde{G}$ of G which is a product of a torus T^n and simply-connected simple Lie groups. Suppose X is a non-trivial retract of $BG_p^\wedge$ with retraction $r : BG_p^\wedge \longrightarrow X$ and inclusion $i : X \longrightarrow BG_p^\wedge$ so that $r \cdot i \simeq 1$. We will find a retract $\widetilde{X}$ of $B\widetilde{G}_p^\wedge$ associated with X . First suppose that G is semi-simple. By a Postnikov system for the 1-connected space X , we obtain a map $X \longrightarrow K(\pi_2 X, 2)$ whose induced homomorphism on the second homotopy groups is an isomorphism. Let $\widetilde{X}$ be the homotopy fibre of the map $X \longrightarrow K(\pi_2 X, 2)$. One can show that the 2-connected space $\widetilde{X}$ is a retract of $B\widetilde{G}_p^\wedge$. Suppose next that G is not semi-simple so that the covering $\widetilde{G}$ contains T^n with $n \geq 1$. We notice that $\pi_2(B\widetilde{G})$ is a free abelian group and that $Bq^* : \pi_2(B\widetilde{G}) \longrightarrow \pi_2(BG)$ is a monomorphism. Hence Bq^* is represented by a square matrix $Q \in Mat_n(\mathbb{Z})$ with $detQ \neq 0$, since $n \geq 1$. Suppose $f = i \cdot r : BG_p^\wedge \longrightarrow BG_p^\wedge$. Since $\pi_2(BG_p^\wedge)$ is a finitely generated module over the principal ideal domain $\mathbb{Z}_p^\wedge$, we can find a positive integer m such that the abelian group $p^m f_*(\pi_2(BG_p^\wedge))$ is torsion-free. Let $Q_0 = (detQ) \cdot Q^{-1} \in Mat_n(\mathbb{Z})$ and let q_0 be the endomorphism of $\widetilde{G} = T^n \times (\widetilde{G}/T^n)$ represented by $p^m Q_0 \times 1$. The composite homomorphism $q_1 = q_0 \cdot q : \widetilde{G} \longrightarrow G$ induces a fibration $B\widetilde{G}_p^\wedge \longrightarrow BG_p^\wedge \longrightarrow K(A, 2)$ where A is a suitable finite abelian p–group. By construction $(Bq_1)_* : \pi_2(B\widetilde{G}) \longrightarrow \pi_2(BG)$ is a homomorphism of multiplication by $p^m \cdot detQ$. Consequently if α denotes the map $BG_p^\wedge \longrightarrow K(A, 2)$, the composite map $\alpha \cdot i \cdot r \cdot (Bq_1)_p^\wedge : B\widetilde{G}_p^\wedge \longrightarrow K(A, 2)$ is null homotopic. Let D be the image of $\pi_2(X)$ under $(\alpha \cdot i)_*$. Then D is a subgroup of A and the universal coefficient theorem shows that the map $\alpha \cdot i : X \longrightarrow K(A, 2)$ factors through $K(D, 2)$. Suppose $\widetilde{X}$ is the homotopy fibre of the induced map $X \longrightarrow K(D, 2)$. One can find maps to make the following diagram homotopy commutative:

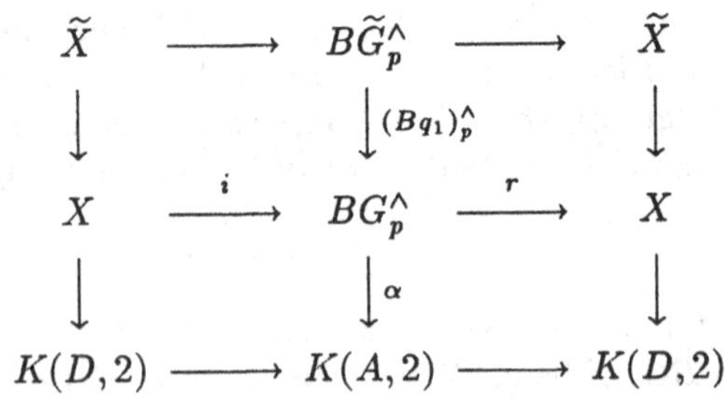

From this diagram one can see that the space $\widetilde{X}$ is a retract of $B\widetilde{G}_p^\wedge$. Notice that the loop space $\Omega\widetilde{X}$ is a retract of the H-space $\widetilde{G}_p^\wedge$. Hence $\widetilde{G}_p^\wedge \simeq \Omega\widetilde{X} \times Y$ for some space Y .

We claim that $\widetilde{X}$ is homotopy equivalent to a classifying space $B\widetilde{K}_p^\wedge$

where $\widetilde{K}$ is a product of a torus and some simply-connected simple Lie groups. If Y is contractible, the retraction $\Omega\widetilde{r} : G_p^\wedge \longrightarrow \Omega\widetilde{X}$ is an equivalence. Hence $\widetilde{r}$ is also an equivalence and therefore $\widetilde{X} \simeq B\widetilde{G}_p^\wedge$. Next, if Y is not contractible, we prove the desired result by induction on the number of factors of $\widetilde{G}$. Notice the restriction $\Omega r|Y$ is null homotopic. Since Y is a retract of $\widetilde{G}_p^\wedge$, the cohomology ring $H^*(Y;\mathbb{Q}_p^\wedge)$ can be regarded as a subalgebra of $H^*(\widetilde{G};\mathbb{Q}_p^\wedge) = \bigwedge_{\mathbb{Q}_p^\wedge}(x_1,\dots,x_k)$. A result of Wilkerson [14] shows that cancellation holds for $G_p^\wedge$; this H-space factors into a finite product of indecomposable H-spaces with the uniqueness of the factorization up to order. Hence we may choose some of x_i's so that $H^*(Y;\mathbb{Q}_p^\wedge) \cong \bigwedge_{\mathbb{Q}_p^\wedge}(x_{i_1},\dots,x_{i_s})$. Recall $H^*(B\widetilde{G};\mathbb{Q}_p^\wedge) = \mathbb{Q}_p^\wedge[y_1,\dots,y_k]$ where each y_i corresponds to x_i under the transgression. Since both $\Omega\widetilde{X}$ and Y are not contractible, we can find x_j , one of the generators of $H^*(\widetilde{G};\mathbb{Q}_p^\wedge)$, such that $x_j \neq x_{i_\alpha}$ for any $1 \leq \alpha \leq s$. Consequently $y_j \neq y_{i_\alpha}$ for any $1 \leq \alpha \leq s$ and there is a subgroup S^1 of one of the factor groups of $\widetilde{G}$ such that $\widetilde{r}|BS^1$ is rationally null homotopic. Let $\widetilde{i}$ be the inclusion $\widetilde{X} \longrightarrow B\widetilde{G}_p^\wedge$ and let $f = \widetilde{i} \cdot \widetilde{r}|BS^1$. Since $H^*(f;\mathbb{Q}_p^\wedge) = 0$, the map f is induced by a trivial homomorphism, [10]. Consequently $f \simeq 0$. We can find a factor subgroup H_i of $\widetilde{G}$ with $S^1 < H_i$ and $\widetilde{r}|BH_i \simeq 0$, since $\widetilde{G}$ is a product of a torus and simply-connected simple Lie groups. Therefore $\widetilde{X}$ is a retract of $B(\widetilde{G}/H_i)$. By the hypothesis of induction, we prove the desired result.

We recall that A is the fundamental group of the homotopy fibre of the map $(Bq_1)_p^\wedge : B\widetilde{G}_p^\wedge \longrightarrow BG_p^\wedge$ and that D is a retract of the group A . Notice that A is a finite abelian p-group. We have the following homotopy commutative diagram:

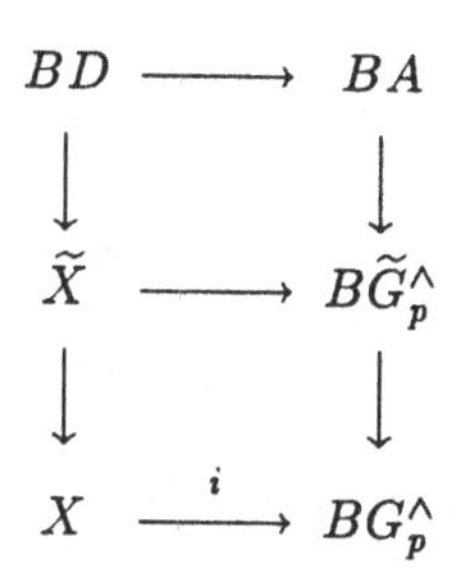

$$\begin{array}{ccc} BD & \longrightarrow & BA \\ \downarrow & & \downarrow \\ \widetilde{X} & \longrightarrow & B\widetilde{G}_p^\wedge \\ \downarrow & & \downarrow \\ X & \xrightarrow{i} & BG_p^\wedge \end{array}$$

Here notice that the map $BD \longrightarrow \widetilde{X} \simeq B\widetilde{K}_p^\wedge$ is induced by an injective homomorphism from D into $\widetilde{K}$, since this map is finite. Applying Lemma

4 to the map $B\widetilde{K}_p^\wedge \longrightarrow X$, we can show $X \simeq BK_p^\wedge$ if $K = N_{\widetilde{K}}Z(\widetilde{K},p)/D$.

Proof of Corollary. Since G is semi-simple, we use the universal covering $\widetilde{G} \longrightarrow G$. At $p = 2$, we notice that $Z(H,2) = Z(H)$ for any simply-connected simple Lie group H . If a space is a retract of BG, its 2-completion is a retract of $BG_2^\wedge$. Along the lines of the proof of the Theorem, we see that the finite 2-group D is included in the center of $\widetilde{K}$. Hence $N_{\widetilde{K}}Z(\widetilde{K},2) = \widetilde{K}$ and $K = \widetilde{K}/D$. Since $\widetilde{K}$ is a product of simply-connected simple Lie groups, the Lie group K is compact, connected, and semi-simple.

References

1 A. Borel and J. de Siebenthal, *Les sous-groupes fermés de rang maximum des groupes de Lie clos*, Comm. Math. Helv. 23 (1949), 200–221.

2 W. G. Dwyer and C. W. Wilkerson, *Spaces of null homotopic maps*, Preprint.

3 W. G. Dwyer and C. W. Wilkerson, *A cohomology decomposition theorem*, Preprint.

4 E. M. Friedlander and G. Mislin, *Locally finite approximation of Lie groups II*, Math. Proc. Camb. Phil. Soc. 100 (1986), 505–517.

5 J. Harper, *Regularity of finite H-spaces*, Illinois J. Math. 23 (2) (1979), 330–333.

6 K. Ishiguro, *Classifying spaces and p-local irreducibility*, J. of Pure and Applied Algebra 49 (1987), 254–258.

7 K. Ishiguro, *Classifying spaces of compact simple Lie groups and p-tori*, to appear in the Proceedings of the 1990 Barcelona Conference.

8 P. G. Kumpel, *Lie groups and products of spheres*, Proc. Amer. Math. Soc. 16 (1965), 1350–1356.

9 C. A. McGibbon *Stable properties of rank 1 loop structures*, Topology 20 (1981), 109–118.

10 D. Notbohm, *Abbildungen zwischen klassifizierenden Raume*, Dissertation, University of Göttingen 1988.

11 D. Notbohm, private communication.

12 J.-P. Serre, *Groupes d'homotopie et classes de groupes abeliens*, Annals of Mathematics 58 (1953), 258–295.

13 R. Steinberg, *Torsion in reductive groups*, Advances in Mathematics 15 (1975), 63–92.

14 C. W. Wilkerson, *Genus and cancellation*, Topology 14 (1975), 29–36.

Mathematics Department, Hofstra University, Hempstead, NY 11550, USA

and

SFB 170, Mathematisches Institut, 3400 Göttingen, Germany.

On the dimension theory of dominant summands

JOHN MARTINO AND STEWART PRIDDY

Section 0. Introduction.

Let G be a finite group and let p be a prime. The p-local stable decompositon

$$BG = X_1 \vee X_2 \vee \cdots \vee X_N$$

of the classifying space of G is known for a variety of specific groups [MiP2, HK, Mt] and a general decomposition procedure is available for all finite groups [MP, BF]. Information about the summands such as their cohomology or connectivity [MiP, FS, CW] can sometimes be determined but it is often quite difficult to obtain. The purpose of this paper is to establish some dimension theoretic results for the Poincaré series

$$PS(X,t) = \sum_{k\geq 0} \dim_{\mathbf{F}_p}(H^k X)t^k$$

of the summands X of BG. Here and throughout the paper cohomology will be taken with coefficients in the prime field $\mathbf{F}_p$.

We say that a power series $f(t) \in Z[[t]]$ has dimension $n, d(f(t)) = n$, if it has a pole of order n at $t = 1$. For a space or spectrum X, the *dimension* of X is defined by

$$d(X) = d(PS(X,t)).$$

We have two main results concerning dimension. If P is a finite p-group we recall that the principal dominant summand X_0 of BP is the dominant summand associated with the trivial representation (see Section 1). If G has P as a Sylow p-subgroup then X_0 is also a summand of BG. Let $n =$ rank of G, that is, the dimension of the largest elementary abelian p-subgroup of G.

THEOREM A. *X_0 has dimension n.*

Our approach to this theorem and the next is via the algebra $D(P)$ of Dickson invariants for the regular representation of P. This is a polynomial algebra of Krull dimension n over which H^*BP is a finitely generated module.

THEOREM B. *Suppose for some $k > 0$, $D(P)$ acts freely on $H^k(BP)$. Then every dominant summand of BP has dimension n.*

The hypothesis of Theorem B means $D(P)x \subset H^*(BP)$ is a free $D(P)$ module for all nontrivial $x \in H^k(BP)$.

COROLLARY C. *If $H^*(BP)$ is free over $D(P)$, then every dominant summand of BP has dimension n.*

Let $H(P)$ denote the subalgebra of H^*BP consisting of even dimensional elements or just $H^*(BP)$ if $p = 2$. Since $H(P)$ is a commutative algebra, it is more convenient for some purposes. Also $H^*(BP)$ is finitely generated module over $H(P)$ thus there is no loss of generality for dimension theory.

We now recall that a commutative algebra is called *Cohen-Macaulay* if its depth equals its Krull dimension, that is, it has a regular sequence [Mu] of length n. A recent result of D. Rusin [R] states that $H(P)$ is Cohen-Macaulay if as a $D(P)$-module the generators of $D(P)$ form a regular sequence. In particular, if $H(P)$ is Cohen-Macaulay then it is free over $D(P)$. Thus we have

COROLLARY D. *If $H(P)$ is Cohen-Macaulay then every dominant summand has dimension n.*

QUESTION. *Theorems A and B lead us to ask if every dominant summand of BG has dimension n. At present we know of no counter example and have been unable to settle this question either way.*

The paper is organized as follows: Section 1 contains some preliminary notions about dimension theory and the Dickson invariants. The proofs of Theorems A and B are given in Sections 2 and 3 respectively. Section 4 is devoted to some related results.

Theorem A answers a question raised by Haynes Miller as to the dimension of the principal dominant summand.

Section 1. Preliminaries.

In this section we establish the preliminary notions which will be needed in the later sections. We begin with a brief sketch of Nishida's theory of dominant summands [N]. For BP the classifying space of a p-group P, Lewis, May and McClure [LMM], using Carlsson's solution of the Segal Conjecture [C], have shown that the ring of stable homotopy classes of self-maps $\{BP, BP\}$ is generated as a free module over the p-adic integers $\widehat{\mathbf{Z}}_p$ by maps of the form $BP \xrightarrow{tr} BQ \xrightarrow{Bh} BP$, where $Q \leq P$, $tr : BP \to BQ$ is the reduced transfer (if $Q = P$ then tr is understood to be the identity) and $h : Q \to P$ is an outer homomorphism. If

$$1 = \sum e_i \in \{BP, BP\}$$

is a decomposition of one into primitive, orthogonal idempotents then $BP \simeq \vee e_i BP$, where each $e_i BP = Tel(BP \xrightarrow{e_i} BP)$ is the infinite mapping telescope, is a stable splitting into indecomposable summands. Define $J \subseteq \{BP, BP\}$ to be the ideal generated by maps of the form $BP \xrightarrow{tr} BQ \xrightarrow{Bh} BP$ where $Q \lneqq P$. Then by Nishida [N]

$$\{BP, BP\}/J \cong \widehat{\mathbf{Z}}_p Out(P)$$

where $Out(P)$ is the outer automorphism group of P. If $1 = \sum \widetilde{e}_j \in \widehat{\mathbf{Z}}_p Out(P)$ is a decomposition into primitive orthogonal idempotents, then each $\widetilde{e}_j = e_{jo} + \cdots + e_{jk}$ can be further decomposed into primitive, orthogonal idempotents in $\{BP, BP\}$ where $e_{jo} \notin J$ and $e_{ji} \in J$ for $i = 1, \ldots, k$.

We call $e_{j_0}BP$ a *dominant summand.* Each $e_{ji}BP$, $i = 1, \ldots, k$ is stably homotopy equivalent to a dominant summand of some BQ, $Q \lneqq P$ since $e_{ji} \in J$. The stable homotopy type of a summand is independent of the choice of idempotents. We also call $\widetilde{e}_j BP \cong e_{j_0}BP \vee (e_{j1}, BP \vee \cdots \vee e_{jk}BP)$ a *superdominant summand.* Since each $\widetilde{e}_j$ corresponds to an indecomposable representation of $\widehat{\mathbf{Z}}_p Out(P)$, we have a correspondence between irreducible representations and dominant summands. We call the dominant summand corresponding to the trivial representation the *principal dominant summand.*

We now develop the main tool we will use to analyze the dimension of a dominant summand. For a finite field k of characteristic p the $GL_n(k)$-invariants of the polynomial algebra $k[x_1, \ldots, x_n]$ were found by Dickson [Dk] to be a polynomial subalgebra of the same Krull dimension n. In [W], it is shown for $k = \mathbf{F}_p$ that the x_i are the images of the Chern classes $\{c_{p^n - p^i} | i = 0, \ldots, n-1\} \subseteq H^*(BU(p^n))$ under the regular representation $T^n \to U(p^n)$ where T^n is a p-torus (or elementary abelian p-group) of $\mathbf{F}_p$-dimension n.

We generalize this definition to all finite p-groups. The *rank* of a p-group P is the maximum $\mathbf{F}_p$-dimension of p-tori in P. Let P be a p-group of order p^r and rank n. Consider the regular representation $reg_P : P \to \sum(P) \subseteq U(p^r)$ where $\sum(P)$ is the group of set bijections of P.

Definition 1.1. The *Dickson invariants* $D(P) = reg_P^* C(n) \subseteq H^*(BP)$ where $C(n) = \mathbf{F}_p[c_{p^s(p^n - p^i)} | i = 0, 1, \ldots, n-1] \subseteq H^*(BU(p^r))$ and $p^s = [P : T]$ for T a maximal p-torus of rank n.

We now develop some basic properties of $D(P)$.

PROPOSITION 1.2. *$D(P)$ is a polynomial algebra of Krull dimension n, where n is the rank of P.*

PROOF: Let T be a maximal p-torus of rank n and $[P : T] = p^s$. $P \cong$

$p^sT = T\coprod\cdots\coprod T$ (p^s copies) as a T-set. Thus

$$\begin{array}{ccc} P & \xrightarrow{reg_P} & \Sigma(P) \\ \nwarrow incl & & \nearrow p^s reg_T \\ & T & \end{array} \quad \text{commutes.}$$

So $incl^* reg_P^* C(n) = p^s reg_T^* C(n) \approx \mathbf{F}_p[(c_{p^n-p^i})^{p^s} | i = 0, \ldots, n-1]$. ∎

PROPOSITION 1.3. *If $x \in D(P)$ and $f \in Out(P)$ then $f^*(x) = x$.*

PROOF: This is immediate from the commutative diagram

$$\begin{array}{ccc} P & \xrightarrow{reg_P} & \Sigma(P) \\ \uparrow f & & \uparrow c_f \\ P & \xrightarrow{reg_P} & \Sigma(P) \end{array}$$

where c_f is conjunction by f in $\Sigma(P)$.

COROLLARY 1.4. *If $\tilde{e} \in \widehat{\mathbf{Z}}_p Out(P)$ then $H^*(\tilde{e}BP) = \tilde{e}^* H^*(BP)$ is a finitely generated $D(P)$-module.*

For X a dominant summand of BP, we now want to establish that our definition of dimension $d(X)$ is well-defined. We use

THEOREM 1.5 (HILBERT, SERRE) [AM]. *Let $A = \oplus A_n$ be a Noetherian graded ring generated as an A_0-algebra by the homogeneous elements $x_1, \ldots, x_s$ of positive degree $k_1, \ldots, k_s$ respectively. If M is a finitely generated graded module over A then*

$$PS(M,t) = \frac{f(t)}{\prod_{i=1}^{s}(1-t^{k_i})}$$

where $f(t) \in \mathbf{Z}[t]$.

PROPOSITION 1.6. *If X is a summand of BP then $PS(X,t)$ is a rational function.*

PROOF: By induction on the order of P we may assume that X is a dominant summand. Let $\tilde{e} \in \widehat{\mathbf{Z}}_p Out(P)$ be a primitive idempotent corresponding to X then by Corollary 1.4 and Theorem 1.5 $PS(\tilde{e}BP,t)$ is a rational function. But $\tilde{e}BP = X \vee Y_1 \vee \cdots \vee Y_k$ where the Y_i's are non-dominant summands. Each Y_i is a dominant summand of BQ for some $Q \lneq P$. By the inductive hypothesis $PS(Y_i, t)$ is a rational function. Since $PS(\tilde{e}BP, t) = PS(X, t) + \sum_{i=1}^{k} PS(Y_i, t)$ we have that $PS(X, t)$ is a rational function. ∎

Thus $d(X)$ is a well-defined number. There is an immediate upper bound for $d(X)$.

THEOREM 1.7 (QUILLEN) [Q2]. *$d(BP)$ = Krull dimension of $H(P)$ = rank of P. So $d(X) \leq$ rank of P for every summand X of BP.*

Another theorem of Quillen's will prove useful in the following sections.

THEOREM 1.8 (QUILLEN) [Q1]. *$H^*(BP) \xrightarrow{\prod incl^*} \prod H^*(BT)$ where the product ranges over maximal p-tori, is a monomorphism modulo nilpotent elements.*

Section 2. The proof of Theorem A.

We begin by recalling the definition of the principal dominant summand X_0 of BP. Let $e \in \mathbf{F}_pOut(P)$ be a primitive idempotent with augmentation 1, i.e. $\varepsilon(e) = 1$, where $\varepsilon : \mathbf{F}_pOut(P) \to \mathbf{F}_p$ is the usual augmentation defined by $\varepsilon(g) = 1$ for all $g \in Out(P)$. Let $\tilde{e} \in \hat{\mathbf{Z}}_pOut(P)$ be a primitive idempotent lifting e, then $\tilde{e}$ decomposes in $\{BP, BP\}$ into a sum of primitive idempotents

$$\tilde{e} = e_0 + e_1 + \cdots + e_k$$

with $X_0 \simeq e_0BP$ the principal dominant summand and e_iBP non-dominant summands for $i \geq 1$. The p-local homotopy type of X_0 is independent of these choices.

PROOF OF THEOREM A: By Proposition 1.3 the ring $D(P)$ of Dickson invariants consists of $Out(P)$ invariants, hence $e^*D(P) = D(P)$ and so

$$D(P) \subset H^*eBP = e^*(H^*BP).$$

Thus the theorem follows provided

$$\dim e_i^*(D(P)) < n$$

for all $i \geq 1$. By the Segal Conjecture [LMM] each e_i, $i \geq 1$, is a $\widehat{\mathbf{Z}}_p$ linear combination of maps of the form

$$f : BP \xrightarrow{tr} BQ \xrightarrow{Bh} BP$$

for various subgroups $Q \leq P$ and homomorphisms h. Thus the theorem follows from

LEMMA 2.1. *If $Q \lneqq P$ or if h is not injective then $f^*(D(P))$ has dimension $< n$.*

PROOF: $\dim f^*D(P) \leq \dim H^*BQ \leq n$. Hence, we are reduced to assuming $\dim H^*BQ = n$. By Theorem 1.7, the p-rank of Q is also n.

Case 1. $Q \lneqq P$ and h is injective: Let $[P : Q] = p^s$, $s > 0$. Then $P \approx p^sQ = Q \coprod \cdots \coprod Q$ (p^s copies) as a Q set. Thus

$$(2.2)\qquad \begin{array}{ccc} P & \xrightarrow{reg_P} & \Sigma(P) \\ & {}_{i_{Q,P}}\nwarrow \quad \nearrow_{p^s reg_Q} & \\ & Q & \end{array} \quad \text{commutes.}$$

Similarly for $Q' = h(Q)$. Let $h' = p^s h : p^sQ \to p^sQ'$. Then h' is a permutation of P and we claim that the diagram

$$(2.3)\qquad \begin{array}{ccc} Q & \xrightarrow{p^s reg_Q} & \Sigma(P) \\ {\scriptstyle h}\downarrow & & \downarrow{\scriptstyle c_{h'}} \\ Q' & \xrightarrow{p^s reg_{Q'}} & \Sigma(P) \end{array}$$

commutes where $c_{h'}$ is conjugation by h'. For $x \in Q$, $y \in p^sQ'$,

$$(c_{h'} \circ p^s reg_Q(x))(y) = h(p^s reg_Q(x)(h^{-1}(y)))$$

$$= h(x.h^{-1}(y)) = h(x).y = p^s reg_{Q'} h(x)(y)$$

Now we claim that

$$h^* i^*_{Q',P} D(P) = i^*_{Q,P} D(P). \tag{2.4}$$

To see this recall from (1.1) that $D(P) = reg^*_P C(n)$, thus $i^*_{Q',P} D(P) = i^*_{Q',P} reg^*_P C(n) = p^s reg^*_{Q'} C(n)$ by (2.2). Hence by (2.3),

$$h^* i^*_{Q',P} D(P) = h^* p^s reg^*_{Q'} C(n) = (p^s reg^*_Q) c^*_{h'} C(n)$$

$$= (p^s req_Q)^* C(n) = i^*_{Q,P} reg^*_P C(n) = i^*_{Q,P} D(P)$$

. Hence $f^* D(P) = tr^* h^* i^*_{Q',P} D(P) = tr^* i^*_{Q,P} D(P) = [P : Q] D(P) = 0$ and so certainly $\dim f^* D(P) < n$.

Case 2. h is not injective: Let $T \xrightarrow{j} Q$ be a p-torus of rank n. Then T is a maximal p-torus and it follows easily that $hj(T)$ has rank $< n$. Hence $j^* h^* D(P)$ has dim $< n$. However by Theorem 1.8,

$$\prod j^* : H^* BQ \to \prod H^* BT$$

is injective modulo nilpotent elements, where the union is taken over all maximal p-tori $T < Q$. Hence $\dim h^* D(P) < n$ and so $\dim f^* D(P) < n$.

Section 3. The proof of Theorem B.

Let $\widetilde{e} \in \widehat{\mathbf{Z}}_p Out(P)$ be a primitive idempotent and let X be the corresponding dominant summand. We can write $\widetilde{e} = e_0 + e_1$ where e_0, e_1 are orthogonal idempotents in $\{BP, BP\}$, $X = e_0 BP$, and $e_1 \in J$. To prove Theorem B, it suffices to show that there is a copy of $D(P)$ in $H^*(X)$ as a vector space.

Now e_1, as is any element of $\{BP, BP\}$, is a linear combination of maps of the form $BP \xrightarrow{tr} BQ \xrightarrow{Bh} BP$. We write $e_1 = \sum_\alpha Bh_\alpha tr + \sum_\beta Bk_\beta tr$

where the h_α's are monomorphisms and the k_β's are not. Since $e_1 \in J$ for $h_\alpha : Q \to P$ a monomorphism we must have $Q \lneqq P$.

Since $\{X, X\}$ is a local ring [Mg] and X is a dominant summand the map

$$X \xrightarrow{i} BP \xrightarrow{\sum Bh_\alpha tr} BP \xrightarrow{\pi} X$$

is nilpotent (i, π are inclusion and projection maps respectively). Thus, for any $y \in H^k(X)$, $(\sum Bh_\alpha tr)^*(y) = z \neq y$. Let $w \in D(P)$ be the top-most Dickson invariant, that is the element which restricts to the image of $(c_{p^n-1})^{p^s}$ on a maximal p-torus of rank n. Since the image of c_{p^n-1} vanishes on any p-torus of rank less than n [W], for any $k_\beta : Q \to P$ which is not a monomorphism $Bk_\beta^*(w) \in H^*(BQ)$ is not detected on the cohomology of p-tori and by Quillen's Theorem 1.8, $Bk_\beta^*(w)$ is nilpotent. Therefore, for some high power of w, $Bk_\beta^*(w^t) = 0$ for every k_β. Let $\bar{w} = w^t$.

Since $\tilde{e} \in \hat{\mathbf{Z}}_p Out(P)$ we know $\tilde{e}^*(\bar{w}y) = \bar{w}\tilde{e}^*(y) = \bar{w}y$ by Corollary 1.4. Moreover

$$e_1^*(\bar{w}y) = (\sum Bh_\alpha tr)^*(\bar{w}y) + (\sum Bk_\beta tr)^*(\bar{w}y) = \bar{w}(\sum Bh_\alpha tr)^*(y) + 0$$

by Frobenius reciprocity since $Bh_\alpha^*(\bar{w}) = incl_Q^*(\bar{w})$ by (2.4). Thus $e_1^*(\bar{w}y) = \bar{w}z$ implies

$$e_0^*(\bar{w}y) = \tilde{e}^*(\bar{w}y) - e_1^*(\bar{w}y) = \bar{w}(y - z) \neq 0$$

since $y - z \neq 0$ and $D(P)$ acts freely on $y - z$ by hypothesis.

Let $u = \bar{w}(y - z)$ and $v \in D(P)$. $\tilde{e}^*(vu) = v\tilde{e}^*(u) = ve_0^*(u) = vu$, but $\tilde{e}^*(vu) = e_0^*(vu) + e_1^*(vu) = e_0^*(vu) + v(\sum Bh_\alpha tr)^*(u)$ by Frobenius reciprocity since $(\sum Bk_\beta tr)^*(u) = 0$. Thus $\tilde{e}^*(vu) = e_0^*(vu) + ve_1^*(u) = e_0^*(vu)$ since $e_1^*(u) = 0$. Hence $e_0^*(vu) = ve_0^*(u)$ for all $v \in D(P) \Rightarrow D(P)u \subseteq H^*(X)$. ∎

Section 4. Addendum.

Finally we include a couple of results on superdominant summands along the lines of Theorems A and B.

PROPOSITION 4.1. *Let Z be a superdominant summand of BP, P a finite p-group. Then $d(Z) \geq$ rank of center of P.*

PROOF: The depth of $H(P)$ is the maximum length of a regular sequence. By Duflot [Df] d=depth of $H(P) \geq$ rank of center of P. Let $u_1, \ldots, u_d$ be the bottom-most Dickson invariants, i.e. the elements which restrict to the images of $(c_{p^r - p^i})^{p^s}$, $i = n-d, \ldots, n-1$ on a p-torus. By Rusin [R], $u_1, \ldots, u_d$ form a regular sequence in $H(P)$. From which it follows that for $Z = \tilde{e}BP$, $\tilde{e} \in \hat{\mathbf{Z}}_p Out(P), H^*(\tilde{e}BP)$ is a free module over $\mathbf{F}_p[u_1, \ldots, u_d]$ which implies $d(Z) \geq d$. ∎

An element $z \in H^n(Z)$ is called *nilpotent* if $Sq^{2^k n} \ldots Sq^n z = 0$ for some k (or $P^{p^k n} \ldots P^n \beta^\epsilon z = 0$ for $p > 2$, where $\varepsilon = 0, 1$ and $z \in H^{2n-\epsilon}(Z)$).

PROPOSITION 4.2. *Let Z be a superdominant summand of BP, P a finite p-group. If $z \in H^*(Z)$ is not nilpotent then $d(Z) \geq$ minimum rank of maximal p-tori of P.*

PROOF: Since z is non-nilpotent, for some maximal p-torus T the restriction of z to $H^*(T)$ does not vanish (Theorem 1.8). Since $incl_T^*(D(P))$ has the same dimension as $D(T)$, $d(Z) \geq d(D(P)z) \geq d(D(T)incl_T^*(z)) =$ rank of T. ∎

REFERENCES

[AM] M. Atiyah and I. MacDonald, "Introduction to Commutative Algebra," Addison-Wesley, Reading, Massachusetts, 1969.

[BF] D. Benson and M. Feshbach, *Stable splittings of classifying spaces of finite groups*, Topology (to appear).

[C] G. Carlsson, *Equivariant stable homotopy and Segal's Burnside ring conjecture*, Annals. of Math **120** (1984), 189–224.

[CW] D. Carlisle and G. Walker, *Poincaré series for the occurence of certain modular representations of $GL(n,p)$ in the symmetric algebra*, (preprint).

[Df] J. Duflot, *Depth and equivariant cohomology*, Comment. Math. Helvetici **56** (1981), 627–637.

[Dk] L. Dickson, *A fundamental system of invariants of the general modular group with a solution of the four problem*, Trans. A.M.S. **12** (1911), 75–98.

[FS] V. Franjou and L. Schwartz, *Reduced unstable A-modules and the modular representation theory of the symmetric groups*, (preprint).

[HK] J. Harris and N. Kuhn, *Stable decompositions of classifying spaces of finite Abelian p-groups*, Math. Proc. Camb. Phil. Soc. **103** (1988), 427–449.

[LMM] G. Lewis, P. May and J. McClure, *Classifying G-spaces and the Segal Conjecture*, Current Trends in Algebraic Topology, CMS Conference Proc. **2** (1982), 165–179.

[Mg] H. Margolis, "Spectra and the Steenrod Algebra," North-Holland, Amsterdam, 1983.

[Mt] J. Martino, *Stable splittings and the Sylow 2-subgroups of $SL_3(F_q)$, q odd*, Ph. D. thesis, Northwestern University 1988.

[MP] J. Martino and S. Priddy, *Classification of BG for groups with dihedral or quaternionic Sylow 2-subgroups*, J. of Pure and applied Algebra (to appear).

[MP2] ——————, *The complete stable splitting for the classifying space of a finite group*, Topology (to appear).

[MiP] S. Mitchell and S. Priddy, *Stable splittings derived from the Steinberg module*, Topology **22** (1983), 285–298.

[MiP2] ——————, *Symmetric product spectra and splittings of classifying spaces*, Amer. J. Math. **106** (1984), 219–232.

[Mu] H. Matsumura, "Commutative Algebra," Benjamin/Cummings, Reading, Massachusetts, 1980.

[N] G. Nishida, *Stable homotopy type of classifying spaces of finite groups*, Algebraic and Topological Theories (1985), 391-404.

[Q] D. Quillen, *The spectrum of an equivariant cohomology ring I*, Annals. of Math. **94** (1971), 549–572.

[Q2] ————, *The spectrum of an equivariant cohomology ring II*, Annals. of Math. **94** (1971), 573–602.

[R] D. Rusin, *The depths of rings of invariants and cohomology rings*, preprint.

[W] C. Wilkerson, *A primer on the Dickson invariants*, Proc. of Northwestern Homotopy Theory Conference, Comtemporary Math. **19** (1982), 421–434.

Yale University
New Haven, CT 06520

Northwestern University
Evanston, IL 60208

Printed in the United States
By Bookmasters